WITHDRAWN
University of
Illinois Library
at Urbana-Champaign

Plasma Processing
and Synthesis of Materials

MATERIALS RESEARCH SOCIETY SYMPOSIA PROCEEDINGS   VOLUME 30

ISSN 0272 - 9172

Volume 1—Laser and Electron-Beam Solid Interactions and Materials Processing,
  J.F. Gibbons, L.D. Hess, T.W. Sigmon, 1981
Volume 2—Defects in Semiconductors, J. Narayan, T.Y. Tan, 1981
Volume 3—Nuclear and Electron Resonance Spectroscopies Applied to Materials Science,
  E.N. Kaufmann, G.K. Shenoy, 1981
Volume 4—Laser and Electron-Beam Interactions with Solids, B.R. Appleton,
  G.K. Celler, 1982
Volume 5—Grain Boundaries in Semiconductors, H.J. Leamy, G.E. Pike,
  C.H. Seager, 1982
Volume 6—Scientific Basis for Nuclear Waste Management, S.V. Topp, 1982
Volume 7—Metastable Materials Formation by Ion Implantation, S.T. Picraux,
  W.J. Choyke, 1982
Volume 8—Rapidly Solidified Amorphous and Crystalline Alloys, B.H. Kear,
  B.C. Giessen, M. Cohen, 1982
Volume 9—Materials Processing in the Reduced Gravity Environment of Space,
  G.E. Rindone, 1982
Volume 10—Thin Films and Interfaces, P.S. Ho, K.N. Tu, 1982
Volume 11—Scientific Basis for Nuclear Waste Management V, W. Lutze, 1982
Volume 12—In Situ Composites IV, F.D. Lemkey, H.E. Cline, M. McLean, 1982
Volume 13—Laser-Solid Interactions and Transient Thermal Processing of Materials,
  J. Narayan, W.L. Brown, R.A. Lemons, 1983
Volume 14—Defects in Semiconductors II, S. Mahajan, J.W. Corbett, 1983
Volume 15—Scientific Basis for Nuclear Waste Management VI, D.G. Brookins, 1983
Volume 16—Nuclear Radiation Detector Materials, E.E. Haller, H.W. Kraner,
  W.A. Higinbotham, 1983
Volume 17—Laser Diagnostics and Photochemical Processing for Semiconductor Devices,
  R.M. Osgood, S.R.J. Brueck, H.R. Schlossberg, 1983
Volume 18—Interfaces and Contacts, R. Ludeke, K. Rose, 1983
Volume 19—Alloy Phase Diagrams, L.H. Bennett, T.B. Massalski, B.C. Giessen, 1983
Volume 20—Intercalated Graphite, M.S. Dresselhaus, G. Dresselhaus, J.E. Fischer,
  M.J. Moran, 1983
Volume 21—Phase Transformations in Solids, T. Tsakalakos, 1984
Volume 22—High Pressure in Science and Technology, C. Homan, R.K. MacCrone,
  E. Whalley, 1984
Volume 23—Energy Beam-Solid Interactions and Transient Thermal Processing,
  J.C.C. Fan, N.M. Johnson, 1984
Volume 24—Defect Properties and Processing of High-Technology Nonmetallic Materials,
  J.H. Crawford, Jr., Y. Chen, W.A. Sibley, 1984

MATERIALS RESEARCH SOCIETY SYMPOSIA PROCEEDINGS   VOLUME 30

Volume 25—Thin Films and Interfaces II, J.E.E. Baglin, D.R. Campbell, W.K. Chu, 1984

Volume 26—Scientific Basis for Nuclear Waste Management VII, G.L. McVay, 1984

Volume 27—Ion Implantation and Ion Beam Processing of Materials, G.K. Hubler, O.W. Holland, C.R. Clayton, C.W. White, 1984

Volume 28—Rapidly Solidified Metastable Materials, B.H. Kear, B.C. Giessen, 1984

Volume 29—Laser-Controlled Chemical Processing of Surfaces, A.W. Johnson, D.J. Ehrlich, H.R. Schlossberg, 1984

Volume 30—Plasma Processing and Synthesis of Materials, J. Szekely, D. Apelian, 1984

Volume 31—Electron Microscopy of Materials, W. Krakow, D. Smith, L.W. Hobbs, 1984

Volume 32—Better Ceramics Through Chemistry, C.J. Brinker, D.R. Ulrich, D.E. Clark, 1984

Volume 33—Comparison of Thin Film Transistor and SOI Technologies, H.W. Lam, M.J. Thompson, 1984

Volume 34—Physical Metallurgy of Cast Iron, H. Fredriksson, 1985

MATERIALS RESEARCH SOCIETY SYMPOSIA PROCEEDINGS   VOLUME 30

Plasma Processing and Synthesis of Materials

Symposium held November 1983 in Boston, Massachusetts, U.S.A.

EDITORS:

J. Szekely
Massachusetts Institute of Technology, Cambridge, Massachusetts, U.S.A.

D. Apelian
Drexel University, Philadelphia, Pennsylvania, U.S.A.

NORTH-HOLLAND
NEW YORK · AMSTERDAM · OXFORD

This work was supported by the U.S. Army Research Office under Grant Number DAAG29-83-M-0368. The views, opinions, and/or findings contained in this report are those of the author(s) and should not be construed as an official Department of The Army position, policy, or decision, unless so designated by other documentation.

This work relates to the Department of the Navy Grant N00014-83-G-0102 issued by the Office of Naval Research. The United States Government has a royalty-free license throughout the world in all copyrightable material contained herein.

©1984 by Elsevier Science Publishing Co., Inc.
All rights reserved.

This book has been registered with the Copyright Clearance Center, Inc. For further information, please contact the Copyright Clearance Center, Salem, Massachusetts.

Published by:
Elsevier Science Publishing Company, Inc.
52 Vanderbilt Avenue, New York, New York 10017

Sole distributors outside the United States and Canada:
Elsevier Science Publishers B.V.
P.O. Box 211, 1000 AE Amsterdam, The Netherlands

Library of Congress Cataloging in Publication Data

Main entry under title:

Plasma processing and synthesis of materials.
   (Materials Research Society symposia proceedings, ISSN 0272-9172; v. 30)
   Includes indexes.
   1. Plasma engineering—Congresses. 2. Chemistry, Organic—Synthesis.
   3. Materials—Congresses. I. Szekely, Julian, 1934-   . II. Apelian,
   Diran. III. Series.
TA2005.P53 1984   660.2'9   84-10288
ISBN 0-444-00895-0

Manufactured in the United States of America

TABLE OF CONTENTS

Preface xi

An Overview of Plasma Processing* 1
    J. Szekely, Massachusetts Institute of Technology

Plasma Generation* 13
    E. Pfender, University of Minnesota

Diagnostics Under Thermal Plasma Conditions* 37
    P. Fauchais, J.F. Coudert, A. Vardelle, M. Vardelle
    and J. Lesinski, Université de Limoges, France

Modeling of Plasma Processes* 53
    M. Boulos, University of Sherbrooke, Canada

Thermal Plasma Melting/Remelting Technology* 61
    W.C. Roman, United Technologies Research Center

Plasmas in Extractive Metallurgy* 77
    W.H. Gauvin and H.K. Choi, McGill University
    and Hydro-Quebec Research Institute

Rapid Solidification by Plasma Deposition* 91
    D. Apelian, Drexel University

Research Needs In Arc Technology* 101
    J.V.R. Heberlein, Westinghouse R&D Center

The Production of Metastable Metallic Particles Directly
From the Mineral Concentrate by In-Flight Plasma Reduction 117
    J.J. Moore, K.J. Reid and J.M. Sivertsen,
    University of Minnesota

Interaction of Coal Particles Injected into Argon
and Hydrogen Plasmas 127
    K. Littlewood, The University of St. George's
    Square, United Kingdom

Design and Use of An Efficient Plasma Jet Reactor for
High Temperature Gas/Solid Reactions 133
    F.W. Giacobbe and D.W. Schmerling,
    Cardox Corporation

Heat Transfer Analysis of the Plasma Sintering Process 141
    E. Pfender and Y.C. Lee, University of Minnesota

In-Flight Metal Extraction in a Novel Plasma Reactor 151
    K.J. Reid[+], J.J. Moore[+] and J.K. Tylko[++],
    [+]University of Minnesota and [++]Plasmatech Inc.

Plasma-Melted and Rapidly Solidified Powders 163
    R.F. Cheney, GTE Products Corporation

The Effect of Structure on the Thermal
Conductivity of Plasma Sprayed Alumina 173
    H.C. Fiedler, General Electric Company

---
*Invited Paper

Novel RF-Plasma System for the Synthesis of Ultrafine,
Ultrapure SiC and Si_3N_4   283
    G.J. Vogt, C.M. Hollabaugh, D.E. Hull
    and L.R. Newkirk, Los Alamos National Laboratory

Use of Optical Emission Spectroscopy as a Diagnostic
Technique for Plasma Deposition of Hydrogenated
Amorphous Silicon and Carbon   291
    K.J. Kampas, Brookhaven National Laboratory

Plasma Arc Carbide Coatings on Titanium   297
    R.D. Shull, P.A. Boyer, L.K. Ives and K.J. Bhansali,
    National Bureau of Standards

Author Index   303

Subject Index   305

PREFACE

In recent years there has been a rapidly growing interest in plasma chemistry and in engineering physics as applied to plasma phenomena as evidenced by the great success of recent symposia held in Zurich, Edinburgh and Montreal. Plasma applications in materials processing is a significant, expanding sub-set of these activities, with a rather special flavor of its own, because in material systems the consideration of processing, structure and properties are closely interwoven. As this field is maturing it was worthwhile to organize a symposium, which brought together plasma specialists and materials scientists and engineers who share an interest in the plasma processing of materials.

In organizing this symposium we sought to achieve multiple objectives. First of all, through the plenary sessions we attempted to provide an up-to-date survey of the current state of the art; the material contained in these papers should serve as a useful introduction to plasma processing. We also sought to open up lines of communication between those active in the plasma field and the materials scientists, and finally wished to provide a coherent forum for reporting the important advances that are being made by the members of the materials community in the plasma processing field.

In organizing these sessions we wish to thank the plenary speakers who travelled long distances to deliver their important contribution; thanks are also due to the authors and the other participants in the symposium, and last but not least we wish to express our sincere appreciation to the very competent staff of the Materials Research Society who assisted us and alleviated the many chores which exist in organizing such a symposium.

Finally thanks are due to the organizing committee and the session chairmen, their help was invaluable. Special thanks go to: Dr. S. Fishman, ONR; Dr. P. Parish, ARO; Dr. W.C. Roman, United Technologies; Dr. Don Polk, ONR; Mr. R.W. Smith, General Electric; Dr. H.F. Winters, IBM. In addition, Dr. Bernard Kear is acknowledged for his vision and his strong support of this symposium.

Julian Szekely
Professor and Associate Director of
 the Materials Processing Center
Massachusetts Institute of Technology
Cambridge, MA 02140

Diran Apelian
Professor and Head
Department of Materials Engineering
Drexel University
Philadelphia, PA 19104

Plasma Processing
and Synthesis of Materials

AN OVERVIEW OF PLASMA PROCESSING

JULIAN SZEKELY
Department of Materials Science and Engineering and Center for
Materials Processing, Massachusetts Institute of Technology,
Cambridge, Massachusetts 02139

ABSTRACT

   An overview is presented of the plasma processing of materials. The principal components of this overview include the definition and classification of plasmas of interest in materials processing, the methods of plasma generation and the basic engineering principles that govern plasma phenomena. Here emphasis is placed on both plasma theory and on the experimental verification of the models. This is followed by a brief description of principal plasma applications, including thermal plasmas used in melting, refining, plasma deposition and plasma synthesis and low pressure plasmas used in synthesis and in the processing of electronic materials.

   The review is concluded by a brief discussion of future prospects for plasma technology, where the main developments envisioned are in the area of improved coatings, the processing of electronic materials, melting and refining of refractory metals and superalloys and finally the possibility of more widespread use for plasma systems in extractive metallurgy.

INTRODUCTION

   The purpose of this paper is to present a broad overview of the current and anticipated role of plasma technology in materials processing. The main thrust of these comments is to introduce the reader to plasma phenomena, indicating the level of basic understanding that exists and stressing the particular plasma properties that may make this technology attractive for materials processing applications.

DEFINITION OF PLASMAS

   A plasma is a gas of sufficient energy content, such that a significant fraction of the species present are ionized and hence are conductors of electricity. The ionized nature is an important facet of plasma behavior, because as sketched in Fig. 1, plasmas are sustained, by "Joule Heating" through the passage of a current through the medium.

   As illustrated in Fig. 2, plasmas may be generated either by the passage of a current between electrodes, by induction, or through the combination of these.

   A common feature of all these systems is that the passage of a current through the conducting medium, i.e. the partially ionized gas gives rise to both Joule Heating which sustains the plasmas, and also the interaction of this current with the magnetic field will give rise to an electromagnetic force field, which in turn may generate flow.

   At this stage it is important to draw a distinction between two types of plasmas, namely, high pressure, thermal or equilibrium plasmas and low pressure, non-equilibrium plasmas, which include glow discharges. For high temperature plasmas, e.g. encountered in welding arcs, arc furnaces, plasma torches, and the like, local thermodynamic equilibrium is attained, that is the electrons and the ions are at the same energy level or temperature.

Under these conditions the system may be described in terms of a single temperature.

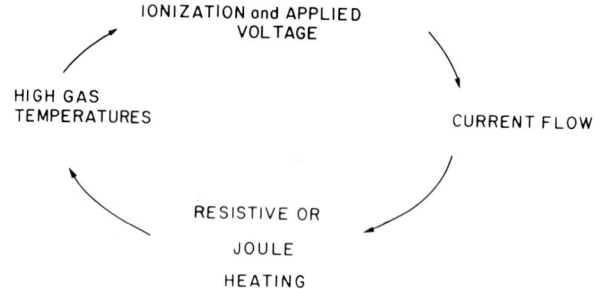

Fig. 1 Plasma as a self-sustaining electric discharge

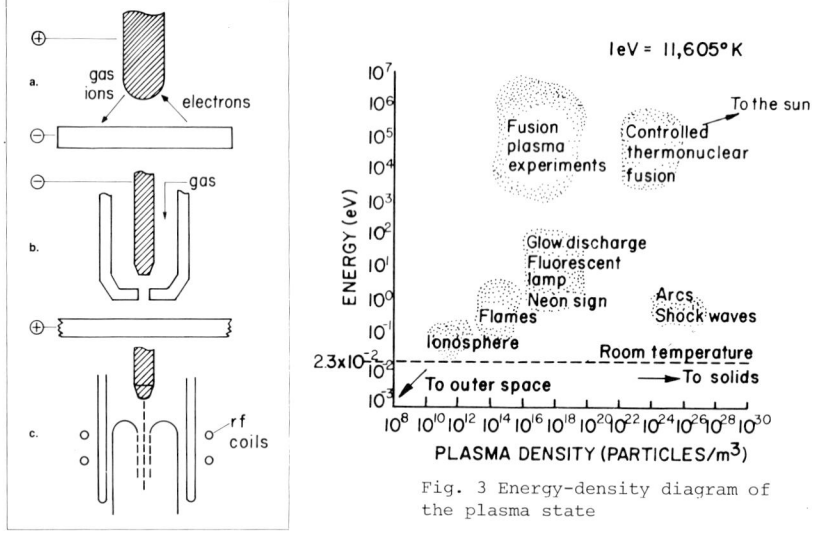

Fig. 2 The three principal methods for generating thermal plasmas: a) plasmas between electrodes, b) plasmas through electrodes, c) the induction-coupled torch

Fig. 3 Energy-density diagram of the plasma state

   In contrast for low pressure plasmas, such as encountered in glow discharges, in plasma etching, in plasma assisted chemical vapor deposition, the collision between the ions and the electrons is much less frequent; under these conditions the energy level (i.e. temperature) of the electrons is much higher than that of the ions. This distinction is illustrated in Fig. 3 and 4, where Fig. 3 shows a broad "road map" of the plasma state, while Fig. 4 illustrates the difference between the "ion temperatures" and the "electron temperatures" that will exist at low pressures.

Thermal plasmas have been used extensively in a broad range of applications, including welding, plasma spraying, in arc furnace technology and the like and our understanding of the fundamentals of these systems has advanced considerably in recent years.

In contrast the application of non-equilibrium, "cold" plasmas is more recent, but these systems are gaining quite widespread use in the processing of electronic materials, e.g. plasma etching, and plasma assisted chemical vapor deposition.

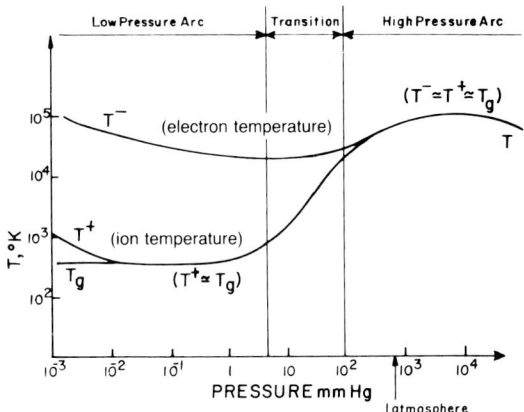

Fig 4 Low pressure versus high pressure arc, and approximate definition of local thermodynamic equilibrium

BASIC UNDERSTANDING OF PLASMA PHENOMENA

One may consider two basic approaches regarding the development of an understanding of plasma phenomena.

One of these would be that of a physicist, concerned with the spectra of energy distribution of the charged particles in plasma systems, that is the ions and the electrons.

From a theoretical point of view, the basic description of a plasma lies in the kinetic theory of matter. One defines a function of position, velocity, and time, $f(r,v,t)$, such that f, dr, dv is the probability of finding particles within the six dimensional volume element dr, dv, centered at the point (r,v) in coordinate and velocity space. Observable properties of the plasma can then be obtained from this function, known as the distribution function, by taking various velocity moments of f.

The equation determining the distribution function is called the kinetic equation:

$$\frac{\partial f}{\partial t} + \underset{\sim}{v} \cdot \underset{\sim}{\nabla}_r f + \frac{\underset{\sim}{F}}{m} \cdot \underset{\sim}{\nabla}_v f = \left(\frac{\partial f}{\partial t}\right)_c \tag{1}$$

where ∇_r is the gradient with respect to position

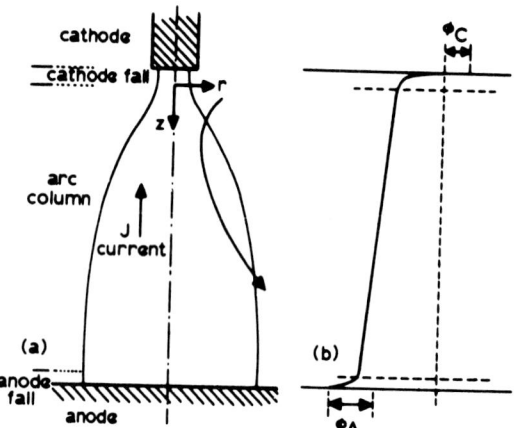

Fig. 5 Schematic representation of an arc and of the potential distribution along the axis of the arc

V_v is the gradient with respect to velocity and the term $(\partial f/\partial t)_c$ is due to collisions.

This kinetic equation is well known as the Boltzmann equation.

Rather more progress has been made in recent years regarding the quantitative description of thermal equilibrium plasmas, through the application on continuum electrodynamics.

Let us consider an arc struck between two electrodes, as sketched in Fig. 5. It is seen that the passage of the current from the anode to the cathode will give rise to a magnetic field, B_θ, which in turn interacts with the current to product an electromagnetic force field.

It has been shown that these systems may be represented by writing down the conventionally employed MHD equations. (1-5) viz:

$$\nabla \cdot \underset{\sim}{v} = 0 \qquad (2)$$

(equation of continuity)

$$\bar{\rho}(\underset{\sim}{v} \cdot \nabla =- \nabla P - \nabla \tau + \underset{\sim}{J} \times \underset{\sim}{B} \qquad (3)$$

(equation of motion)

$$\bar{C}_p \bar{\rho}(\underset{\sim}{v} \cdot \nabla T) = \nabla \cdot (K_{eff} \nabla T) + \underset{\sim}{J} \cdot \underset{\sim}{E} - S_R$$
$$+ \frac{5}{2} \frac{k_B}{e} \underset{\sim}{J} \cdot \nabla T \qquad (4)$$

(thermal energy balance equation)

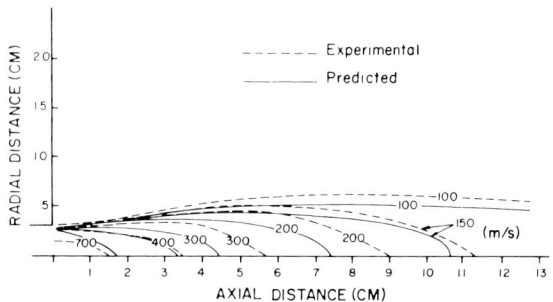

Fig. 6(a) Comparison between predicted and experimental velocity fields for nitrogen/hydrogen plasma

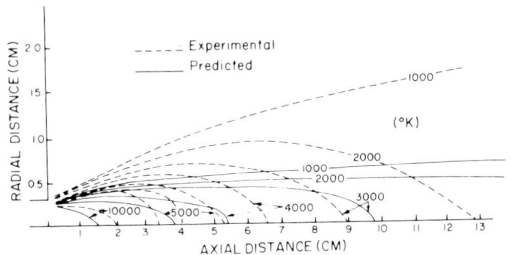

Fig. 6(b) Experimental and predicted temperature fields for nitrogen/hydrogen plasma

where
- $\bar{\rho}$ = average density
- $\underset{\sim}{v}$ = velocity vector
- ∇p = pressure gradient
- $\underset{\sim}{\tau}$ = total stress tensor (sum of viscous and turbulent stresses)
- $\underset{\sim}{J} \times \underset{\sim}{B}$ = Lorentz force
- K_{eff} = effective thermal conductivity, incorporating the turbulent components
- k_B = Boltzmann's constant
- e = electric charge
- T = temperature of plasma
- $\underset{\sim}{J} \cdot \underset{\sim}{E}$ = Joule Heating
- S_R = radiation loss per unit volume
- $\bar{C}_p$ = average specific heat of plasma

The appropriate subsidiary relationships required, to complete the statement of the problem will have to include Maxwell's equations in order to compute the JxB term, expressions for the temperature dependence of the fluid density, viscosity, thermal conductivity and most important of all, the temperature dependence of the extent of ionization.

Plasma Spray Processing of Ceramic Oxides 181
    N.R. Shankar, H. Herman[+] and R.K. MacCrone[++],
    [+]State University of New York at Stony Brook and
    [++]Rensselaer Polytechnic Institute

An Experimental Study of Powder Melting During
Low Pressure Plasma Deposition 187
    M. Paliwal and D. Apelian, Drexel University

Melting of Powder Particles in a Low Pressure Plasma Jet 197
    D. Wei[+], D. Apelian[+], S.M. Correa[++] and M. Paliwal[+],
    [+]Drexel University and [++]General Electric Company

Microprocessor Control of the Spraying of Graded Coatings 207
    R. Kaczmerek[+], W. Robert[+], J. Jurewicz[+],
    M.I. Boulos[+] and S. Dallaire[++], [+]University of
    Sherbrooke (Canada) and [++] National Research
    Council (Canada)

Consolidation of Nickel Base Superalloys Powder
by Low Pressure Plasma Deposition 217
    R.W. Smith, L.G. Peterson and W.F. Schilling,
    General Electric Company

Method for Producing Prealloyed Chromium Carbide/Nickel/
Chromium Powders and Properties of Coatings Created Therefrom . . 225
    D.L. Houck and R.F. Cheney, GTE Products Corporation

Microstructure and Phase Composition of Sputter-
Deposited Zirconia-Yttria Films 235
    R.W. Knoll and E.R. Bradley,
    Pacific Northwest Laboratory

Multiple Arc Discharges for Metallurgical
Reduction of Metal Melting 245
    J.E. Harry and R. Knight, University of
    Technology, United Kingdom

On the Allowance for the Temperature Dependence of Plasma
Properties for Selection of Dimensionless Numbers to
Correlate Characteristics of Electric Arcs 255
    V.A. Vashkevich, S.K. Kravchenki, T.V. Laktyushina
    and O.I. Yasko, Luikov Heat and Mass Transfer
    Institute, USSR

Non-Equilibrium Modeling and Dissipative Structures
in Solid Material Plasma Interactions 263
    Yu.L. Khait, Ben Gurion University of the
    Negev, Israel

Surface Treatment of the Glass Fibers in Low
Pressure Microwave Plasma 271
    R. Parosa, Technical University of Wrocław, Poland

A Mass Spectrometric System for the Study of
Transient Plasma Species in Thin Film Deposition 277
    N.P. Johnson[+], A.P. Webb[++] and D.J. Fabian[++],
    [+]University of Strathclyde (Scotland) and
    [++]Simon Fraser University (Canada)

These equations are of course mutually coupled, because of the interdependence of the electric conductivity, the temperature and the fluid velocity.

Allowance may also be made for particle - plasma interactions (6-8) to calculate the time - temperature history of solid particles injected into the plasma, of importance in plasma reactors, and also for plasma - particle - surface interactions, crucial in plasma deposition.

Figs. 6, 7 and 8 illustrate how well these techniques may be used to describe the behavior of real practical systems. Many other excellent examples are available in the papers presented at this symposium. (9-13) and at other recent meetings. (14)

In contrast to the reasonability good understanding that is being developed for thermal, equilibrium plasmas, our knowledge of the low pressure, non-equilibrium plasmas is much less complete.

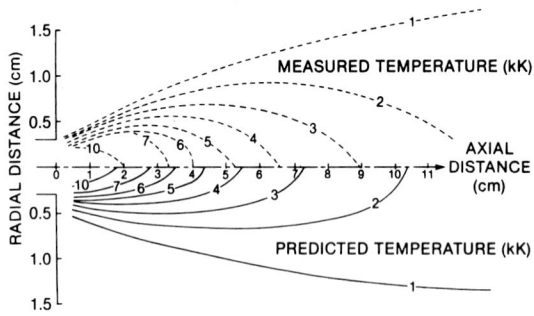

Fig. 7 Measured and predicted temperature contours

Fig. 9 shows a schematic sketch of a glow discharge system, used in plasma assisted chemical vapor deposition, or plasma etching. In these low pressure systems the ion temperatures are relatively low, in the range of 0.1 eV, while the electron temperature may be significantly higher, in the range of 1-10eV. In thermal plasmas no complex molecules or structures can survive, indeed this is an important principle utilized in plasma extractive metallurgy. In contrast in non-equilibrium plasmas the collision of the high energy electrons with molecular species gives rise to activated complexes, which upon reacting with each other or with solid surfaces, may result in interesting and useful solid products. This principle is being utilized in plasma synthesis, in plasma etching and in plasma assisted chemical vapor deposition.

In the quantitative representation of these reactions the electron energy distributions must be calculated by manipulating some form of the kinetic equation and this energy distribution function may then be related to the kinetic coefficients characterizing the formation (and destruction) of the activated species. As discussed by Winters (15), Bell (16), Kushner (17), Garscadden (18) and others (19-21), appropriate master equations must be written down to represent the formation, destruction and transfer of each of the species within the system.

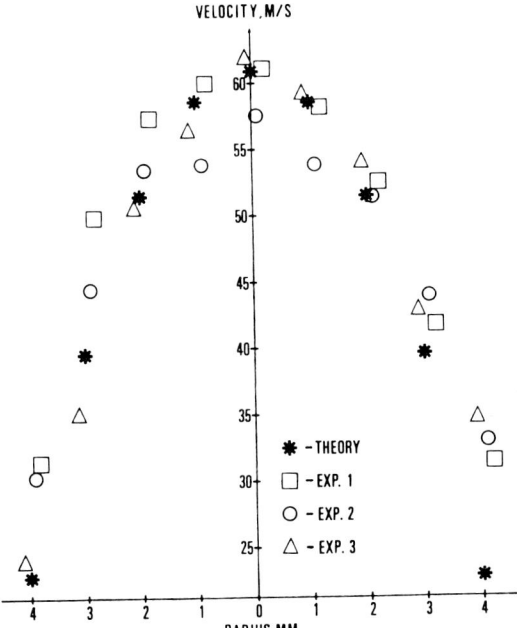

Fig. 8 Radial distribution of axial velocity in an atmospheric argon arc. I=70A, ṁ=0.2 g/s. Scattering centers: 8μm alumina. Observation station: 5.5 cm below cathode

As a corollary of this approach the system must be sampled for these activated species.

As an illustration, Fig. 10 shows the computed composition of the various species leaving the discharge region of a plasma reactor, as computed by Kushner, while Fig. 11, also taken from the work of Kushner indicates how the bias field may be related to the etch rate and the selectivity in a silicon/silica etching system. A similar approach is being applied at present to the plasma assisted DVD of silicon. (22)

PLASMA APPLICATIONS IN MATERIALS PROCESSING

It would be tempting to say that the previously described science base has played a key role in the realization of existing plasma applications. However, up to the present the actual industrial plasma applications are owed far more to ingenuity, inventiveness, intuition and possibly sheer luck than to a solid science base. The rapid evolution of the science base for plasma processing will, however play a dominant role in the development of a new generation of plasma applications and in the refinement of the existing technologies. In view of the very extensive treatment of this topic by papers that will be given in this symposium, only a very broad outline will be given here.

Thermal Plasmas

The main practical applications of thermal plasmas at present is in the areas of:

(i)   Melting and Refining
(ii) Plasma Deposition
(iii) Plasma Consolidation and Synthesis - very interesting potential in the area of -
(iv) Extractive Metallurgy.

In all these applications thermal plasmas may be regarded as a well focussed form of thermal energy, derived from an electrical source, available at high temperature levels, say above 2,000-3,000K. As far as the energy delivery rates are concerned single guns or torches may range in power from about 10 KW to say 5 MW.

As will be discussed in greater detail by Roman, (23) existing plasma melting and refining operations on an industrial scale are dominated by the use of this technology for the processing of titanium (scrap or sponge), with important efforts being initiated for the melting of steel, especially stainless steel. In this regard the use of plasmas in titanium processing is an established technology, while the use of plasma furnaces (replacing electric arc furnace installations as illustrated in Fig. (12) may well be realized on a more widespread scale in the near future.

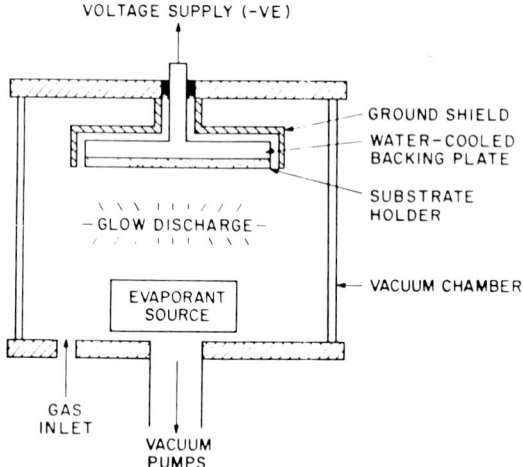

Fig. 9   Sketch of a typical glow discharge system

Plasma Spraying; to form coatings as thermal barriers has been practiced extensively during the past two decades, with the scientific interest being concentrated on characterization of the deposits, rather than on the process itself. Recent research is paving the way toward the extension of this technology to result in the build-up of massive deposits approaching near net shape and also the formation of more complex, sophisticated coatings, possibly through in situ plasma synthesis.

Plasma synthesis and Consolidation; involves the production of sub-micron ceramic powders, spheroidization, and the densification of powders and conglomerates. In virtually all these applications plasma synthesis and consolidation competes with conventional technologies, and has to offset the enhanced cost by yielding products of superior quality.

Plasma Extractive Metallurgy; should be appealing in situations where the reaction would have to be carried out at very high temperatures (refractory materials) in arc furnaces, or where the process would require a complex sequence of intermediate steps (e.g. titanium or magnesium production). At present an important additional limitation is posed on plasma processing by the fact that the maximum power attainable by a single gun is less than about 5 MW.

If we define extractive metallurgy as the smelting of ores (as distinct from the melting of scrap) bona fide, existing large scale applications do not exist at the present time, if we discount the plasma component of SKF's PLASMARED system. Interesting applications are now being pursued, through the pilot plant stage for the smelting of ferroalloys, and also in the smelting of ferrous ores and the recycling of metallic waste. Not fully confirmed reports suggest that the construction of some full scale plasma processing facilities is being contemplated at several locations. The use of this technology for the smelting of aluminum, magnesium or titanium ores, while appealing in principle, still lies in the future.

Low Pressure Plasmas

Low pressure plasmas, through the selective generation of the desired

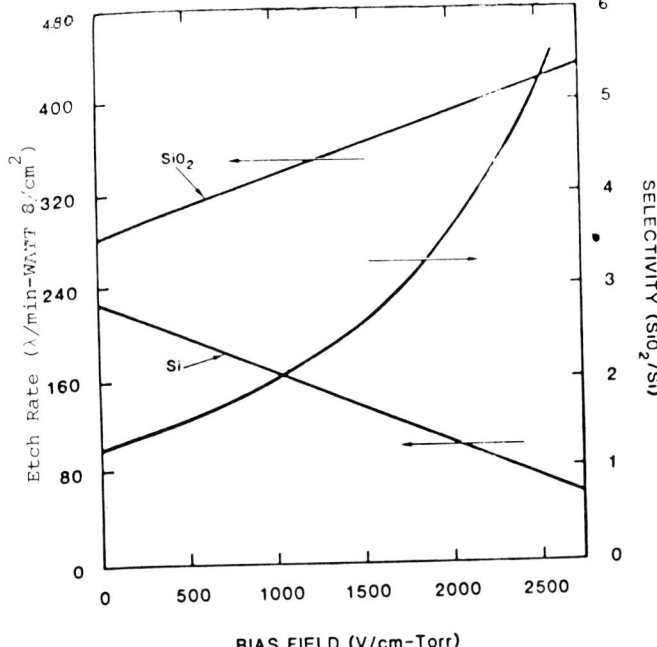

Fig. 9   Sketch of a typical glow discharge system

activated species offer very attractive and cost effective (compared to laser processing) means for the preparation of sophisticated structures under carefully controlled conditions.

Present uses include:
- surface cleaning
- chemical alteration of surfaces
- etching (which represents perhaps the most important present day application)
- plasma assisted film deposition and
- texturing.

This is an extremely interesting technology, employed largely on an empirical basis at present, which will benefit a great deal from the rapidly growing science base.

FUTURE PROSPECTS FOR PLASMA TECHNOLOGY IN MATERIALS PROCESSING

In a sense plasma technology in materials processing may be likened to the awakening of a sleeping giant. The essentially empirical application of high and low pressure plasma systems primarily to surface modifications has received a major impetus for further growth, primarily by the demand for much more sophisticated materials which plasma systems have a unique capability of delivering in a cost effective manner. The rapidly evolving science base for these operations, as evidenced by both the quantity and the quality of the papers that are appearing on this subject in symposia and also in the archieval journals should provide a much better perspective for these developments.

The important growth areas in plasma applications are likely to be in

- significantly improved coatings for thermal barriers and corrosion resistance aided by in situ composites and in situ synthesis

- the production of electronic materials through the more sophisticated low pressure, non-equilibrium plasmas, the application of plasma etching and plasma assisted chemical vapor deposition

- the use of plasma guns for the melting and refining of refractory metals, superalloys and high grade steels and finally

- the emergence of plasma technology as a vital component of extractive metallurgy.

As will be described in the subsequent papers, the evolving science base, through modelling, diagnosis and most important, their interaction will have to play a vital role in this development.

ACKNOWLEDGMENTS

The author wishes to thank his collegues in the plasma field, notably Professors Fauchais, Pfender, Apelian, Boulos and Gauvin and Drs. Mogab, Winters, Kear and Thomson, for helpful discussions. Thanks are also due to the US Department of Energy and the US Army Research Office for the support of this plasma research under grants, DE-AC02-78ER04799 and DAAG29-83-K-0022.

REFERENCES

1. M.I. Boulos, Proceedings of this Symposium.
2. J. Szekely, Proceedings of 7th International Conference on Vacuum Metallurgy, pp 1028, Tokyo, Japan (1982)
3. J. McKelliget, J. Szekely, M. Vardelle and P. Fauchais, Temperature

J. Plasma Chemistry and Plasma Processing, 2, (3), 315, (1982).

4. S.M. Correa, M.I. Boulos, M.I. Boulos eds. 6th International Symposium on Plasma Chemistry Universite de Sherbrooke and McGill University Publishers, pp 77, 1983.
5. D. Celmenti and D.M. Beneson, M.I. Boulos eds. 6th International Symposium on Plasma Chemistry Universite de Sherbrooks and McGill University Publishers, pp 126, 1983.
6. J.A. Lewis, W.H. Gauvin, A.I.Ch.E. Jnl., 19, (5), 982-990, (1973).
7. D. Bhattacharyya, W.M. Gauvin, A.I.Ch.E. Jnl., 21, (5), 879-885, (1975).
8. N.N. Sayegh, W.H. Gauvin, A.I.Ch.E. Jnl., 25,(3), 522-534, (1979).
9. E. Pfender and Y.C. Lee - Proceedings of this Symposium.
10. D. Apelian, Proceedings of this Symposium
11. R. Kaczmarek, W. Robert, J. Jurewitz and M.I. Boulos, Proceedings of this symposium.
12. P. Fauchais, Proceedings of this Symposium.
13. Y.L. Khait, Proceedings of this Symposium.
14. M.I. Boulos and R.J. Muntz eds. 7th International Symposium on Plasma Chemistry, Universite de Sherbrooke and McGill University Publishers, 1983.
15. C.J. Mogab and H.F. Winters, in Report of the National Materials Advisory Board on Plasma Processing, NRC, 1984.
16. A.T. Bell, Fundamentals of Plasma Chemistry.
17. M. Kushner, J. Appl. Phys. 53, 4, pp 2923, (1982).
18. A. Garscadden, in M.I. Boulos and R.J. Muntz eds., 6th, International Symposium on Plasma Chemistry, pp 388, (1983).
19. G. Turban, Y. Catherine and B. Grolleau, Plasma Chemistry and Plasma Processing, 2, No. 1 pp 61, (1982).
20. J.C. Knights, J.P.P. Schmitt, J. Perrin and G. Guelachvilli, J. Chem. Phys. 76, 7, pp 3424, (1982).
21. J.J. Wagner and S. Veprek, Plasma Chemistry and Plasma Processing, 3, No. 2, 219, (1983).
22. J. Szekely, R. Reif and A. Balazs - to be published.
23. W. Roman, Proceedings of this Symposium

PLASMA GENERATION

Dr. E. Pfender*
*University of Minnesota, Dept. of Mech. Eng., 111 Church St.
S.E., Minneapolis, Minnesota, 55455

1. INTRODUCTION

In general, a plasma consists of a mixture of electrons, ions, and neutral species. Although there are free electric charges in a plasma, negative and positive charges compensate each other, i.e. overall a plasma is electrically neutral, a property which is known as quasi-neutrality. In contrast to an ordinary gas, the free electric charges in a plasma give rise to high electrical conductivities which may even surpass those of metals. A hydrogen plasma, for example, at atmospheric pressure heated to temperatures of 10^6 K, has the same electrical conductivity as copper at room temperature. As the plasma temperature increases, the electrical conductivity increases beyond that of copper. Plasma temperatures of the order of 10^6 K and above are typical for thermonuclear fusion experiments.

Plasmas may be generated by passing an electric current through a gas. Since gases at room temperature are excellent insulators, a sufficient number of charge carriers have to be generated to make the gas electrically conducting. This process is known as electrical breakdown and there are many possibilities to accomplish this breakdown. Breakdown of the originally nonconducting gas establishes a conducting path between a pair of electrodes. The passage of an electrical current through the ionized gas leads to an array of phenomena known as gaseous discharges. Such gaseous discharges are the most common, but not the only means for producing plasmas. For certain applications plasmas are produced by electrodeless rf discharges, by microwaves, by shock waves, by laser or high energy particle beams. Finally plasmas may also be produced by heating of gases (vapors) in a high temperature furnace. Because of inherent temperature limitations, this method is restricted to metal vapors with low ionization potentials.

For the following considerations plasmas produced by electrical discharges will be discussed. In principle, such plasmas are divided into two types. The first is the "hot" or "equilibrium" plasma which is characterized by an approximate equality between heavy particle and electron temperatures, i.e. the thermodynamic state of the plasma approaches equilibrium or, more precisely, local thermodynamic equilibrium (LTE). Such plasmas are known as <u>thermal plasmas</u>. LTE comprises not only kinetic equilibrium ($T_e = T_h$; T_e = electron temperature, T_h = heavy particle temperature) but also chemical equilibrium, i.e. particle concentrations in a LTE plasma are only a function of the temperature. Typical examples of thermal plasmas are those produced in high intensity arcs and plasma torches or in high intensity rf discharges.

The second type of plasma is known as "cold" or "<u>non-equilibrium</u>" plasma. In contrast to thermal plasmas, cold plasmas are characterized by high electron temperatures and rather low "sensible" temperatures of the heavy particles ($T_e \gg T_h$). Plasmas produced in various types of glow discharges, in low intensity rf discharges, and in corona discharges are typical examples of such non-equilibrium plasmas.

Thermal as well as non-equilibrium plasmas cover a wide range of temperatures and electron densities. Fig. 1 shows as a survey the approximate range of electron temperatures and electron densities of natural and man-made plasmas. The electron

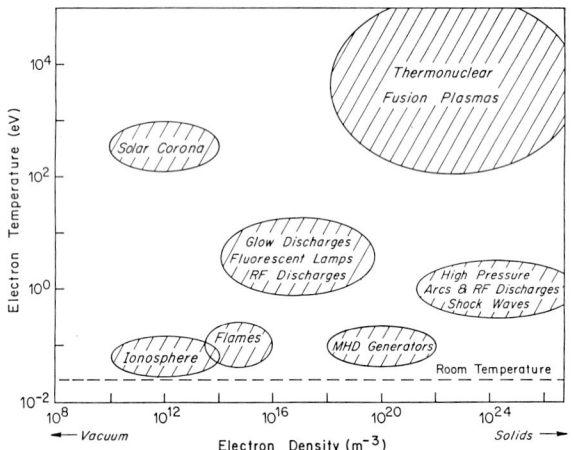

FIG. 1. Classification of natural and man-made plasmas.

temperatures are given in units of electron volts (1 eV corresponds to an electron temperature of approximately 7,740 K).

As a general criterion for the existence of a hot or a cold plasma, the ratio of E/p or E/n has been proposed (E = electrical field strength, p = pressure, n = particle number density). This criterion reflects the energy exchange process between electrons and heavy particles in a plasma. The collisional "coupling" and the associated energy exchange between electrons and heavy particles is enhanced by high particle densities or high pressures. High electric fields tend to increase the excess energy of the electron gas, therefore, thermal plasmas are characterized by small values of E/p or E/n. For cold plasmas, values of these parameters are higher, usually by several orders of magnitude. Fig. 2 shows the seperation of electron and heavy particle temperatures in an electric arc as a function of the pressure (100 kPA ≈ 1 atm).

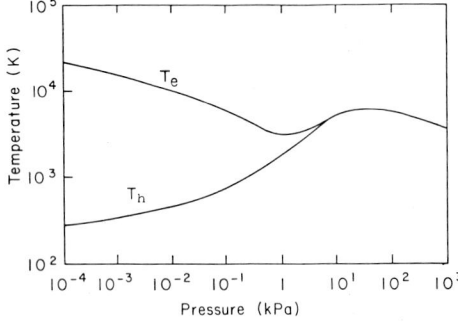

FIG 2. Seperation of electron temperatures and heavy particle temperatures at low pressures.

Because of space limitations, this survey will be restricted to the most common electrically generated steady state thermal plasmas. Non-thermal plasmas, transient plasmas, and those which are not produced by electrical discharges (laser beams, high energy particle beams, shock waves, heating in a furnace) will not be included in this survey.

A large number of older references related to thermal arcs are listed in a survey on "Electric Arcs and Thermal Plasmas" [1]. Newer references are given in a review on "Electric Arcs and Arc Gas Heaters" [8] by the author of this review.

2. THERMAL PLASMAS

In this survey, thermal plasmas will be considered, i.e. plasmas which approach a state of Local Thermodynamic equilibrium (LTE). The conditions which govern LTE in a steady state, optically thin (absorption is negligible) plasma produced from an atomic gas may be summarized as follows:

(a) The different species which constitute a plasma have a Maxwellian distribution of their velocities or kinetic energies. In terms of kinetic energies this requirement may be expressed by

$$f(E_\nu) = \frac{2E_\nu^{\frac{1}{2}}}{(\pi)^{\frac{1}{2}} (kT)^{3/2}} \exp(-E_\nu/kT) \qquad (1)$$

where $E_\nu = \frac{1}{2} m_\nu v_\nu^2$ represents the kinetic energy of particles of species ν (ν = electrons, ions, neutral atoms) in the plasma, k is the Boltzmann constant and T is the absolute temperature.

(b) E/p or E/n is sufficiently small and the temperature is high enough so that $T_e = T_h$ (T_e = electron temperature, T_h = heavy particle temperature). This requirement may be expressed by

$$\frac{T_e - T_h}{T_e} = \frac{m_h}{24 m_e} \frac{(\lambda_e eE)^2}{(kT_e)^2} >> 1 \qquad (2)$$

where m_h, m_e, λ_e, and E represent the mass of the heavy particles, the electron mass, mean free path of the electrons and the electric field strength, respectively.

(c) Collisions are the dominating mechanism for excitation and ionization. This requirement is met for sufficiently high electron densities ($n_e \gtrsim 10^{16} cm^{-3}$). In this situation the population of excited states follows as Boltzmann distribution, i.e.

$$\frac{n_{r,s}}{n_r} = \frac{g_{r,s}}{Z_r} \exp(E_{r,s}/kT) \qquad (3)$$

and the electron density is described by the Eggert-Saha equation, i.e.

$$\frac{n_{r+1} n_e}{n_e} = \frac{2 Z_{r+1}}{Z_r} \cdot \frac{(2\pi m_e kT)^{3/2}}{h^3} \exp(E_{r+1}/kT) \qquad (4)$$

In these equations $n_{r,s}$ represents the number density of excited atoms or ions in the quantum state s, n_r is the total number density of particles of species r, Z_r is their partition function, $E_{r,s}$ is the energy of the s-th quantum state, $g_{r,s}$ is the statistical weight of this state, E_{r+1} is the energy required for producing an (r+1)-times ionized atom from an r-times ionized atom, and h is Planck's constant.

(d) Spatial variations (gradients) of the plasma properties

must remain sufficiently small so that

$$\tau_{diff} \gg \tau_{relax} \qquad (5)$$

for all species in the plasma. In Eq. (5) τ_{diff} represents the diffusion time of a particle between two given locations in the plasma and τ_{relax} represents the corresponding relaxation time.
   In order to meet the requirement of LTE, all the temperatures appearing in Eqs. (1-4) must be the same, i.e. kinetic temperatures, excitation temperatures, and ionization temperatures must be identical.
   Further details on LTE and deviations from LTE may be found in the literature [1-6].

2.1 High intensity arcs

   For the sake of simplicity only d.c. arcs will be considered, although a.c. arcs are in wide use in actual applications.
   The potential distribution in high as well as in low intensity arcs shows a peculiar behavior as indicated in Fig. 3.

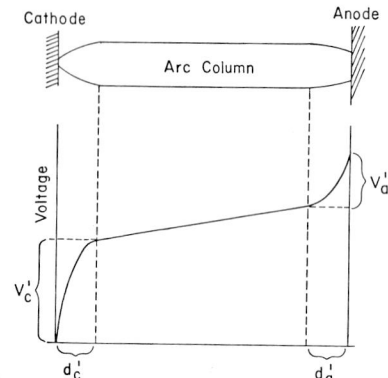

FIG. 3. Typical potential distribution along an arc.

Steep potential drops in front of the electrodes and relatively small potential gradients in the arc column suggest dividing the arc into three parts: the cathode region, the anode region, and the arc column. The latter represents a true plasma which will approach a state of LTE in a high intensity arc.
   A high intensity arc represents a simple and rather straight forward means for producing a thermal plasma. Such an arc is defined as a discharge operated at current levels I >50A and pressures p > 0.1 atm. In contrast to low intensity arcs, high intensity arcs are characterized by strong macroscopic flows induced by the arc itself [7,8]. Any variation of the current-carrying cross section of the arc leads, via the interaction of the arc current with its own magnetic field, to a pumping action as sketched in Fig. 4. At sufficiently high currents (I > 100A) and axial current density variations, flow velocities of the order of 100 m/s are produced. The cathode jet phenomenon is a typical example. Its effect on the arc will be discussed in section 2.1.1.5
   Temperatures and charged particle densities which are the most important properties of an arc plasma may vary over a wide range. These properties are determined by the arc parameters including the arc geometry. Table I provides an indication of temperatures and electron densities for different types

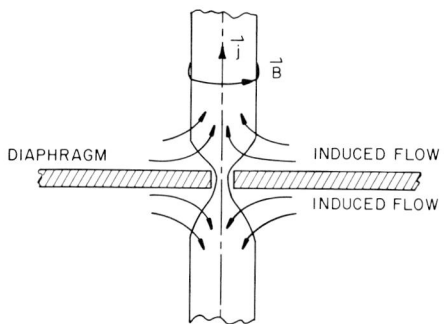

FIG. 4. Schematic of the pumping action induced by arc constriction.

of arcs. Some of them will be discussed in the following sections.

TABLE 1: SURVEY OF ARC TEMPERATURES AND ELECTRON DENSITIES

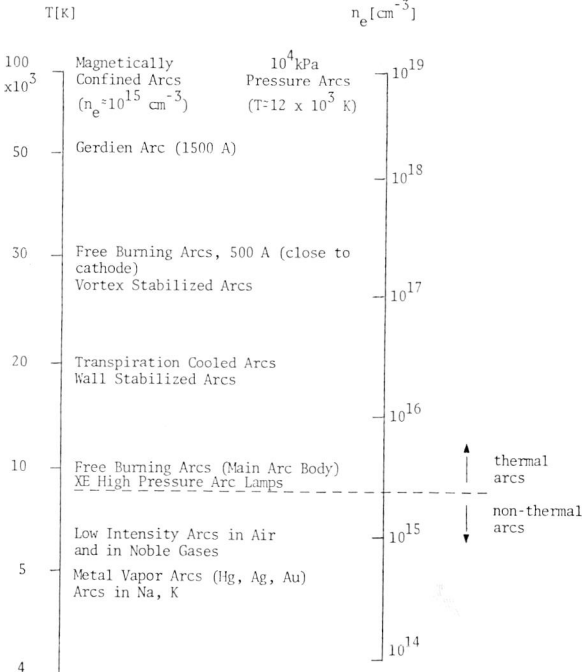

2.1.1 Classification of high intensity arcs

In view of arc applications, a classification in terms of stabilization of the arc column appears to be useful. There is a direct link between the method of stabilizing the arc column and the options available for the design of arc devices.

Most electric arcs require for their stable operation some

kind of stabilizing mechanism which must be either provided externally or which may be produced by the arc itself. The term "stabilization" as applied in this paragraph refers to a particular mechanism which keeps the arc column in a given, stable position, i.e. any accidental excursion of the arc from its equilibrium position causes an interaction with the stabilizing mechanism such that the arc column is forced to return to its equilibrium position. This stable position is not necessarily a stationary one; the arc may, for example, rotate or move along rail electrodes with a certain velocity. Stabilization implies in this situation that the arc column can only move in a well defined pattern, controlled by the stabilizing mechanism.

2.1.1.1. Free-burning arcs

As the name implies no external stabilizing mechanism is imposed on the arc in this case; but this does not exclude that this arc generates its own stabilizing mechanism. Although high intensity arcs may be operated in the free-burning arc mode, they are frequently classified as self-stabilized arcs (see part 2.1.1.5. of this paragraph) if the induced gas flow, due to the interaction of the self-magnetic field and the arc current, is the dominating stabilizing mechanism. Therefore, free-burning high intensity arcs to which the described conditions apply will be discussed in part 2.1.1.5.

Arcs operated at extremely high currents (up to 100 kA) known as Ultra-High Current Arcs, should also be mentioned in this category. Although most experiments in this current range utilize pulsed discharges, the relatively long duration ($\approx$ 10ms) of the discharge justifies classifying them as arcs. There is considerable interest in such arcs in connection with melting and steelmaking, utilization in chemical arc furnaces and high power switchgear. Visual observations of ultra-high current arcs in arc furnaces reveal a rather complex picture of large, grossly turbulent plasma volumes, vapor jets emanating from the electrodes, and parallel current paths with multiple, highly mobile electrode spots. In this situation there is no evidence for any dominating stabilizing mechanism. Induced gas flows and vapor jets exist simultaneously, interacting with each other in a complicated way. Depending on the polarity of the arc and the electrode materials, stable vapor jets have been observed which are able to stabilize the arc column. Thus, the generation of vapor jets by the arc represents another possible mechanism for self-stabilization of arcs.

In a comprehensive survey Edels [9] gave a description of the characteristic features and properties of ultra-high current arcs, including 120 pertinent references. Continuing studies in this area are mainly concerned with radiation properties and flow fields in such arcs.

2.1.1.2 Wall-stabilized arcs

The principle of wall-stabilization of arcs has been known for more than 75 years, introduced in connection with arc lamps. A long arc enclosed in a narrow tube with circular cross section will assume a rotationally symmetric, coaxial position within the tube. Any accidental excursion of the arc column towards the wall will be compensated by increasing heat conduction to the wall which reduces the temperature and, therefore, the electrical conductivity at this location. In short, the arc will be forced to return to its equilibrium position. In this situation, increased thermal conduction and the associated secondary effects represent the stabilizing mechanism.

In order to cope with the extremely high wall heat fluxes experienced with high intensity arcs enclosed in small diameter tubes, metal tubes have been introduced as arc vessel, consisting of a stack of insulated, water-cooled disks (usually Cu). This arrangement is known as the wall-stabilized, cascade arc which has been extensively used as a basic research tool.

Fig. 5 shows a cut-away view of a typcial wall-stabilized

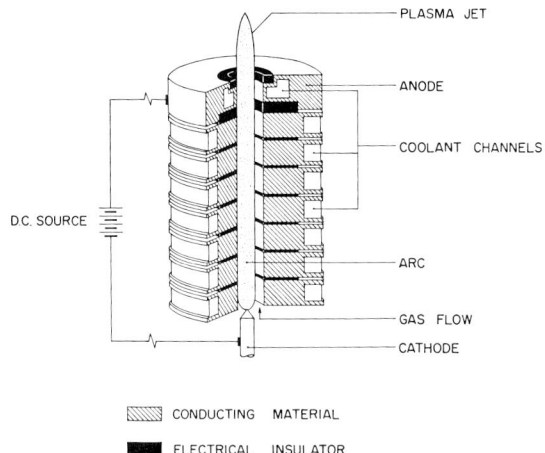

FIG. 5. Cut-away view of a wall-stabilized, cascaded arc.

arc. Due to the much higher electrical conductivity of metals compared with that of the arc column, segmentation of the tube enclosing the arc is necessary because a continuous metal tube would cause a double arc (arcing from the cathode to the metal tube and from the metal tube to the anode) seeking the path of least resistance.

The maximum possible temperature or enthalpy attainable in a constricted, wall-stabilized arc is limited by the highest permissible heat flux hwich the wall is able to withstand. Sophisticated water-cooling arrangements permit wall heat fluxes up to 2×10^5 kW/m^2.

2.1.1.3 Vortex-stabilized arcs

The principle of vortex-stabilization of arcs has been already reported around the turn of the last century [10]. In the case of vortex- or whirl-stabilization the arc is confined to the center of a tube in which an intense vortex of a gas or liquid is maintained. Centrifugal forces drive the cold fluid towards the walls of the arc chamber which, in this way, is thermally well protected. In addition to the circumferential component of the vortex flow, there is also an axial component superimposed which supplies continously cold fluid.

A well-known example of a vortex-stabilized arc is the so-called Gerdien arc [11] which is schematically illustrated in Fig. 6. In this case the stabilizing fluid is water, and the arc plasma is generated from water vapor in the core of the vortex. Because of the extreme cooling of the arc fringes, the

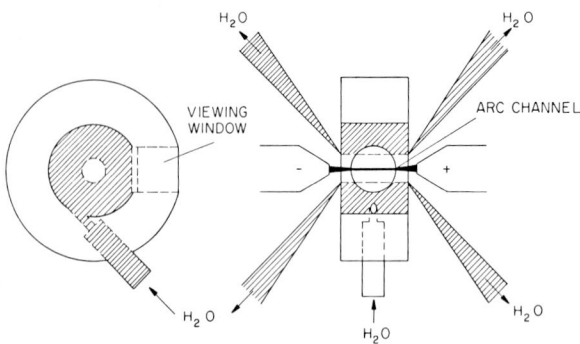

FIG. 6. Schematic of water-vortex-stabilized (Gerdien) arc.

power dissipation per unit volume and the associated arc temperatures reach much higher values than feasible in wall-stabilized arcs. Arc temperatures in excess of 50,000 K have been found using an arc current of 1,450 A and confining the arc to a diameter < 2.3 mm.

For actual applications of vortex-stabilized arcs various gases and gas mixtures are utilized as working fluids. Fig. 7

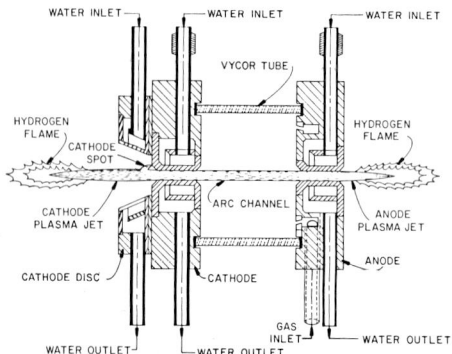

FIG 7. Schematic of a gas vortex-stabilized arc.

shows a schematic of a gas vortex-stabilized arc arrangement developed for the generation of fully ionized atmospheric pressure hydrogen plasmas. Both electrodes in this arrangement are water cooled. The working fluid enters tangentially at the anode through small orifices. The vortex generated in this way confines the arc to the center of the arc chamber reducing the arc diameter to approximately 2-3 mm. The intense convective cooling of the arc fringes due to the vortex flow around the arc enhances the power dissipation per unit length of the arc column, which, in turn, results in high axis temperatures. In the previously mentioned vortex-stabilized hydrogen arc, axis temperatures close to 25,000 K have been reached.

2.1.1.4 Electrode-stabilized arcs

In extremely short arcs (electrode gaps in the order of 1 mm) the behavior of the arc is determined by the vicinity of the electrodes. The arc column as such does not exist any more. The remaining arc consists of the non-uniform electrode regions, which, in contrast to a fully developed arc column, reveal strong axial gradients of the plasma properties [12]. These regions may be considered as thermal boundary layers which, in this situation, may even partially overlap. The contour of the remaining part of the arc approaches the shape of an ellipse with the electrode roots as focus points.

2.1.1.5 Self-stabilized arcs

The transition from a low intensity to a high intensity arc, which occurs at atmospheric pressure above currents of 50 A, manifests itself in a drastic change of the stability of the arc column. Below 50 A the arc column is subject to irregular motion induced by free convection effects. For arc currents in the range from 50 to 100 A the column becomes suddenly motionless and stiff with a visually well defined boundary. The cathode jet phenomenon which has been mentioned previously gives rise to this transition. The maximum velocity in the jet is related to total current and current density at the cathode, vis.

$$v_{max} \sim (Ij)^{\frac{1}{2}} \qquad (6)$$

As soon as this velocity substantially exceeds those induced by free convection effects (in the order of 1 m/s), the described transition will occur. The current at which this transition takes places depends on the conditions in the cathode region (current density and variation of current density in axial direction). The phenomenon is usually reversible, i.e. by lowering of the current a transition to the free convection dominated low intensity arc occurs.

The well-known bell shape of the free burning high intensity arc is observed when the cathode jet impinges on an anode normal to the cathode axis. Fig 8 shows calculated and measured isotherms of such an arc [13].

2.1.1.6 Forced convection stabilized arcs

By superimposing an axial flow to an otherwise unstable arc, stability can be achieved. In this case convective heat transfer from the arc to the cold gas shroud surrounding the arc plays essentially the same role as conduction in the case of a wall-stabilized arc.

2.1.1.7 Magnetically stabilized arcs

Since an arc is an electrically conducting medium it will interact not only with its own magnetic field, but also with externally applied magnetic fields. Over the past 15 years this interaction attracted increasing interest because of its potential for many arc applications. Magnetically influenced or stabilized arcs are extensively used in the development of arc gas heaters for material processing, in circuit breakers, in arc furnaces, etc. According to the existing literature the interaction of arcs with magnetic fields may be divided into the following categories:

(a) Magnetic stabilization of arcs in cross flow.

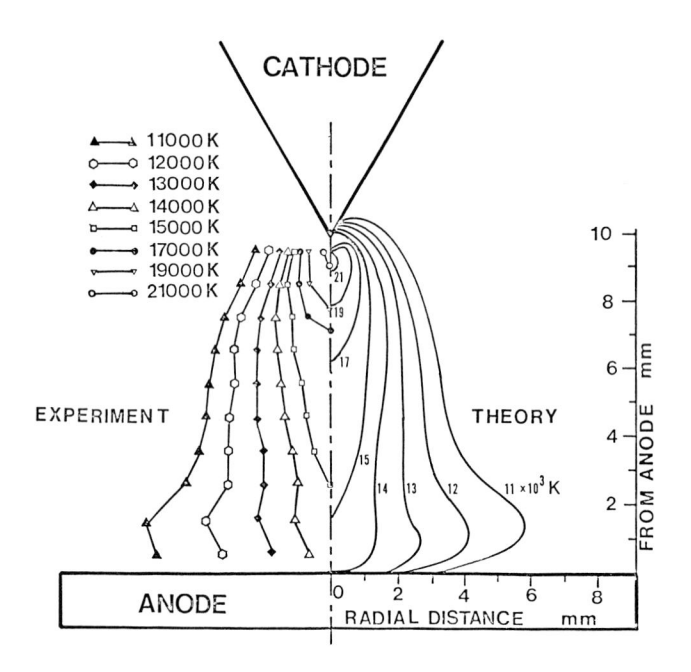

(b) Magnetically deflected arcs.
(c) Magnetically driven arcs.

The first category refers to arcs exposed to strong cross flows so that the arc column bends in the downstream direction if the electrode roots are fixed. At the same time arc length and arc voltage drop increase and for a sufficiently strong flow the required arc voltage may surpass the available voltage, i.e. the arc extinguishes. In order to stabilize an arc in this situation, a magnetic field may be applied so that the drag force exerted on the arc by the flow is balanced by the $\vec{j} \times \vec{B}$ force. Experimental studies on magnetically balanced arcs in cross flow indicate that the magnetic field strength required for balancing the arc is proportional to c^2 where c is the gas speed, i.e. the arc behaves like a solid body as far as aerodynamic drag is concerned. The cross section of the arc assumes the shape of an ellipse with the major axis normal to the flow. Without the balancing $\vec{j} \times \vec{B}$ force, the major axis of the ellipsoidal cross section of the arc is in flow direction. These findings have been confirmed by a number of other studies including detailed analyses. Many refined experimental studies have been reported including balanced arcs in supersonic flow.

The second category deals with magnetically deflected arcs and the secondary effects induced by this deflection. Although the applied magnetic fields exert a strong influence on the arc, the primary stabilizing effect of an arc enclosed in a tube is due to confining walls. In this sense, this type of arc may also be classified as a magnetically influenced, wall-stabilized arc.

The interaction of curved arcs with their own magnetic field can produce similar effects as observed in magnetically deflected arcs.

Magnetically driven arcs are also classified under magnetically stabilized arcs as previously explained. Such arcs have been extensively studied, from a more basic point of view in connection with the phenomenon of retrograde motion. With applications for arc gas heaters in mind, the coaxial, magnetically rotated arc has been of great interest for efficient heating of gases to well-controlled temperature levels in the range from 3×10^3 to 6×10^3 K.

2.1.2 Arc characteristics

The characteristics of low current arcs (I < 50 A) are usually falling provided that the arc can freely expand with increasing current and that there is no severe influence on the arc by evaporation of electrode material. This applies, for example, to free-burning arcs in air or other gases as long as the arc remains relatively short (L ≤ 20 cm). Due to free convection effects, the shape of the column of long arcs changes continuously accompanied by severe voltage fluctuations. Stabilization of the arc in this situation becomes a necessity as discussed in the previous paragraph. Arc stabilization may change the energy balance of the arc drastically so that the resulting characteristic may be rising.

In contrast to low current arcs, high intensity arcs frequently show rather flat or even rising characteristics. The falling trend of a characteristic is a result of the increasing arc conductance with increasing current caused either by an increase of the electrical conductivity (temperature) or the arc diameter, or both. In high intensity arcs this general trend may be overbalanced by disproportionally high losses from the arc to which the arc responds with an increase of the field strength.

The foregoing discussion illustrates that no general prediction about the characteristics of arcs can be made because external conditions as for example stabilizing walls, magnetic of gas dynamic fields, etc. may reverse trends.

Experiments showed that the characteristic of a free-burning high intensity argon arc with a water-cooled anode can be represented by [14]

$$V_{arc} = 4.3 \ I^{0.25} L^{0.3} \qquad (7)$$

I [A], L [cm]

for the following parameter range

$$200 \leq I \leq 2,300 A$$

$$0.5 \leq L \leq 3.15 \ cm$$

For this arc, the characteristic shows a slightly rising trend whereas the characteristics of wall-stabilized arcs also in argon as shown in Fig. 9 indicate at low currents the typical falling trend of low current arcs, changing to a positive slope at higher currents. The falling trend of a characteristic may continue to relatively high currents in gas vortex-stabilized arcs as shown in Fig. 10. The importance of pressure variations on the characteristic is demonstrated in Fig. 11 for a wall-stabilized arc in argon.

Operation of an arc with falling characteristic requires electrical stabilization which will be discussed in the following paragraph.

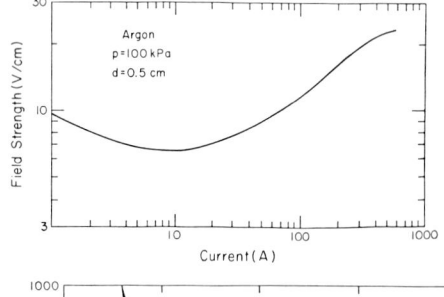

FIG. 9. Measured characteristics of a wall-stabilized argon arc.

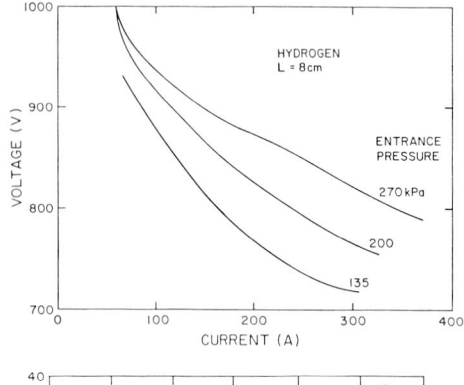

FIG. 10. Measured characteristics of a vortex-stabilized hydrogen arc.

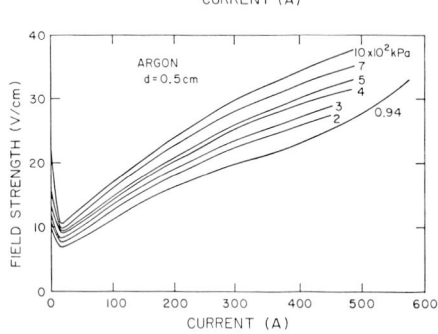

FIG. 11. The infuence of pressure on the characteristics of a wall-stabilized argon arc.

2.1.3 Electrical stability

Arcs with rising characteristics do not require any precautions in the electric circuit in contrast to arcs with falling characteristics. For the following discussion only arcs with falling characteristics will be considered. Figure 12b shows a typical arc characteristic with load line according to the circuit diagram in Fig. 12a. The load line intercepts the characteristic in point A and B but only point A is an electrically stable point of operation. This fact is expressed by the Kaufman stability criterion which states that for a stable point of operation

$$\frac{dv}{di} + R > 0 \tag{8}$$

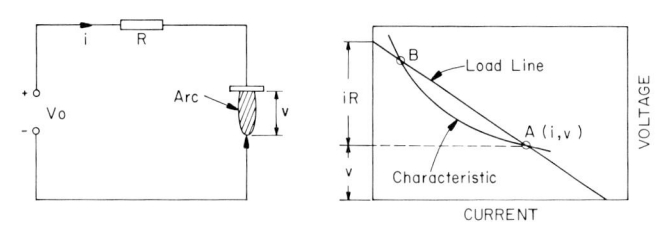

FIG. 12. Electric stability of an arc with falling characteristic.

$$\text{or} \quad R > \left|\frac{dv}{di}\right| \tag{9}$$

According to this inequality the load line must intercept the characteristic from above for producing a stable point of operation. In a more detailed electrical stability analysis the arc inductance and capacitance must be taken into account. A calculation based on small disturbances from the equilibrium state leads to two conditions for arc stability

$$\frac{dv}{di} + R > 0$$

and

$$\frac{1}{L}\frac{dv}{di} + \frac{1}{RC} > 0 \tag{10}$$

The first condition is identical with the Kaufman criterion whereas the second criterion establishes an upper limit for the load resistance

$$R < \frac{L}{C\left|\frac{dv}{di}\right|} \tag{11}$$

L represents the arc inductance which is assumed to be in series with an induction-free arc and C the capacitance of the arc configuration which is assumed to be in parallel with a capacity-free arc configuration. The second criterion is not critical for high current arcs because L is an increasing function of the current, C is relatively small and $|dv/di|$ does not assume extremely large values in high current arcs.

2.2 Thermal RF-Discharges

An rf-discharge may be maintained either by capacitive or inductive coupling with the power source. In the case of capacitive coupling the high frequency electric field is responsible for maintaining the discharge. For this reason this type of discharge is known as the E-discharge. In contrast, an inductively coupled discharage is maintained by the time-varying magnetic field, denoted as H-discharge.

For the generastion of thermal plasmas, the H-discharge is by far more important because the E-discharge which relies on displacement currents for establishing a closed electrical circuit, requires extremely high frequencies for producing a thermal plasma.

Since the appearance and the properties of the plasma produced in a high-pressure ($p \geq 100$ kPa) H-discharge resemble that of an electric high intensity arc, this type of discharge is sometimes referred to an as "induction arc" or an "electrodeless arc".

One of the most important advantages of both types of rf-discharges is the absence of contaminants from the electrodes in the plasma because the electrodes or the magnetic field coil are not in physical contact with the plasmas. This makes rf-discharges particularly suitable for plasma chemistry and plasma processing applications for which a clean plasma is essential. Unfortunately, the coupling between power supply and plasma is less efficient than in the case of electric arcs.

In the following, only the H-discharge will be discussed with exception of some comparisons with the E-discharge.

2.2.1 Basic considerations

For the following considerations it will be assumed that the plasma in the tube sketched in Fig. 13 is already esta-

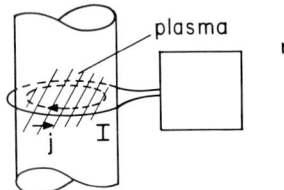

FIG. 13. Schematic of an H-discharge with single-loop coil.

blished. The time-varying magnetic field caused by the rf-current flowing in the coil induces an electric ring field which in turn, drives a current of density $\vec{j}$ since the plasma is a conducting medium. This induced current consists of closed loops directed opposite to the primary current in the coil. The current distribution and the associated temperature distribution show off-axis peaks as indicated in Fig. 14.

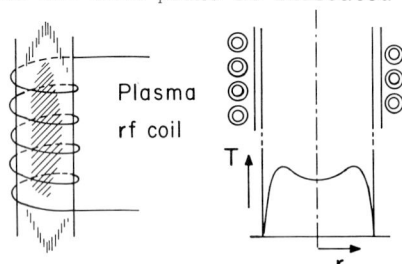

FIG. 14. Appearance of the plasma and schematic temperature distribution in an H-discharge.

The off-axis peak of the current density is primarily due the skin effect. Since the plasma has a relatively high electrical conductivity, the alternating magnetic field cannot penetrate the plasma, especially at very high frequencies. This fact is usually expressed by the skin depth δ which is a measure for the penetration of the rf-field into the plasmas. The skin depth is given by

$$\delta = \left(\frac{2}{\sigma \omega \mu}\right)^{\frac{1}{2}} \qquad (12)$$

where σ represents the electrical conductivity of the plasma, $\omega = 2\pi f$ the radian frequency of the power supply and μ is the permeability of the plasma. Although Eq. 12 holds strictly only for plane materials with uniform electrical conductivity, this equation is useful for examining trends. An increase of the frequency as well as of the electrical conductivity reduces the skin depth.

The non-uniform current density distribution with an off-axis peak close to the walls of the discharge vessel gives rise to enhanced heat dissipation (j^2/σ) close to the wall. This effect in combination with radiative cooling of the plasma causes an off-axis peak in the temperature distribution as indicated in the sketch of Fig. 14.

The H-discharge may be operated in a wide frequency range from approximately 100 kHz to almost 100 MHz. It has been suggested to divide this frequency range (or wavelength range: $\lambda = c/f$ where c is the speed of light) into the following intervals assuming a characteristic length of L = 1m.

(1) $\frac{\lambda}{L} \gg 100$ (f << 3 MHz) low frequency discharge

(2) $100 > \frac{\lambda}{L} > 10$ (3 MHz < f < 30 MHz) high frequency discharge

(3) $\frac{\lambda}{L} \ll 10$ (f >> 30 MHz) ultra high frequency discharge

As the dissipated power in the plasma increases, the plasma approaches LTE, i.e. the state of a thermal plasma. Fig. 15 shows qualitatively how the power dissipation per unit volume varies as a function of the frequency. The figure shows also a comparison with the E-discharge.

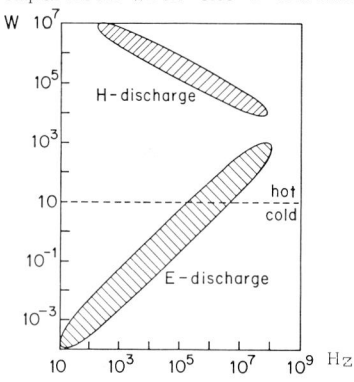

FIG. 15. Power dissipation in the H- and E-discharge.

Stability considerations discussed in a review by Eckert [15] seem to indicate that the skin effect has a stabilizing influence on the H-discharge provided that the ratio $R/\delta \geq 1.75$ where R is the discharge tube radius and δ the skin depth (Eq. 12).

2.3 Plasma Torches and Plasma Jets

In contrast to current-carrying discharges, plasma jets represent field-free plasmas which are characterized by a more or less rapid decay of the temperature and the associated plasma state. Such plasma jets may be produced by high intensity d.c. or a.c. arcs or by high intensity rf-discharges in so-called plasma torches. The art of designing a plasmas torch lies mainly in finding ways of transferring the energy dissipated in the arc or in the rf discharge in the desired way to the flowing gas within the restraints imposed by physical laws.

The following considerations will be restricted to plasma jets generated by means of d.c. arcs because the underlying physical principles are the same for d.c. and a.c. arcs as well as for rf-discharges.

The attractiveness of plasma jets for actual and potential applications is due to the ease with which high temperature levels can be produced, far beyond the levels feasible in conventional combustion processes and without the contamination due to combustion products. Although the application of electric arcs for heating gases to high temperature levels has been known for more than 75 years, the development of plasma torches is still - to a large degree - an empirical science. With a very few exceptions plasmas torches have defied a comprehensive analytical treatment. Even in extremely simple, laboratory - type arc configurations, the analytical treatment may be faced with unsurmountable problems. In particular, there are three areas in which basic knowledge is still lacking:
   (a) Interaction of arcs with magnetic and/or flow fields.
   (b) Effects in the electrode regions and at the electrodes.
   (c) Thermodynamic state of the arc plasma (deviations from LTE).

Additional complexities introduced by the "mission-oriented" design and secondary effects during operation of actual plasma torches aggravates the situation further. In spite of these problems and the lack of theoretical guidance, experimental ingenuity produced astonishing results over the past 25 years.

Plasmas torches have been designed for a wide spectrum of applications including chemical and material processing. They are successfully employed in extractive metallurgy, for welding, cutting, spraying, surfacing, spheroidizing; and they are also used in space-related applications, for example, reentry simulation and ablation studies. The latter application provided a strong impetus for research in the field of plasma torches, especially during the late fifties and the sixties. The power levels at which plasma torches are operated varies from a few kW (typical for laboratory type plasma torches) to several MW or tens of MW. Chemical processing, for example, frequently requires heating of large gas volumes to modest enthalpy (temperature) levels whereas aerospace applications need extremely high enthalpy levels. The first type of plasma torch (low enthalpy device) may operate as a highly efficient energy conversion device in contrast to high enthalpy plasma torches which are inherently inefficient.

The increasing interest in arc heater technology and its enormously increasing activities starting in 1955 are reflected by the vast number of papers published during the past 25 years. Developments up to 1960 are summarzied in [16], covering arc gas heater applications for chemical synthesis, refractory processing, with special emphasis on reentry simulation and space propulsion. The increasing interest in reentry simulation and space propulsion in the early sixties is documented by a number of technical reports [17-22] and by two AGARD volumes devoted to arc heaters and MHD accelerators [23]. Aerospace related developments in the U.S. continued to dominate the scene in the later part of the sixties [24-32], leveling off in the early seventies [33-37].

The development and the application of arc gas heaters and plasma torches in areas which are not aerospace related are difficult to assess because many promising developments in industry are considered as proprietary. The patent literature, however, indicates that there has been increasing activity in this field over the past 25 years. The various designs and claims about the performance of plasma torches in the patent literature will not be discussed in this review.

Most of the available, pertinent literature on arc gas heaters and plasma torches in non-aerospace fields has been published abroad [38-45].

2.3.1 Classification and performance of plasma torches.

In a plasma torch electric energy is converted into thermal and kinetic energy as shown schematically in Fig. 16. The pre-

ENERGY CONVERSION PROCESS

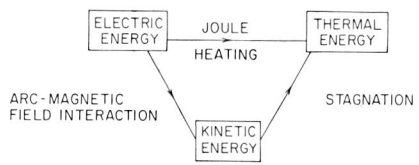

FIG. 16. Schematic of the energy conversion process in an arc gas heater.

valent energy conversion path is usually from electric energy over joule heating to thermal energy. Only in cases of strong interaction of the arc with its own or an externally applied magnetic field is an appreciable fraction of the available electric energy used for acceleration of the plasma. By stagnation, the kinetic energy may be also converted into thermal energy.

The principally different requirements of the performance capabilities of plasma torches used for reentry simulation and related tasks and those desired for chemical and material processing suggest dividing plasma torches into two broad categories. Aerospace applications require plasma torches designed for extremely high enthalpy and high impact pressure levels. The efficiency in such applications is of minor importance as long as the necessary investment for the power supply remains within reasonable limits. Testing (for example reentry simulation) lasts usually only for seconds or at most for minutes.

In contrast, applications in chemical engineering and in material processing are mainly concerned with relatively low enthalpy levels and pressure levels around atmospheric pressure, but the efficiency is probably the most important consideration because such torches should run continuously for weeks or even months. Since this type of torch is of primary interest in this review, the following discussion will be restricted to low enthalpy plasma torches.

The previous classification does not imply that arc gas heaters which fall between these two extremes are non-existant or technically not viable. In fact, a number of existing plasma torches fall somewhere between these two extremes. Plasma torches, for example, in which the plasma column is forced through a constrictor either attached or separate from the anode, belong in this category.

The relatively modest, but uniform enthalpies needed in certain applications (for example in chemical processing) suggest a distribution of the heat dissipation over larger volumes than provided by the natural size of an arc. This goal may be reached in two different ways, either by rotating an arc or by expanding the column of an arc. Both approaches have been pursued and will be briefly discussed at the end of this paragraph. A plasma torch is probably the simplest tool for generating a high temperature plasma. Fig. 17 shows schematically the

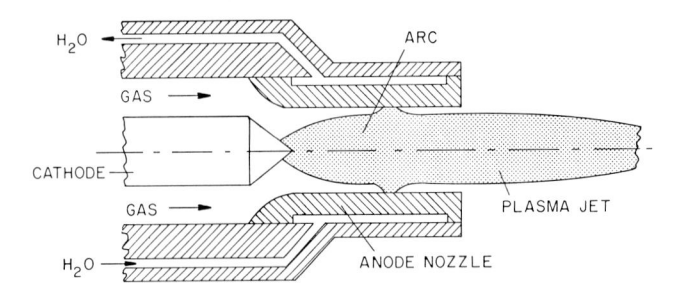

FIG. 17. Typical electrode configuration of an arc plasma torch.

essential components of a plasma torch. The arc is initiated between the tip of the cathode and the water-cooled anode. The working gas is introduced either axially of with an additional swirl component. The latter improves arc stability in the vicinity of the cathode and rotates the anode root. The gas heated by the arc emanates as a plasma jet from the torch orifice.

The plasma jet may be laminar or turbulent depending on the chosen arc parameters. In the laminar regime (relatively small mass flow rates) the torch operates quietly and the highly luminous plasma jet may reach lengths up to 30 cm in atmospheric air at sufficiently high power levels. Transition to turbulence (relatively high mass flow rates) is characterized by increasing audible noise and decreasing length of the plasma jet. The maximum temperature in the plasma jet is a function of the operating parameters and may vary from 8,000 to 20,000 K close to the nozzle orifice. Since the plasma jet is a field-free plasma, the plasma temperature decays rapidly with increasing distance from the nozzle orifice, especially when turbulent mixing enhances the energy exchange between plasma jet and the surrounding atmosphere. A reduction of the energy exchange with the ambient atmosphere (for example by lowering the ambient pressure) produces substantially longer plasma jets.

The flow velocities in the jet may range from almost zero to sonic velocities. If the conventional anode nozzle is replaced by a supersonic nozzle, a supersonic plasma flow may be produced.

Plasma torches have been designed for power levels from a few kW up to approximately 1 MW. Thermal efficiencies are in the range from 30 to 90%. Usually inert gases or their mixtures are used as working fluids. Hydrogen, ammonia, hydrocarbons, oxygen and other corrosive gases can also be used as working fluids if the necessary precaution is taken to protect the electrodes. Frequently, corrosive gases are introduced into the plasma flow downstream of the anode orifice to circumvent electrode problems.

For certain applications a magnetic field coil surrounds the anode which distributes the anode heat load over a larger area and in this way the lifetime of the anode is increased and the level of contamination of the plasma is reduced.

For applications which require high specific heat fluxes (for example welding or cutting) it is customary to transfer the arc to the work piece. In this transferred mode the work piece is the anode and the nozzle serves as arc constrictor.

In the following figures some typical data are shown, obtained with a small, modified, commercial plasmas torch [46]. Fig. 18 shows a cross-section of the torch without the gas flow passage. The inside nozzle diameter is 6.3 mm. Velocity and

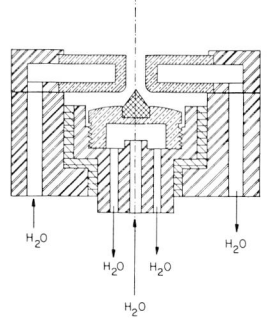

FIG. 18. Actual design of an arc plasmas torch (schematically).

enthalpy distributions at a distance of 5 mm from the nozzle are shown in Fig. 19 and Fig. 20, respectively. For these

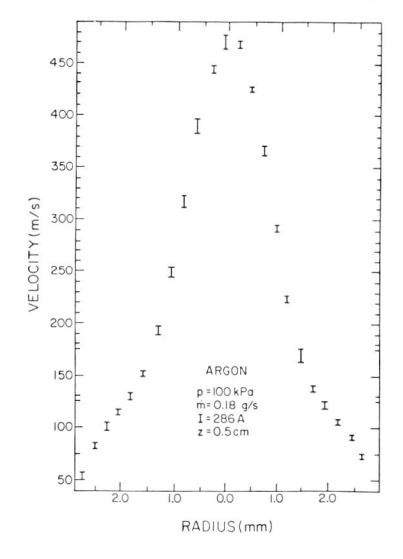

FIG. 19. Velocity distribution in a plasma jet.

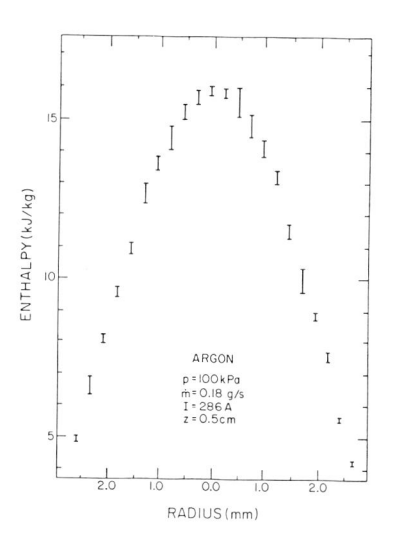

FIG. 20. Enthalpy distribution in a plasma jet.

measurements an enthalpy probe [47] was used. The vertical bars in these figures reflect the experimental error. The corresponding temperature distribution, measured spectrometrically, is shown in Fig. 21. For comparison, temperature distributions at lower currents are also included in this figure. For the highest current (286 A) the power input to this torch is approximately 5.5 kW and the efficiency is close to 50%.

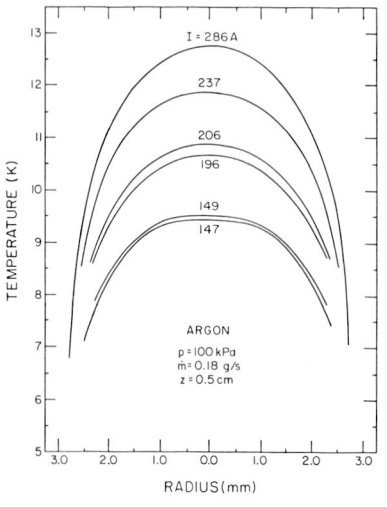

FIG. 21. Temperature distribution in a plasma jet.

Due to the relatively small mass flow rate the plasma jet is laminar at this current level.

In the late fifties and early sixties the magnetically spun arc operated in a coaxial electrode configuration attracted great interest. A sketch of such a torch is shown in Fig. 22 [18]. The magnetic field coil wrapped around the anode produces

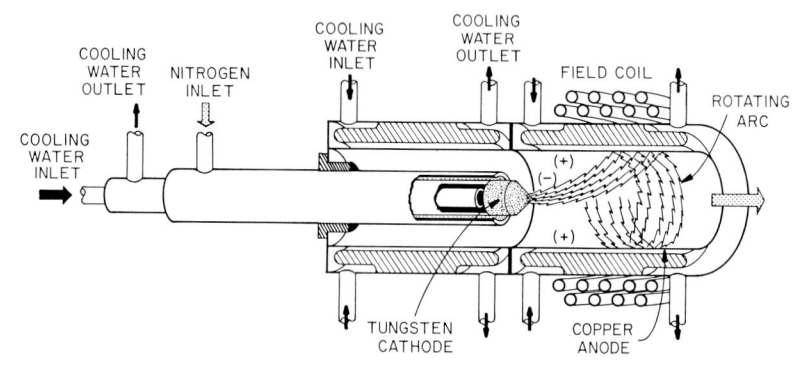

FIG. 22. Schematic of a magnetically spun arc torch.

a magnetic field with a component perpendicular to the current flow in the vicinity of the anode. The $\vec{j} \times \vec{B}$ force in this region drives the arc in azimuthal direction and a constant velocity of this magnetically spun arc is established by a balance between magnetic driving force and drag force acting on the column due to its motion relative to the surrounding gas. It was anticipated that this device would provide an efficient and almost uniform heating of large gas volumes blown through the coaxial gap between the electrodes by spinning the arc with sufficiently high velocities. It was further argued that with increasing spinning velocities an increasing fraction of the annular gas volume would be ionized and eventually the arc would fill the entire gap between the electrodes optimizing the energy exchange between plasma and cold gas. Unfortunately, the expected high enthalpy levels could not be confirmed, because the arc continues to rotate as a rather constricted spoke, resulting in relatively poor energy exchange between arc and cold gas.

Nevertheless, this arrangement offers four features which render it very attractive as arc torch for chemical processing.
- (a) Rotation of the arc distributes the anode heat load and reduces anode erosion in this way.
- (b) The enthalpy level of the heated gas can be easily controlled by monitoring arc current and spinning velocity.
- (c) Relatively large volumes of gas can be heated to fairly uniform temperatures.
- (d) Rotation of the arc increases the potential drop so that more power can be delivered to the gas at a given current.

Another possible arrangement of a rotating arc device is shown in Fig. 23. In this case the electrodes consist of two concentric metal rings so that the arc current flows essentially in radial direction providing a strong interaction with the applied magnetic field.

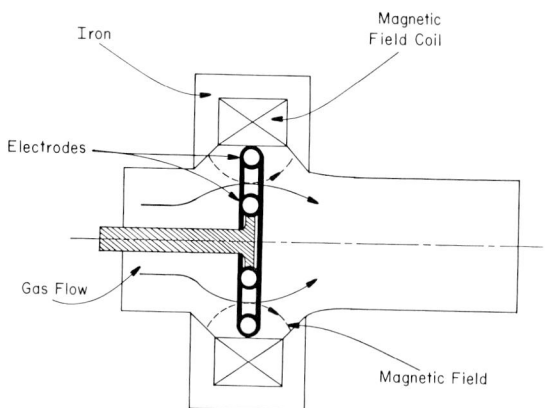

FIG. 23. Concentric electrode arrangement for a magnetically spun arc torch.

REFERENCES

1. W. Finkelnburg, H. Maecker, "Electrische Bögen und thermisches Plasma" Encyclopedia of Physics, Vol. XXII, 254 (1956), Springer-Verlag. Germany.
2. H.R. Griem, Plasma Spectroscopy, McGraw-Hill Book Company, New York, 1964.
3. R.H. Huddlestone and S.L. Leonard, Editors Plasma Diagnostic Techniques, Academic Press, New York and London, 1965.
4. W. Lochte-Holtgreven, Editor Plasma Diagnostics, North-Holland Publishing Company, Amsterdam, 1968.
5. H.W. Drawin, High Temperature-High Pressures, Vol. 2, 359 (1970).
6. M. Mitchner and C.H. Kruger Jr., Partially Ionized Gases, John Wiley and Sons., New York 973.
7. H. Mäcker, Z. Physik, 141, 198 (1955).
8. E. Pfender, "Electric Arcs and Arc Gas Heaters", Ch. 5 in Gaseous Electronics, Vol. 1, 291 (1978), ed. M.N. Hirsh and H.J. Oskam, Academic Press, New York, 1978.

9. H. Edels, "Properties of the High Pressure Ultra High Current Arc", Proc. of the Eleventh Internat. Conf. on Phen. in Ionized Gases, Invited Papers, 9, Prague Czechoslovakia, Czechoslovak Academy of Sciences, Institute of Physics, 18040 Prague 8, Na Slovance 2, CSSR, 1973.
10. O. Schoenherr, Elektrotechn. Z. 30, 365 (1909).
11. H. Gerdien and A. Lotz, Wiss. Veröff, Siemens-Konz. 2, 489 (1922).
12. G. Ecker, "Electrode Components of the Arc Discharge", Erg. D. ExaktenNaturwiss., Bd.33, 1 (1961), Springer-Verlag, Germany.
13. K.C. Hsu, K. Etemadi, and E. Pfender, J. Appl. Phys. 54, 1293 (1983).
14. R.C. Eberhart and R.A. Seban, Int. J. Heat Mass Transfer, 9, 939 (1966).
15. H.U. Eckert, High Temp. Sci., Vol. 6, No. 2, 99 (1974).
16. R.R. John and W.L. Bade, "Recent Advances in Electric Arc Plasma Generation Technology", ARS Journal, No. 31, 4 (1961)..
17. R.C. Eschenbach et al., "Performance Improvement of Air Heaters for Aerodynamic Wind Tunnels", Air Force Flight Dynamics Lab. Rept. 65-87, Linde Company, Division of Union Carbide, 1965.
18. D.R. Boldman, C.W. Shepard, and J.C. Fakan, "Electrode Configurations for a Wind-Tunnel Heater Incorporating the Magnetically Spun Electric Arc", NASA TN D-1222, 1962.
19. G.L. Cann et al., "Thermal Arc Jet Research", Aeronautical Systems Division, Technical Documentary Report No. ASD-TDR-63-632, 1963.
20. C.E. Shepard, V.R. Watson, and H.A. Stine, "Evaluation of a Constricted-Arc Supersonic Jet", NASA TN D-2066, 1964.
21. G.L. Cann, R.D. Buhler, R.L. Harder, and R.A. Morre, "Basic Research on Gas Flows Through Electric Arcs-Hot Gas Containment Limits", ARL 64-69, 1964.
22. G.L. Marlotte, G.L. Cann and R.L. Harder, "A Study of Interactions Between Eletric Arcs and Gas Flows", ARL Report 68-0049, 1968.
23. AGARDograph 84, Part 1 and 2, 1964.
24. R.C. Eschenbach et al., "Performance Improvement of Air Heaters for Aerodynamic Wind Tunnels", Air Force Flight Dynamics Lab. Rept. 65-87, Linde Company, Division of Union, Carbide, 1965.
25. V.R. Watson, "Comparison of Detailed Numerical Solutions with Simplified Theories of the Constricted-Arc Plasma Generator", Proc. of the 1965 Heat Transfer and Fluid Mechanics Institute, Stanford Univ. Press, 24 (1965).
26. J.W. Vorreiter and C.E. Shepard, "Performance Characteristics of the Constricted-Arc Supersonic Jet", Proc. 1965 Heat Transfer and Fluid Mechanics Institute, Stanford Univ. Press, 42 (1965).
27. C.E. Shepard, J.W. Vorreiter, H.A. Stine and W. Winovich, "A Study of Artificial Meteors as Ablators", NASA TN D-3740, 1967.
28. C.E. Shepard, D.M. Ketner and J.W. Vorreiter, "A High Enthalpy Plasma Generator for Entry Heating Simulation", NASA TN D-4583, 1968.
29. G.L. Marlotte, G.L. Cann, and R.L. Harder, "A Study of Interactions Between Electric Arcs and Gas Flows", ARL Report 68-0049, 1968.
30. J.C. Beachler, "Design and Shakedown Operation of the Air Force Flight Dynamics Laboratory's 2 Ft (4 Megawatt) Electro-Gasdynamic Facility", Air Force Flight Dynamics Laboratory

Rept. 68-3, Wright-Patterson Air Force Base, Ohio 1968.
31. R.T. Smith and J.L. Folek, "Operating Characteristics of a Multi-Megawatt Arc Heater Used with the AFFDL Fifty Megawatt Facility", Proc. of the 15th Annual Tech. Meeting, Inst. of Environ. Sciences, 281, 1969.
32. R. Richter, "Ultra-High Pressure Arc Heater Studies", AECD TR 69-180, Arnold Engineering Develop. Center, Arnold Air Force Station, Tennessee, 1969.
33. R.L. Harder and G.L. Cann, AIAA Journal, Vol. $\underline{8}$, No. 12, 2220 (1970).
34. R. Richter, "Ultra-High Pressure Arc Heater Studies, (Phase III)", AECD TR 70-106, Arnold Engineering Develop. Center, Arnold Air Force Station, Tennessee, 1970.
35. C.E. Shepard, AIAA Journal, Vol. $\underline{10}$, No. 2, 117 (1972).
36. G.L. Cann, "An Experimental Investigatiaon of a Vortex Stabilized Arc in an Axial Magnetic Field", ARL 73-0043, 1973.
37. J.H. Painter, "High-Pressure Arc Heater Electrode Heat Transfer Study", AIAA Paper N. 74-731, AIAA/ASME Thermophysics and Heat Transfer Conf., 1974.
38. B. Gross, G. Grycz and K. Miklossy Plasma Technnology, Iliffe, London & SNTL, Prague, 1969.
39. U. Landt, "Entwicklungen auf dem Gebiet der anorganischen Plasmachemic Teil 1; Reaktionene im Plasmastrahl", Chemie Ing.-Techn. 42, Jahrg. Nr. 9/10, 617 (1970).
40. I.G. Sayce, "Plasma Processes in Extractive Metallurgy", Advan. Extr. Met. Refining, Proc. Int. Symp. 241, 1971.
41. M.L. Thorpe, "High Temperature Technology and its Relationship to Mineral Exploitation", Advan. Extr. Met. Refining, Proc. Int. Symp. 275, 1971.
42. A.S. Anshakov, M.F. Zhukov, and A.N. Timoshevsky, "Arc Dynamics in Arc Heater Tunnel", Proc. of the Eleventh Int. Conf. on Phen. in Ionized Gases, Prague, Czechoslovakia, Contributed Papers, 225, Czechoslovak Academy of Sciences, Inst. of Phys., 18040 Prague 8, Na Slovance 2, CSSR, 1973.
43. A.G. Shashkov and O.I. Yas'ko, IEEE Transaction on Plasma Science, Vol. $\underline{PS-1}$, No. 3, 21 (1973).
44. A.S. Shaboltas and O.I. Yas'ko, J. Eng. Phys., Vol. $\underline{19/6}$, 1529 (1974).
45. L.I. Sharakhovskii, J.Eng. Phys., Vol. $\underline{20/2}$, 222 (1974).
46. C. Boffa and E. Pfender, "Enthalpy Probe and Spectrometric Studies in an Argon Plasma Jet", HTL TR No. 73, University of Minnesota, 1968.
47. J. Grey and P.F. Jacobs, AIAA Journal, Vol. $\underline{5}$, No. 1, 84 (1967).

DIAGNOSTICS UNDER THERMAL PLASMA CONDITIONS

P. FAUCHAIS, J.F. COUDERT, A. VARDELLE, M. VARDELLE, J. LESINSKI
Equipe Thermodynamique et Plasmas, Laboratoire Ceramiques Nouvelles
LA 320 Université de Limoges France

ABSTRACT

A number of diagnostics problems in the low temperature plasma generators have been analysed with special attention to the plasma spraying process, and a general approach to these problems has been proposed. The advantages, difficulties and limitations of optical nonintrusive methods have been discussed. Examples based on authors experience during last few years, are given.

I - INTRODUCTION

Since a few years the interest for the use of thermal plasmas in industry is really increasing. At the beginning plasma spraying at atmospheric pressure was the only technique used in industry /1/, specially to spray refractory metals and ceramics. It has then be extended to vaccuum spraying mainly in aeronautic and nuclear industries /2/ for alloys such as MCrAlY and companies, owing to the excellent quality of the deposits (better than the cast alloys) forsee the possibility to realize massive pieces with that technique. If a decade ago, Westinghouse /13/ was the only company developping for sale plasma generators for gas heating at a level of 2 MW about , today Plasma Energy Corp. /4/ and Accurex in U.S.A. /5/, SNIAS-Jeumont Scheider in France /6/, Hüls in B.D.R. /7/, SKF in Sweden /8/ offer such devices. Also the plasma furnaces are now at the level of pilot plants with SKF in Sweden for sponge iron, Middelburg in South Africa (process of Foster Wheeler /10/) for ferrochromium extraction as well as Mintech /11/ with its own technique and VEB Edelstahlwerke in DDR /12/ (commercialized by Voest Alpine in Western countries) for the refusion of iron pellets. In such a development, scientific research lags behind industry probably because industries, at such levels of investments and of power, were not willing to develop sophisticated and expensive methods of measurements (some of which being still under development). If a lot of plasma diagnostics started about 20 years ago, at that time it was a very laborious and tedious work to get reasonable amount of data and some measurements were simply impossible because in laboratories, there were no desk computers, lasers, OMA, interference filters, holographic gratings ...That is why it is only since a few years that such measurements have been gathered on experiments with thermal plasma at a power level of a few tens of kW such as spraying devices or RF generators. We will then present these measurements techniques through the experience we have developed with plasma spraying generators at atmospheric pressure. In such a device we can distinguish five zones (fig. 1).

```
                                            coating
     torch      plasma                      ▨▨        substrate
   ┌─────┐                                  ▨▨      ╱
   │     │─ ─  ──·∵⋅∴∵⋅∴∵⋅∴∵⋅∴·─ ─ ─ ─ ─    ▨▨ ─ ─ ─
   └─────┘                                  ▨▨      ╲

   │    I    │      II      │     III          │ IV │      V
   ────────────────────────────────────────────────────────────────────
    voltage    temperature   surface temperature      temperature
    current    velocity      velocity                 thermal conductivity
    flow rates enthalpy      (size)                   surface quality

                                                 temperature
```

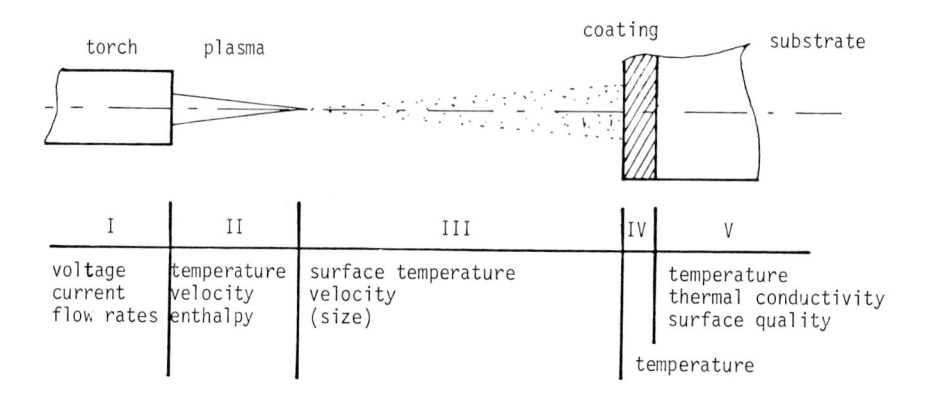

FIG. 1

In table I, we have summarized what we are able now to measure by optical measurement and in this paper we will comment successively measurements of the plasma and of the particles in the first three zones described in table I.

TABLE I. The measurable parameters in the different zone of a plasma jet used for spraying at atmospheric pressure.

Zone I :
inside the
generator :
Arc characteristics, gas and water mass flow rates, energy balance correlated to electrodes dimensions, plasma gas nature, flow rate and injection

Zone II :
Plasma core

Measured parameters for pure plasma
By emission spectroscopy : excited species and electron densities, excitation and rotation temperatures, equilibrium conditions. Conditions of measurement : cylindrical symmetry
By laser anemometry of small particles : plasma gas velocity with difficulties and a low precision

Measured parameters for plasma and carrier gas
By emission spectroscopy : same parameters as for pure plasma but the carrier gas injection has to be symmetrical too - determination of the equilibrium perturbation and of the temperatures and populations distributions
By laser fluorescence : relative penetration of the carrier gas if it contains a tracer

Measured parameters for plasma and particles
For the plasma : if the injection is symmetrical the parameters modification , the vaporization of certain species through the creation of new excited species
For the particles :
. by laser anemometry : their velocity with a precision seriously lowed for diameters smaller than 10 μm
. by laser fluxmetry : particles trajectories (down to 5 μm)

Zone III :
plasma plume
For the plasma : with difficulties, by emission spectroscopy rotation temperatures (down to 3000 K), kinetic temperature through probes (thermocouples, melting points...)
For the particles : as in zone II velocity and flux + :
. surface temperature by statistical pyrometry in flight (questionnable precision)
. size evolution of the particle by laser techniques (low precision)

Zone IV :
Sprayed deposits
While spraying : temperature of the surface (pyrometry) and of the interface with the substrate (thermocouples)
After spraying : mechanical and structural properties, thermal properties, porosity

Zone V : deposit-
substrate interface :   Adhesion, boundings ...

II - OPTICAL DIAGNOSTIC FOR TEMPERATURES MEASUREMENTS

II - 1 - Introduction

Generally speaking the optical techniques are based on the analysis of the light emitted by the plasma jet. Their main interest is that they are non pertubating methods, but on another hand, according to the high complexity of the medium under analysis, many problems may occur and have to be discussed.

First of all, we may distinguish two kinds of complementary techniques, which are to be chosen according to the region of the plasma we have to look at.

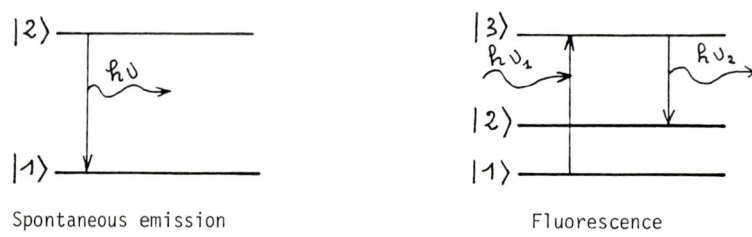

Spontaneous emission   Fluorescence

FIG. 2

In the core of the plasma, the collisional excitation of the various chemical species is so strong that the populations of the radiative quantum states are high enough to emit radiation escaping from the medium. In that case the medium is a self emitting one, and its analysis may be achieved using the classical methods of emission spectroscopy /12/. But if we want to look at the regions where the temperatures are below 3 000 K, (these regions are of primary importance for plasma spraying, for exemple) the jet is no more bright enough and the emission spectroscopy is no longer an usefull technique. So, if we want to look at some low populated radiative states, we have to increase their population by the absorption of the light of an external source. Its spectral power has to be sufficiently high, as in the case of the pulsed tunable dye lasers which are now commercially available. The use of such new devices leads to various techniques, one of them being the laser induced fluorescence (LIF).

II - 2 - Emission spectroscopy

II - 2 - 1 - Abel's inversion

For the optical diagnostics, we dispose of the image of the plasma on the entrance slit of a monochromator and the portion of spectrum falling on the exit plane is then detected using photomultipliers or vidicon detectors. What is observed in reality is the intensity of the light integrated all along the line-of-sight with the contributions of the local volumic emission coefficients, which may be represented, in the case of an axially symmetric plasma, by a given function of the radius. From the line-of-sight intensity contour available from experiment, we may deduce the radial values of the volumic emission coeffcients, which are the only quantities of interest, by using the so called Abel's inversion /13 - 16/. When the plasma is not axially symmetric, the problem becomes a two or three dimensionnal one and the intensity measurements should be done through several directions of the line of sight /17 - 18/.

II - 2 - 2 - Light scattering

When observing the bright part of the plasma we have to deal with some difficulties arising from the fact that the intensity of the light emitted from the regions close to the axis of the jet are more than three decades over the light emitted from the edge, i.e. few millimeters from the axis.

As a consequence of these high gradients of light intensity, the experimental data have to be corrected for the effect of light scattering by the optical system. This is done by conveniently placing diaphragms to reduce the scattered light, or by using a scattering function deduced from the signal given by a source with a well known intensity distribution /19/.

II - 2 - 3 - Automatization of the measurements

A number of methods have been developped to get the line of sight intensity contour. Their differences lie in the way by which the scanning of the jet is done. As a consequence of the high gradient of the jet brightness, the scanning of the line-of-sight intensity contour has to be done with the spatial resolution of about a tenth of millimeter.

A first example is given in /20/ where the plasma torch is moved in a direction perpendicular to the line-of-sight. The displacement is realised by stepping motors and the position of the analysed point is deduced from the signal given by a magnetic captor.

In the method described in /21/ the scanning of the jet is achieved by using a rotating mirror. Another mean proposed by /22/ is to unfold the light beam coming from the plasma by moving an opaque screen placed between the source and the collection optics.

The method we have chosen is to use an optical multichannel analyser (O.M.A.) as the detection system. The spectral image of the jet falling on the detector is scanned by a reading electron beam, the intensity of which is a measure of the number of photons impinging the target consisting of a photodiodes matrix /27 - 28/.

A stepwise scanning of the image in then obtained from which we can deduce the real illumination profile by applying the Shannon theorem and taking into account the O.M.A. apparatus function deduced from previous calibrations. The signal treatment includes the deconvolution by the scattering function and the noise is eliminated by numerical filtering. The resulting smoothed profile of the line of sight intensity is then treated by the classical Abel's inversion method.

II - 2 - 4 - The parameters which can be reached from emission spectroscopy

The intensity of an atomic, ionic or molecular spectral line is proportional to the population of the upper level of the transition and to the energy of quantum. So, when measuring the intensity of the line, one may reach the population of a given quantum state of a given chemical species. On another hand, in the case of local thermodynamic equilibrium (LTE), the calculation of the chemical composition of the plasma and the use of Boltzmann's law allow to relate the different quantum states population to the temperature /12/. As a consequence ,provided the knowledge of transition probabilities and the validity of LTE, the measurements of absolute line intensity may give the temperature of the plasma, through the determination of the volumic emission coefficients of the line after Abel's inversion. As the temperature falls down under approximately 6 000 K the atomic lines become too weak to be usefully detected.

The validity of the LTE model and particularly the Boltzmann distribution of populations of excited levels is related to the electrons concentration, because collisions of all species with electrons are the principal mechanism of equilibrating the energy distribution between quantum states of species in the plasmas (so called collision dominated plasmas).

When electrons concentration is not high enough non equilibrium populations can be observed /23/. In that case it may be shown that the quantum levels of the atoms or ions which are closed to the Ground State are no longer in equilibrium with the rest of the states. Only the levels close to the ionisation limit show an equilibrium distribution which may be related to an excitation temperature. This is the case of partial local thermodynamic equilibrium, where the excitation temperature may be deduced from the so called Boltzmann's plot. Eddy /24/ gives a critical review of spectroscopic diagnostic technics in the case of multi-temperatures plasma. In that case, we may distinguish the electron temperature, the excitation temperatures, the vibrationnal and rotational temperatures and the kinetic temperature related to the translationnal energy of the heavy species. The differences between these various temperatures are indications of the degree of desequilibrium of the plasma.

Usually, the kinetic and rotationnal temperatures are equal, the energy exchange between these degrees of freedom being very efficient as it is caracterized by relaxation times of about nanosecond.

By the use of the Saha's law /24/, the electronic temperature may be obtained if one knows the concentration of the electrons and the populations of some atomic levels close to the ionization limit. The electronic concentration is determined both by continuum radiation analysis and Stark's profile analysis of the atomic lines /23, 25, 28/.

In that case, we have to use a monochromator with a high resolution power and to scan the line spectral profile over a stepwise Set of sampling wavelength. The spatial contour for each spectral intensity has to be inversed by Abel's transform, and then the local values of the spectral emission coefficients of the line have to be treated for the reconstruction of the line shape emitted from different radial locations in the plasma. The full width half maximum (FWHM) of the line is then directly related to the electron concentration /25,26/. In practice, at atmospheric pressure such measurements are limited to electronic concentration higher than about 10^{16} e/cm^3.

The rotational temperature is deduced from the analysis of the spectrum of molecular bands, registered with high resolution in order to separate the different compounds of the rotational lines. Various methods have been proposed including absolute line intensity measurements, relative line intensity measurements and methods the principle of which may be related to the Boltzmann's plot, or where temperature appears like a fitting parameter which optimise the fit of the experimental shape of the band to a numerically simulated one /20, 27, 28/. The measurements on molecular bands allow the knowledge of rotationnal temperatures as low as 3 000 K. This method appears as a complementary one of the diagnostics on atomic lines.

II - 2 - 5 - Laser induced fluorescence

The principle of the method is to populate by optical pumping a quantum level from which other transitions may occur giving rise to an emission of various spectral lines (atomic or molecular) which can be analysed by time resolved spectroscopy. Several papers have been published on the subject regarding the fluorescence of NO or OH in combustion flames /29, 37/.

The advantage of the method is that this technique is a spatially resolved one and, as a consequence, may avoid the Abel's inversion. Moreover, when the allowed transition is coupling two states where the lower is the ground state, one may probe the regions of the jet where temperatures are very low.

Practically, the laser beam which wavelength is tuned on the transition of interest, is focused in the plasma and the fluorescence is observed at right angle. In our case we have probed the NO produced in an argon plasma in which a N_2-O_2 mixture was injected with a very light flow rate. The pumped transition was the (0-0) gamma band at 2 260 Å coupling the X^2 ground state to the first electron excited state, that is the A^2 state. The fluorescence was then observed on the A^2 (v = 0) X^2 (v = 2) transition at 2478 Å.

The pumped transition is coupling in reality rovibronic states and as the rotational relaxation is very fast compared to the time of the laser-plasma interaction, we may assume that the two vibrational levels keep a rotational distribution caracterised by the kinetic temperature of the gas. That is, when analysing the molecular emission of the (0-2) band we may deduce the kinetic temperature of the medium.

Experimentally, we have used a tunable dye laser pumped by a YAG laser delivering 10 pulses per second each of them of 10 nsec duration and about one joule in energy. The light coming out the YAG was then tripled in frequency and entered the dye laser working with coumarine 450. The exit laser beam was then doubled in frequency by using a KDP cristal. The resulting pulses were of about 10 joules at 2 260 Å. The fluorescence light was then detected in the exit plan of the monochromator by using a photomultiplier, the rise time of which was less than 2 nsec. The photomultiplier signal was then sent to a waveform digitizer of 400 MHZ band pass which was connected to a microcomputer.

In order to obtain the rotational temperature, the spectral profile of the molecular band was then fitted by a least square technique /38/.

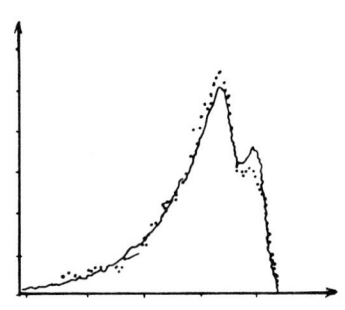

FIG. 3 : Comparison between measured (...) and calculated spectra (0 2) band. The best fit (—) is obtained for Trot = 1170 K.

III - MEASUREMENTS RELATIVE TO PARTICLES

III - 1 - Particles surface temperatures measurements

Papers concerning measurements of the temperature of the particles suspended in high temperature streams can be divided into two groups. The first, comprises papers in which authors measured the mean temperature of the stream of particles. They use different versions of the colour pyrometry method i.e. analyzing the spectral distributions of the thermal radiation of a multiparticle system /39,40,41,42,43/. The second group comprises a number of papers, in which authors measure the temperatures of single particles /44,45,46,47,48,49,50,51,52/ generally by photographic techniques or by analysis of the radiation emitted from the surfaces of single particles.

When photographic detection is employed to determine the total radiation intensity of particles moving so fast that during the exposure time they move a distance much greater than their diameter, the recorded intensity is not only dependent on temperature but also on the ratio of particle diameter to its velocity. English /49/ uses the discrete version of the colour pyrometry method. The radiation of a single particle passing through the slit is analysed by two photomultipliers operating in two spectral regions and two pulses are recorded with the aid of a double beam oscilloscope. The ratio of the amplitudes of these pulses is a function of the temperature of the particle and is independent of its diameter and velocity.

In the already mentioned paper /48/ combusted, pulverized coal has a broad distribution of particle size but an almost uniform temperature. While in plasma spraying, usually the lateral injection of the powder causes a large temperature distribution at a given point, even in spite of a narrow size tolerance of the powders used for plasma spraying. In this case the intensities of the radiation emitted from the surface of single particles of known diameter give informations about their temperature distribution /49,50,51,52/. If the above assumption cannot be applied, one must use the colour pyrometry method in order to eliminate size dependence of signals. Wavelengths must be carefully chosen in function of the spectral distribution of the plasma-particles system. Unfortunately due to a very high plasma radiation, all these methods are unsatisfactory in zone II where plasma particle heat transfer is nearly achieved.

III - 2 - Particles velocity measurements

Optical methods for particle velocity measurements comprise essentially the photographic techniques and the laser doppler anemometry (L.D.A.).

In the photographic methods /53,54,55,56,57,58/ either the image of a particle is followed along its trajectory (at this time, techniques can be classifed as direct photography) or the velocity of the image of a particle is composed with a known velocity (rotating mirror, film motion...).

The major problem appears to be the contrast between the particles and the surrounding gas (overcome in some cases with high intensity sources), chiefly when the particles are "cold".

Photographic methods enable to determine the velocity of single particles, but they are laborious and have a weak spatial resolution. Recent photographic experiments in plasma are rare.

L.D.A. methods /59,60,61/ know actually a tremendous development in plasma processes (see the review paper of Gouesbet /62/).

In L.D.A., the velocity of one particle is measured from the frequency shift due to Doppler effect undergone by a laser beam scattered by the particle. The main advantage of L.D.A. is a high spatial and temporal resolution. The optical configuration most commonly used are the fringe pattern system and the two focus systems (notice that for the last one, the use of laser source is strictly necessary for cold particles). In signals processing systems, one can distinguish between frequency domain (frequency tracker, counter) and correlation methods.

Among these systems, only counters and frequency trackers are able to associate a velocity to a given single particle. They have been successfully used in plasmas /63,64,65,66/. The use of high speed transient recorder with digital computer processing of the stored signal is also very promising /67,68/.

Photon correlation method may be useful in situations where scattering is weak or background noise important /69,70/ and they have been used in plasma situation /71/.

Notice that the total intensity of light scattered by a small particle is a function of its size (cross section) and spatial distribution of the scattered light depends on the ratio of particle diameter to the wavelenght of the incident light. These diffraction patterns have several lobes with the maximums in forward and backward directions and a relative minimum at 90°. So the observation perpendicular to the beam permits to achieve the best spatial resolution but unfortunetaly one should dispose in this case, of powerful laser and of photodetector with high sensitivity.

III - 3 - Particle size measurements

There are a lot of methods available for particle size measurement /72/ (photographic, holographic...) but the only ones which can provide simultaneous informations on particles sizes and velocities are concerned with light scattering ; these methods can be distinguished /73/ in visibility approach /74,75/, absolute intensity methods /76,77,78/ and relative intensity methods /79,80,81,82/.

Until now, these methods have not been developped in plasmas flows as far as we are aware of.

III - 4 - Particle flux measurement

The number of particles travelling at differents points in the plasma jet is generally measured by counting, during a given time, the pulses resulting from the light scattered by the particles passing through a focused laser beam. LDA systems using frequency trackers or counters as signal processors can be extended to particle concentration measurements.

The mean particles trajectories are determined using the maximum of the radial distribution of the particles /83/.

III - 5 - Interpretation of particles parameters measurements

The knowledge of particles parameters (size, surface temperature and velocity) allows a better understanding of plasma particle momentum and energy transfer. For example in plasma spraying, properties of the plasma coating depends largely upon the velocity and the bulk temperature of the particles at their impact on the substrate.

For such measurements, the "good" particle should be inside a given volume in the 3 dimensionnal space : temperature, velocity, diameter. Fig. 4

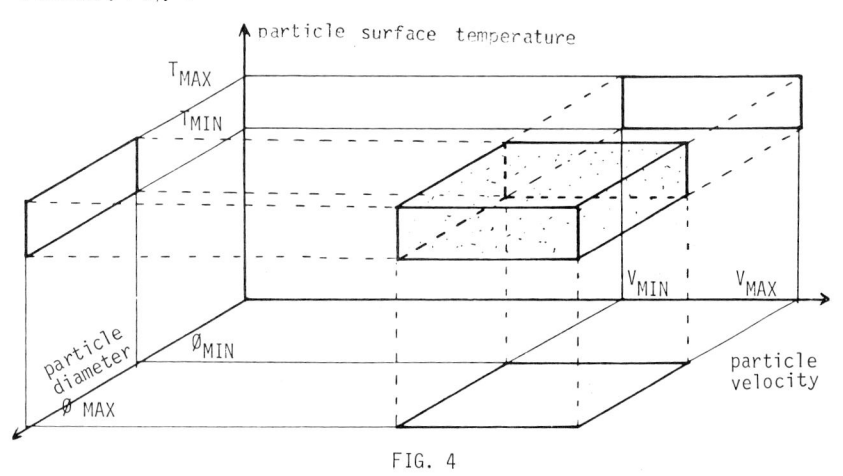

FIG. 4

Each real particle in the jet can be represented in this space by one point. Then, what one has to do is :
  1 - to establish the limits of the "box"
  2 - to put as many particles as possible into the "box".
But of course the parameters of each particle may vary along its

trajectory in the jet and when we are talking about optimas it concerns parameters of the particles just before the impact onto the workpiece and the distribution concerns all particles at that distance regardless their positions relative to the jet axis.

In order to establish the limits of the box, one should analyse the properties of obtained coating relative to the 3 parameters of the particles measured in the region III. In order to put particles into "the box" one should optimise the design of the torch and its operating parameters while measuring 3 parameters of particles in the region 3.The idea to choose the region 3 (or rather the plane just before the impact on the substrate) is based on the fact that in this plane all most important parameters are measurable and relatively easy to interprete.

The proposed procedure, in our opinion is the best practical approach to the optimisation of the process but not to the understanding of the elementary processes which take place in the plasma i.e. in the region II.

To determine the good particles, it is necessary to know for each particle, the surface temperature, the velocity and the size. Notice that these measurements are difficult due to :
- the radiation of the surrounding gas giving rise to background noise
- the small diameters of particles (optical problems)
- the high velocities of the particles
- the necessity of a small measurement volume in order to have one particle in a time in this volume. In practice this means for the real plasma spraying condition that the measuring volume should be smaller than 1 mm^3.

CONCLUSION

We have tried in this paper to summarize our knowledge in the field of optical measurements of thermal plasma and particles. The restrictions we have either for the ranges of temperatures, of particles diameters, of concentrations, or for the precision of our measurements specially for particles surface temperature and diameters are still important and correlated with the location precision requested for the high gradients encountered (up to 4 000 K/mm and 200 m/s/mm). They make the physical understanding of the phenomena involved difficult. However it is clear that such measurements are the only way to improve the use of the plasma generators as well for spraying as for the chemical reactions closely related to the penetration of the reactants in the plasma, to the equilibrium conditions and to the heat and mass transfer. It is also the only mean to correlate with the reality the numerous models actually developped for plasma flows and for heat and momentum transfer between plasma and particles and to get the data for these models. Unfortunately also the high degree of sophistication of the measurements techniques and thus the high price of the equipments does not help a fast develpment of these necessary measurements.

REFERENCES

1. A. VARDELLE, P. FAUCHAIS, M. VARDELLE - Actualité Chimique 10, (1981) 69
2. D. APELIAN, M. PALCIWAL, R.W. SMITH, W.F. SCHILLING - Melting and solidification in plasma spray deposition - internal Report G.E. R and D Schnectady (N.Y.) (1983)
3. M.G. FEY - Electric Arc Plasma Heater for Process Industries, Ind. Heat, June (1976)
4. S.L. CAMATCHO - Tech. Applications Services Corp. Raleigh N.C., U.S.A., "Plasma arc torches for industrial applications" Workshop on Industrial Plasma Developments, Univ. of Sherbrooke, July 21-23 (1983). The corresponding papers have to be asked to the authors.
5. J. HARTMAN - Accurex Corp., Mountain View, CA, U.S.A. cf. /4/
6. M. LABROT - Aerospatiale, Saint Medard en Jalles, France, "High power plasma torch development", cf /4/
7. R. MÜLLER - Chemishe Werke Hüls, Marl, West Germany "The Hüls Arc Process" cf/4/
8. S. SANTEN - SKF Steel Engineering AB, Hofors, Sweden, "Plasma Smelting" cf /4/
9. J.R. MONK - "Application of plasma to metallurgical processes", Vth Int. Symp. on Plasma Chem. (Ed.) R. Waldie Univ. of Edinburgh, GB
10. N.A. BARCZA - Mintek, Randburg, South Africa "Application of transferred arc plasma to the melting of metal fines" cf/4/
11. A.S. BOWDACHYOV, G.I. MEERSON, V.A. KHOTIN, G.N. OKOROKOV - "Design of ceramic crucible plasma furnaces and operating results UIE 9 Cannes 20-24 /10/80 (ed.) CFE, 79 rue Miromesnil, Paris, France
12. First report on measurement of temperature and concentration of excited species in optically thin plasma. I.U.P.A.C. Plasma Chemistry Subcomittee, (eds) P. FAUCHAIS, University of Limoges, France, K. LAPWORTH N.P.L. Teddington G.B. (1979)
13. BARR W.L. - J. Opt. Soc. Am, 52, 885, (1962)
14. CREMERS C.J. and BIRKEBACK R.C. - Appl. Opt. 5, 1057, (1966)
15. DEUTSH M., BENIAMIRY I. - J. Appl. Phys. 54, (1), 137, (1983)
16. DEUTSH M. - Appl. Phys. Lett. 42, (3), 837, (1983)
17. WETZER J.M. - I.E.E.E. Transaction in Plasma Science, Vol. PS-11, N°2, 1983
18. GRAVELLE D., BEAULIEU M., CARLONE C., BOULOS M. - I.S.P.C. 1-4-3, Proceeding p. 108,Montreal July 1983
19. VENABLE W.H., SHUMAKER J.B. - J.O.S.R.T. 9, 1215, (1969)
20. BARONNET J.M. - Thèse d'Etat, Université de Limoges, France Nov(1978)
21. ETEMADI K., PFENDER F. - Rev. Sci. Instrum. 53 (2), 255, (1982)
22. MEHMETOGLY M.T., KITZINGER F., GAUVIN W.H. - Rev. Sci. Instrum. 53, (2), 285, (1982)
23. DRAWIN H.W. - Reactions under plasma conditions, (ed) Venugopalan M., Wiley Interscience (1971)
24. EDDY T.L. - I.E.E.E. Trans. Plasma Science, PS-4, 103-111, (1976)
25. GRIEM H.R.- Plasma Spectroscopy, Mc Graw Hill (1964)
26. GRIEM H.R. - Spectral line broadening by plasmas, Academic Press (1974)
27. COUDERT J.F., BARONNET J.M., CATHERINOT A., FAUCHAIS P. - 7th International Conference on Gas Discharges and their applications, London 31st Aug. 3rd Sept. 1982, Processing, p. 523
28. COUDERT J.F., BARONNET J.M., FAUCHAIS P. - I.S.P.C. 6 A-4-2, Montreal July (1983) Proceedings p. 102
29. DAILY J.W. - Applied Optics, 15 (1976) 955, 16 (1977) 568, 17 (1978) 225
30. CROSLEY D.R. - ed. "Laser probes for combustion chemistry" A.C.S. Symposium Series 134 (1979)

31. LUCHT R.P., LAURENDEAU N.M., App. Opt. 18 (1979) 856, 21 (1982) 3729
32. STEPOWSKY D., COTTERRAN M.J. - J. Chem. Phys. 74 (12) (1981) 6674
33. BRADSHAW J.D., OMENETTO N., BOWER J.N., WINEFORDNER J.D. - App. Opt. 19 (1980) 2709
34. GRIESER D.R., BARNES R.H. - App. Opt. 19 (1980) 741
35. ZIZAK G., BRADSHAW J.D., WINEFORDNER J.D. - App. Opt. 19 (1980) 3631
36. OZAKI T., MATSUI Y., OHSAWA T. - J. Appl. Phys. 52 (4) 1971
37. ALDEN M., EDNER H., HOLMSTEDT G., SVANBERG S., HUGBERG T. - App. Opt. 21 (1982) 1236
38. COUDERT J.F., CATHERINOT A., BARONNET J.M. - I.S.P.C. 6, A-4-1, Montreal July (1983), Proceeding p. 97
39. HANTZSCHE H., Proc. 7th Int. Conf. on Metalisation, London, paper 15 (1973)
40. SHIMANOVICH V.D., SHIPAI A.K., SOLOVIOV B.M., PUZRIAKOV A.F., Proc. Conf. Fizika, Technika Primenenije Nizkotemperaturnoj Plazmy, Alma-Ata, USSR, p. 209 (1970)
41. ANTONOV G.S., LOSKUTOV V.S., SOLOVIOV B.M., SHIMANOVICH F.D., SHIPAI A.K., "Generatory Nizkotemperaturnoj Plazmy" (Energia, Moskow, USSR) p. 490 (1969)
42. NIKOLAIEV A.V., ibidem, p. 579
43. CHARETTE A., MEURUS P., Rev. Int. Hautes Temper. Refract., 17, 45, (1980)
44. GUREVICH M.A., DRESVIN S.V., KALGANOVA J.V., KLUBNIKIN V.S., NIZKOVSKI A.A., "Voprosy Fiziki Nizkotemperaturnoj Plazmy", (Nauka i Technika, Minsk, USSR) p. 580 (1970)
45. DRESVIN S.V., KALGANOVA J.V., Proc. Conf. Fizika, Technika i Primenenije Nizkotemperaturnoj Plazmy, Alma-Ata, USSR, p. 617 (1970)
46. DRESVIN S.V., KALGANOVA J.V., Proc. Conf. Fizika, Technika i Primenenije Nizkotemperaturnoj Plazmy, Alma-Ata, USSR, p. 617 (1970)
47. GUREVICH M.A., STEINBERG V.B., J. Techn. Phys. USSR, XXVII 394 (1958)
48. POPOV A.V., Sixth International Symposium on Combustion, Newhaven, Connecticut (1956)
49. ENGLISH P.E., Acta Imeco, p. 469 (1967)
50. KRUSZEWSKA B., LESINSKI J., Revue de Physique Appliquée, 12, 1209 (1977)
51. VARDELLE A., BARONNET J.M., VARDELLE M., FAUCHAIS P., IEEE transactions on plasma science PS-8, 417 (1980)
52. MEUBUS P., ELAYOUBI, ISPC 6 proceedings edited by M. BOULOS and R.J. MUNZ, Univ. de Sherbrook, CN, 24-28 July (1983)
53. LEMOINE A., LE GOFF P., Rev. Gen. thermique, 85, 33-48 (1969)
54. LEMOINE A., LE GOFF P., Chimie et industrie - Génie chimique, 102, 1304 - 1311 (1969)
55. LEWIS A., GAUVIN W.H., J. of the society of motion picture and television engineers 80, 951 - 958 (1971)
56. WALDIE B., Velocity measurements in and around the coil region of an induction plasma. International round table on study and application of transport phenomena in thermal plasma. Conference Proceedings. Laboratoire des Ultra-réfractaires du C.N.R.S., Odeillo, 12-16 Sept. 1975
57. ANDRE M., les techniques de cinématographie ultra-rapide. Rapport CEA, Centre de Limeil France 1977
58. FRIND G., GOODY C.P., PRESCOTT L.P., Proceedings of 6th international symosium on plasma chemistry edited by M. BOULOS and J. MUNZ, 24-28 july 1983, Montreal pp. 120-125
59. DURST F., MELLING A., WHITFLAW J.H. - Principles and practise of laser doppler anemometry. Academic Press 1976
60. WATRASIEWISZI B.M., BUDD M.J. - Laser doppler measurements, Butterworths and al. 1976

61. DURRANI T.S., GREATED C.A. - laser systems in flow measurements, Plenum Press 1977
62. GOUESBET G., "Particles velocity and diameter measurements under plasma conditions" to be published by IUPAC Plasma Chemistry Subcomittee
63. GOUESBET G., TRINITE M., in Heat and Mass Transfer. 4, 141, (1977)
64. GOUESBET G., TRINITE M., J. Phys. E : Sci. Instr. 10, 1009, (1977)
65. LESINSKI J., MIZERA-LESINSKA B., FANTON J.C. and BOULOS M.I., LDA measurements under plasma conditions. 4th International Symposium on Plasma Chemistry. University of Zurich. Aug. 27/Sept. 1, 1979
66. LESINSKI J., MIZERA-LESINSKA B., JUREWICZ J. and BOULOS M., Particle and gas velocity measurements in a d.c. plasma jet. Presented at the plasma chemical processing session of the 88th AIChE National Meeting, Philadelphia. Pennsylvania. June 8-12, 1980
67. DURST F. and TROPEA C., Digital processing of LDA-signals by means of a transient recorder and a computer. Proceedings of the
68. LDV-conferences, VAESSEN P.H.M., KROESEN G.M.W., SCHRAM D.C., Proceedings of the 6th International symposium on plasma chemistry 24-28 Montreal (1983) paper A-4-7
69. PIKE E.R., p. 246 in Photon Correlation Spectrometry and Velocimetry, Plenum Press, New-York, 1977
70. BIRCH A.D., BROWN D.R. and THOMAS J.R., Journal of Phys. D 8, 438, 1975
71. VARDELLE M., VARDELLE A., BARONNET J.M., FAUCHAIS P., 4th International symposium on plasma chemistry Zurich (1979) proceedings edited by S. VEPRIECK and J. HUTZ, Univ. of Zurich, CH
72. OLDSCHMIDT V.M., "Proceedings of the dynamic flow conference" "Measurements in two phase flow", Institut de mécanique statistique de la turbulence, Marseille 289-319 (1978)
73. GOUESBET G., GREHAN G., 4th international symposium on plasma chemistry Zurich August 27 - Sept. 1st (1979) p. 603 "laser doppler systems for plasma diagnostics : a review and prospective paper"
74. FARMER W.M., Proceedings of the dynamic flow conference, Institut de mécanique statistique de la turbulence, Marseille 373, 396 (1978) "Measurement of particle size and concentration using LDV techniques"
75. DURST F., ELIASON B., Proceedings of the LDA symposium, university of Denmark (1975) "Properties of laser doppler signals and their exploitation for particle size measurements"
76. YULE A.J., CHIGIER N.A., ATAKAN S. and UNGUT A., J. Energy, 1, 4, 220, 1977
77. UNGUT A., YULE A.J., TAYLOR D.S. and CHIGIER N.A., AIAA 16th Aerospace Sciences Meeting, Huntsville, Alabama, January 16-18, 1978
78. CHIGIER N.A., UNGUT A. and YULE A.J., Particle size and velocity measurement in flames by laser anemometer. 17th Symposium (International) on Combustion. Leeds University. August 20-25, 1978
79. WU J.- J. of applied optics 16, 3, 596-600 (1977)
80. CHOU H.P., WATERSON R.M. - LDA symposium, technical university of Denmark, Lyngby, Denmark (1975)
81. GRAVATT C.C. - J. of the air pollut. contr. assoc. 23, 12, (1973)
82. HIRLEMAN E.D. - Optics lett. 3, 19-21, (1978)
83. LOMBARD D., VARDELLE M., FAUCHAIS P. - To be published in Advances in ceramics, volume 4.

MODELLING OF PLASMA PROCESSES

Maher I. Boulos
Department of Chemical Engineering, University of Sherbrooke, Sherbrooke, Québec, Canada, J1K 2R1

ABSTRACT

A review is made of some of the mathematical modelling work carried out over the last few years with the objective of computing the flow and temperature fields for different thermal plasma systems. Typical results are given for d.c. plasma jets, transfered and non-transfered arc plasma reactors and for inductively coupled plasma torches. This is followed by a brief discussion of the important problem of plasma-particle heat transfer.

1. INTRODUCTION

Increasing attention has been given over the last ten years to the mathematical modelling of plasma torches and reactors in view of their use for the thermal treatment of solid particles in order to induce physical and/or a chemical changes. With only a few exceptions, the proposed models are based on the assumption of local thermodynamic equilibrium (LTE). They also dealt mostly with dilute gas-solid systems thus neglecting any plasma-particle interactions effects. This simplified the problem considerably by allowing the reactor modelling to be carried out in two independent steps. The first dealing with the calculation of the flow and temperature fields in the reactor in the absence of solid particles, followed by computations of the trajectories and the temperature histories of single particles as they are injected into the flow.

The first part of this paper will be mainly concerned with models developed for the calculation of the flow and temperature fields in d.c. plasma jets, transfered and non-transfered arc reactors and in inductively coupled r.f. plasmas. This will be followed by a brief discussion of heat transfer to a single particle under plasma conditions.

It should be emphasized that this paper is by no means a comprehensive review of the literature on such a vast subject. The examples given have been chosen merely to highlight the state-of-the-art in this field.

2. MODELLING OF ARCS AND D.C. PLASMA JETS

The principal difference between the modelling of an arc and a d.c. plasma jet is that the former is current carrying while the latter is not. This has direct consequences not only on the assumptions involved but also on the equations to be solved.

2.1 Arcs

In spite of the intensive research effort that has been devoted over the last thirty years to the study of electric arcs, the mathematical modelling of arcs has developed rather slowly due to important difficulties encountered in the analytical description of the electrode regions which are characterized by their particularly small dimensions (of the order of 10^{-2} mm), the presence of steep temperature gradients and important non-equilibrium effects. It is not surprising that the first theoretical analysis of an electric arc which can be traced back to the work of Elenbaas (1) and Heller (2) in 1935 was limited to the fully developed arc column in an asymptotic equilibrium flow regime.

Accurate predictions of the flow and temperature fields in the arc column were later obtained by the numerical solution of the full, two-dimensional system of the mass, momentum and energy conservation equations as reported by Watson and Pegot (3) in 1967. The principal assumptions used can be summarized as follow:
- Axisymmetric, steady state laminar flow
- Local thermodynamic equilibrium (LTE)
- Negligible thermal diffusion, gravity and viscous dissipation effects
- Optically thin plasma
- Thermodynamic and transport properties are only temperature dependent

Typical results were reported for a wall-stabilized nitrogen arc in axial flow with an arc channel radius of 6.35 mm and an arc current of 580 A. It should be noted that the plasma enthalpy increases rapidly in the entrance region, reaches a peak and then levels off towards the fully developed region of the arc. The mass flux profiles on the other hand, reveal that most of the gas by-passes the arc column and flows essentially in the outer annular region surrounding it. Nevertheless, the high centerline temperatures and the correspondingly low gas densities are responsible for the high axial velocities along the axis of the flow as shown.

Further work in this area carried out by Pfender and his collaborators at the University of Minnesota (4-11), was mainly devoted to the solution of the problem of deviations from LTE in the anode region of the arc (6-8) and to the computation of the flow and temperature fields in arc plasma reactors of different configurations under transfered and non-transfered arc conditions (9-11).

Chen and Pfender (8) proposed a two-temperature model for the anode contraction region of a high intensity arc. This necessitated the addition of a charge conservation equation and separate electron and heavy particle energy transfer equations to the standard mass and momentum transfer equations. Computations of the composition of the two-temperature arc plasma was based on Dalton's law and the quasi-neutrality condition of the plasma. The transport properties of the plasma had to be reevaluated as function of the electron and heavy particle temperatures. The effect on the gas viscosity, however, was negligible because it is mainly a function of the heavy particle temperature, but was rather important for the electrical conductivity and the radiation losses which depend primarily on the electron temperature. Electron and heavy particle temperature isotherms given for the anode contraction region of an argon arc at atmospheric pressure and a current of 200 A show considerable deviations from LTE conditions in the immediate vicinity of the anode surface.

The work of Chen et al (9) was mainly concerned with the study of effect of side gas injection from an annular ring on the flow and temperature fields in the arc column. The results obtained using a two-temperature model showed that the arc becomes constricted at the location of gas injection. Enhanced joule heating in the constricted arc path, results in the increase of both the electron and heavy particle temperatures in this region. Virtually no mixing is observed between the two gas streams with the injected cold gas not being able to penetrate the hot arc column. It should be pointed out that this conclusion would not necessarily hold if the injected cold gas was introduced through a series of holes equally distributed around the periphery of the arc confinement chamber. The flow pattern in this latter case is essentially three dimensional and considerably more complex.

Mazza and Pfender (10) on the other hand were concerned with the modelling of an arc plasma reactor which could be used for thermal plasma synthesis. The model developed in this case was based on the assumption that LTE conditions prevailed over the entire calculation domain. The other assumptions involved were essentially the same as those listed earlier in relation to the work of Watson and Pegot (3) with the exception that the boundary layer approximation was applied in this case implying that axial gradients are considered negligible with respect to radial gradients and that axial velocities are assumed to be much larger than radial velocities. Typical examples of the computed velocity and temperature fields in the reactor for an argon plasma operated in the non-transfered and transfered arc modes are given in Figures 1 and 2, respectively. In the non-transfered arc mode, the arc current was 900 A, the reactor diameter was 10 mm and its length was 180 mm. The plasma gas flow rate was 5×10^{-6} kg/s. In the transfered arc mode, on the other hand, the arc current was 150 A, the transfered arc length was 157.5 mm and the injected gas flow rate 5×10^{-4} kg/s, while the quench gas flow rate was 8.0×10^{-4} kg/s. It is interesting to note the important difference between the flow and temperature fields in each case. The transfered arc mode of operation results in an early increase of the centerline arc temperature, followed by a gradual temperature drop further downstream. It is to be noted that the observed sudden increase in the plasma velocity downstream of the confined region of the reactor is due to the important quench gas injection in this case.

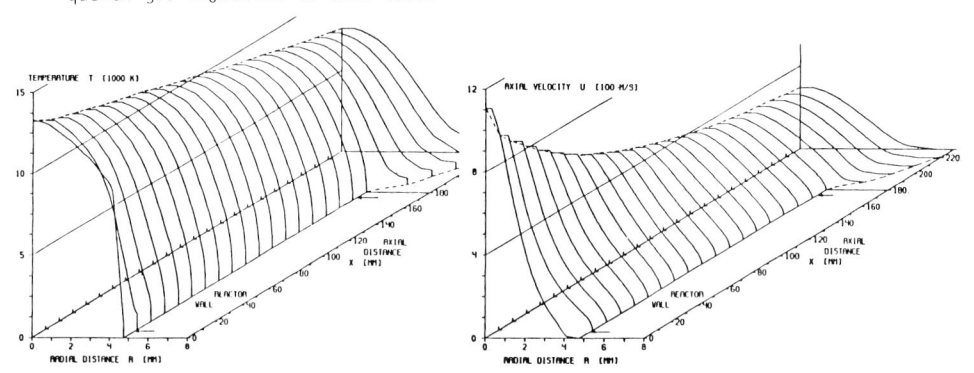

Figure 1. Temperature and velocity fields for a non-transfered arc reactor Argon plasma, I=900 A, $\dot{m}=5.0 \times 10^{-6}$ kg/s (after Mazza and Pfender (10))

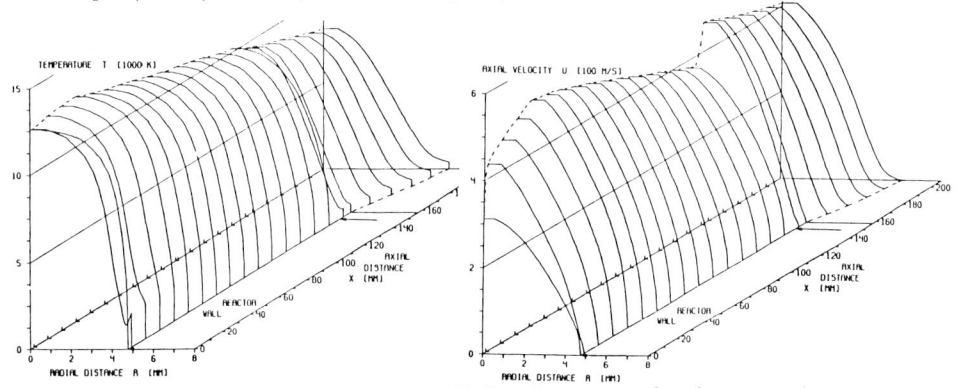

Figure 2. Temperature and velocity fields for a transfered arc reactor, Argon plasma, I=150 A, $\dot{m}=5.0 \times 10^{-4}$ kg/s (after Mazza and Pfender (10))

2.2 d.c. plasma jets

In contrast to electric arcs, d.c. plasma jets are non-current carrying, representing generally turbulent free jets which are being increasingly used in such applications as plasma reactors for chemical synthesis or for the thermal treatment of powder as in plasma spray coating.

The first attempts to model the flow and temperature fields in a d.c. plasma jet can be traced back to the work of Donaldson and Gray (12) and Boulos and Gauvin (13) who solved the simplefied integral boundary layer equations assuming a similarity between the non-dimensional velocity and temperature fields $(u-u_a/u_0-u_a)=(T-T_a/T_0-T_a)$.
The results which compared favorably with measurements of the axial velocity and temperature profiles along the centerline of the jet were later used (13) for the computation of the trajectories and temperature histories of MoS_2 particles injected at different angles and initial velocities in the flow. A statistical analysis of the trajectories and the transformation of different particles clearly indicated that the particle size distribution and injection conditions in the plasma can have an important effect on the overall effeciency of the powder treatment in the plasma reactor.

It is only recently, however, that the problem of modelling of the flow and temperature fields in a turbulent d.c. plasma jet has been fully addressed by McKelliget et al (14) and Correa (15). The approach used in this case is based on the solution of the full two-dimensional continuity, momentum and energy transfer equations simultaneously with the corresponding turbulent kinetic energy equation. The principal assumptions involved (14) can be summarized as follow:
- Axially symmetric, two-dimensional, steady state turbulent flow
- The velocity and temperature profiles at the exit of the plasma torch are known
- The plasma is in local thermodynamic equilibrium (LTE)
- Optically thin plasma
- The specific heat of the plasma is independent of temperature
- The turbulent momentum and energy transport parameters (viscosity and thermal conductivity) are large compared to their molecular components. A constant value of the molecular viscosity was assumed
- In representing the entrainment, the difference between the physical properties of the plasma gas and the surrounding air was not taken into consideration.

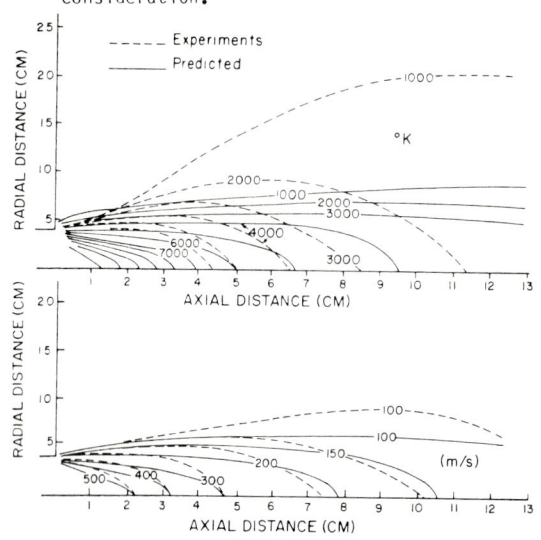

Figure 3. Experimental and predicted velocity and temperature fields for an argon/hydrogen d.c. turbulent plasma jet (after McKelliget et al (14))

In spite of the important simplifying assumptions involved, especially with regard to the transport properties of the plasma, the computed flow and temperature fields given in Figure 3 compared well with the experimental measurements of Vardelle et al (16). The reported measurement in this case were obtained for an argon/hydrogen plasma at atmospheric pressure operated at a power level of 29 kW with a total gas flow rate of 90 l/min and a hydrogen concentration of 16.7% by volume. The exit plasma torch nozzle diameter was 8.0 mm.

3. MODELLING OF THE INDUCTIVELY COUPLED R.F. PLASMA

Since the development of the inductively coupled r.f. plasma torch in the early sixties, special attention has been given to the development of mathematical models for the quantitative description of the phenomena which occurs in the discharge region. These varied from relatively simple one-dimensional models concerned with the calculation of the radial temperature profile in the induction zone to more elaborate two-dimensional models used to calculate the temperature, velocity and concentration profiles in the torch. A general review of the state-of-the-art of induction plasma modelling has been published by Boulos and Barnes (17).

In order to allow for the analytical solution of the governing equations, the one-dimensional models (18-23) had to be kept relatively simple and with few exceptions adopted the following assumptions:

- Local thermodynamic equilibrium (LTE)
- Neglected convective heat tranfer, with the exception of Keefer et al (23)

The proposed one-dimensional models differed, however, in the following:

- The degree to which they took into account the variation of the physical properties of the plasma with temperature (Electrical and Thermal conductivity)
- Whether or not they took into account radiation heat losses from the plasma in the energy equation

While the principal advantage of one-dimensional models is that they offered a relatively simple way of estimating the temperature profiles in the center of the discharge, they suffered from two main limitations. First they provided no information about the temperature field outside the induction zone and second, they could not be used for the calculation of the flow field in the torch.

A number of two-dimensional models (24-32) were consequently developed over the last ten years. These have the following assumptions in common:

- They all assumed local thermodynamic equilibrium (LTE)
- They took into account conductive, convective and radiative heat transfer
- They also took into account the variations of the thermodynamic and transport properties of the plasma with temperature
- They also maintained the one-dimensional electric and magnetic field assumption

They varied however, in the following points:

- Whether, or not, they included the full momentum transfer equations
- Whether they solved the transient or the steady state forms of the equations. In either case they aimed at steady state solutions.

As an example of the capabilities of the developed two-dimensional models, typical results of the flow and temperature fields reported by Mostaghimi et al (32) are given in Figure 4. The corresponding torch geometry and a summary of the operating conditions are given on Figure 5. These were obtained through the solution of the two-dimensional continuity, momentum and energy equations, written in terms of the velocity pressure and enthalpy, simultaneously with the one-dimensional electric and magnetic fields equations written in terms of the axial magnetic field intensity, the azimulthal electric field and the phase difference between them.

Two important aspects are to be noted in the computed flow and temperature fields. The first is the presence of an off axis maximum in the radial temperature profiles in the coil region which is due to the very nature of induction coupling of energy in a cylindrical conductor by which the energy is mostly dissipated in the outer region of the conductor, the thickness of which is governed by the skin depth which is, in turn, a function of the oscillator frequency and the average electrical conductivity of the conductor.

The second point which is to be noted from Figure 4, is the presence of a relatively large cold region in the center of the discharge in the immediate neighbourhood of the central injection probe. The relative size and the temperature field in this region has been found to depend primarily on the dimensions of the probe, r_1 as indicated in Figure 4, and on the injected gas flow rate, Q_1. Since in the event of the use of the inductively coupled plasma for the thermal treatment of materials, the powder is injected in the plasma via the carrier gas stream; it is important that Q_1 be kept at a minimum to avoid local cooling of the center of the plasma and the eventual reduction of the thermal treatment efficiency of the powder.

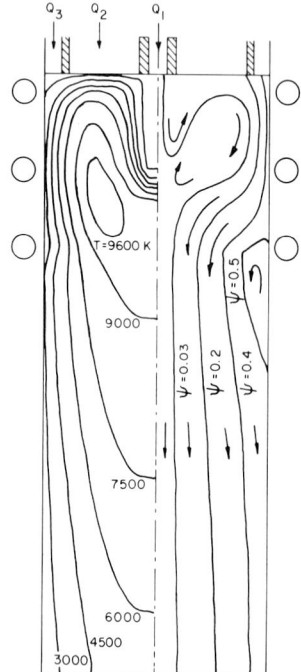

Figure 4. Temperature isotherms and stream lines for an inductively coupled plasma (after Mostaghimi et al (32))

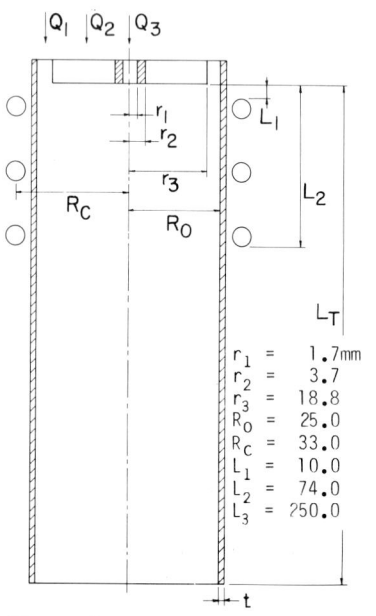

Figure 5. Schematic and dimensions of the induction plasma torch; Argon plasma, f = 3.0 MHz, P = 3.0 kW, Q_1 = 3.0, Q_2 = 3.0, Q_3 = 14.0 ℓ/min (after Mostaghimi et al (32))

r_1 = 1.7mm
r_2 = 3.7
r_3 = 18.8
R_0 = 25.0
R_C = 33.0
L_1 = 10.0
L_2 = 74.0
L_3 = 250.0

4. PLASMA-PARTICLE HEAT TRANSFER

Since the thermal treatment of powders in plasma torches and furnaces represents one of the most important applications of plasma technology, it is not surprising that considerable attention has been given to the important problem of plasma-particle heat transfer. A number of mathematical models have been developed for the thermal treatment of powders in d.c. plasma torches (13,33-37) and a few were applied to the induction plasma (38-39). In either case, the models varied in the assumptions made and whether or not they took into account the effects of internal heat conduction in particles on the overall heat transfer process between the plasma and the particles. Only Fiszdon (35) and Yoshida and Akashi (39) followed the internal heat conduction in the particles while the others assumed the particles to have a uniform temperature.

The specific question of whether, and when, is it necessary to take into account the internal heat conduction in the particles has been studied by Bourdin et al (40) and Chen and Pfender (41). The results reported by Bourdin et al (40) showed that differences as high as 1 000 K could develop between the surface temperature and that of the center of alumina particles as small as 20 µm in diameter when immersed in a nitrogen plasma at 10 000 K. The controlling parameter seems to be the Biot number which is simply the ratio of the thermal conductivity of the plasma to that of the particle (k/k_s). According to Bourdin et al (40), during the transient heating of a particle under plasma conditions, internal heat conduction in the particle should be taken into account if the Biot number is greater than 0.02. The work of Chen and Pfender (41), goes further to indicate that inspite of the differences in initial heating, the analytical expression based on infinite thermal conductivity predict the correct total time for both heating and evaporation even for low-conductivity materials such as alumina.

Chen and Pfender (42-44) also studied the evaporation of single particles under plasma conditions and the behaviour of small particles in a thermal plasma flow. In the latter case they propose a Knudsen number correction of the heat transfer rate to the particles to account for deviations from continuum fluid mechanics which becomes increasingly important for sub-micron particles.

5. CONCLUSION

From the above it can be concluded that while our knowledge of the mathematical modelling of plasma processes have improved considerably over the last few years there are still important areas when much more work needs to be done these can be identified as follows:

- Non-equilibrium phenomena and deviations from LTE
- Modelling of plasmas at reduced pressures (soft vacuum)
- Modelling of plasmas produced from mixed gases
- Plasma-surface interation phenomena
- Plasma-particle interaction under dense loading conditions

Obviously some of these topics would require years of study, but the gained insight in the fundamentals of the processes involved is indispensable to sustain the steady development of plasma technology.

ACKNOWLEDGEMENT

The manuscript was reviewed by Professor E. Pfender whose contribution is gratefully acknowledged.

REFERENCES

1. Elenbaas, W. Physica, 2, 169 (1935).
2. Heller, G., Physica, 6, 389 (1935).
3. Watson, V.R. and E.B. Pegot, NASA, TN, D-4042 (1967).
4. Pfender, E., "Electric arcs and arc gas heaters "Gaseous electronics. Ed. M.N. Hirsh and H.J. Oskam, Academic press, Vol 1, 291-398 (1978).
5. Pfender, E., Pure and Applied Chemistry, 52, 1773 (1980).
6. Dinuleseu, H.A. and E. Pfender, J. Appl. Phys., 51, 3149 (1980).
7. Chen, D.M. and E. Pfender, IEEE, Trans. Plasma Sci., PS-8, 252 (1980).
8. Chen, D.M. and E. Pfender, IEEE, Trans. Plasma Sci., PS-9, 265 (1981).
9. Chen, D.M., K.C. Hsu and E. Pfender, Plasma Chemistry and Plasma processing, 1, 295 (1981).
10. Mazza, A. and E. Pfender, ISPC-6, Montreal, 1, 41 (1983).
11. Young, R.M., Y.P. Chyon, E. Fleck and E. Pfender, Ibid, 1, 211 (1983).
12. Donaldson, C.P. and K.E. Gray, AIAAJ, 4, 2017 (1966).
13. Boulos, M.I. and W.H. Gauvin, C.J.Ch.E., 52, 355 (1974).
14. McKellizet, J., J. Szekely, M. Vardelle and P. Fauchais, Plasma Chemistry and Plasma Processing, 2, 317 (1982).
15. Correa, S.M., ISPC-6, Montreal, 1, 77 (1983).
16. Vardelle, A., J.M. Baronet, M. Vardelle and P. Fauchais, IEEE, Trans. Plasma Sc., PS-8, 417 (1980).
17. Boulos, M.I. and R.M. Barnes,"Induction plasma modelling, state-of-the-art", Developments in Atomic plasma Spectrochemical analysis, R.M. Barnes Ed., Heyden Press p 20 (1981).
18. Freeman, M.P. and J.D. Chase, J.Appl. Phys., 39, 180 (1968).
19. Eckert, H.U., J. Appl. Phys., 41, 1520 (1970).
20. Eckert, H.U., J. Appl. Phys., 41, 1529 (1970).
21. Mensing, A.E. and L.R. Boedeker, NASA CR-1312 (1969).
22. Pridmore-Brown, D.C., J. Appl. Phys., 41, 3621 (1970).
23. Keefer, D.R., J.A. Sprouse, and F.C. Loper, IEEE, Trans. PL. Sc., PS-1, 71 (1973).
24. Miller, R.J. and J.R. Ayen, J. Appl. Phys., 40, 5260 (1969).
25. Barnes, R.M. and R.G. Schleicher, Spectrochim Acta, B-30, 109 (1975).
26. Barnes, R.M. and S. Nikdel, Appl. Spectroscopy, 29, 477 (1975).
27. Ibid, J. Appl. Phys., 47, 3929 (1976).
28. Delettrez, J.A., "A numerical calculation of the flow and electrical characteristics of an argon induction discharge, Ph.D. thesis, U. of Calif., Davis (1974).
29. Boulos, M.I., IEEE Trans. PL. Sc., PS-4, 28 (1976).
30. Boulos, M.I., R. Gagné and R.M. Barnes, C.J.Ch.E., 58, 367 (1980).
31. Mostaghimi, J., P. Proulx and M.I. Boulos, "An analysis of the computer modelling of the flow and temperature fields in an inductively coupled plasmas, to be published (1984).
32. Ibid, "Parametric study of the flow and temperature fields in an inductively coupled r.f. plasma torch, to be published (1984).
33. Bhattacharya, D. and W.H. Gauvin, AIChEJ, 21, 879 (1975).
34. Gal-Or, B., J. of Eng. for power, 102, 589 (1980).
35. Fiszdon, J.K., Int. Heat & Mass Transfer, 22, 749 (1979).
36. Vardelle, M., A. Vardelle, P. Fauchais and M.I. Boulos, AIChEJ, 29, 236 (1983).
37. Wei, D., S.M. Correa, D. Apelian and M. Paliwal, ISPC-6, Montreal, 1, 83 (1983).
38. Boulos, M.I., IEEE, Trans. PL. Sc., PS-6, 93 (1978).
39. Yoshida, T. and K. Akashi, J. Appl. Phys., 48, 2252 (1977).
40. Bourdin, E., P. Fauchais and M.I. Boulos, Int. J. H. & M. T., 26, 567 (1983).
41. Xi Chen and E. Pfender, Plasma Chemistry and Plasma Processing, 2, 293 (1982).
42. Ibid, 2, 185 (1982).
43. Ibid, 3, 97 (1983).
44. Ibid, 3, 351 (1983).

THERMAL PLASMA MELTING/REMELTING TECHNOLOGY

WARD C. ROMAN
United Technologies Research Center
East Hartford, CT 06108

ABSTRACT

An overview is presented on several aspects of thermal plasma melting/remelting technology. Included is a brief description of the fundamentals of thermal plasma, basic types of plasma reactors, examples of unique configurations, highlights of some commercial applications, and recent state-of-the-art advances.

INTRODUCTION

As with many other fields of science, thermal plasma melting and processing technology has experienced some severe oscillations in direction and emphasis over the past fifteen years. The recent energy crisis and stricter environmental regulations have resulted in renewed emphasis on plasma technology. The long range projection on coal and nuclear fuel availability suggests that an "electric energy economy" may be part of the solution to the long-term energy problem. As the U.S. and foreign countries shift more toward electricity as the major energy base, numerous high temperature plasma material processing applications will inevitably result. In some cases, merely a retrofit to an existing process will take place. In others, an updated processing system will replace an obsolete system for generating the same final product at reduced cost and increased efficiency. Finally, in the future completely new plasma processing systems will be initiated to provide new materials and products not defined at this time.

FUNDAMENTALS OF THERMAL PLASMA TECHNOLOGY

Plasma As a Heat Source

In this overview, the word plasma will refer to the gaseous region of an electric discharge that is characterized by intense luminosity occurring simultaneously with the presence of electrons, atoms, and positive and negative ions. In this state the gas which normally is nonconductive possesses a significant electrical conductivity and, therefore, will be in a partially ionized state. In such a plasma discharge, the free electrons gain energy from the applied electric field used and subsequently lose this energy through collisions or recombination with neutral gas particles. Plasmas produced by electric field discharges are normally classified into two types. The first is thermal plasmas also referred to as equilibrium or hot plasmas. These are identified by high gas temperature and an approximate equality between gas and electron temperature. In a thermal plasma, the thermodynamic state of the plasma approaches equilibrium. These are the types discussed in this overview. For reference, the other type of plasma is referred to as nonequilibrium or cold plasma; this type is identified by a low gas temperature but a relatively high electron temperature. Table I shows typical operating regimes that distinguish the thermal plasma arcs and radio frequency (rf) discharges from glow (cold type) discharges.

TABLE I. Typical Operating Regimes of Discharges

Discharge type	Power, Q	Pressure, P	Flow rate, ṁ	Gas temp, Tg
Glow	< 5 kW	≤ 100 torr	≤ mg/s	< 1000 K
RF 1 ≤ freq ≤ 40 MHz	1 kW — 1 MW	10 torr – 1 atm	≤ 50 g/s	1000 — 10,000 K
Arcs	1 kW — 10 MW	100 torr – ^0 atm	≤ 500 g/s	3000 — 30,000 K

TABLE II Types of Thermal Plasmas

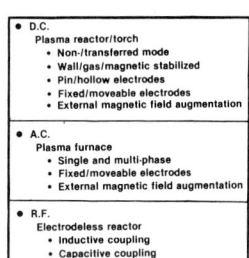

- D.C.
  Plasma reactor/torch
  - Non-/transferred mode
  - Wall/gas/magnetic stabilized
  - Pin/hollow electrodes
  - Fixed/moveable electrodes
  - External magnetic field augmentation
- A.C.
  Plasma furnace
  - Single and multi-phase
  - Fixed/moveable electrodes
  - External magnetic field augmentation
- R.F.
  Electrodeless reactor
  - Inductive coupling
  - Capacitive coupling

Figure 1 illustrates several features that distinguish a thermal plasma arc from other type discharges. In general, the voltage required to sustain an arc discharge is much lower than for other types of discharges. Furthermore, the potential distribution changes rapidly near the electrodes in the area referred to as the cathode and anode fall regions. The cathode fall assumes values of about 10 volts relative to values of greater than 100 volts in glow discharges. Depending on the energy balance of the arc column and its length, the total voltage drop may be relatively high. The specific voltage gradient also depends on the pressure, monatomic or diatomic gas type, metal or liquid vapor environment, etc. For example, a plasma arc column operating in atmospheric pressure air will have a column voltage gradient of about 10 V/cm. If external aerodynamic or magnetic field stabilization is applied to the column combined with higher pressure operation, total arc voltages exceeding 10,000 volts can be reached. Under high pressure operation a significant fraction of the total plasma power is dissipated as radiation (e.g., 15 to 45%).

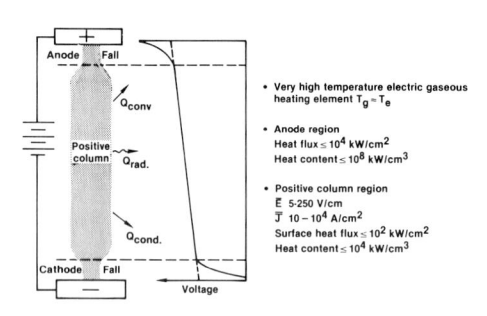

- Very high temperature electric gaseous heating element $T_g \approx T_e$
- Anode region
  Heat flux ≤ 10^4 kW/cm^2
  Heat content ≤ 10^6 kW/cm^3
- Positive column region
  $\bar{E}$ 5-250 V/cm
  $\bar{J}$ $10 - 10^4$ A/cm^2
  Surface heat flux ≤ 10^2 kW/cm^2
  Heat content ≤ 10^4 kW/cm^3

FIG. 1. Thermal Plasma Arc Characteristics

Depending on the particular application different electrode materials are used. In the few cases where electrode contamination is permitted, graphite/carbon electrodes are used. In nonconsumable electrode applications, long lifetime is a prerequisite that must be provided by minimizing electrode erosion (normally via novel water-cooling schemes). The penalty paid in this case is a significant reduction in the overall thermal efficiency. Another technique for reducing electrode erosion is using external magnetic and/or aerodynamic fields to rapidly move the arc attachment points on the electrode surface, thus effectively spreading out the heat load. Depending on the application, either cold or hot-type cathodes are used. Cold cathodes rely primarily on field emission as a source of electrons compared to thermionic emission provided by the hot type. The geometry and shape of the cathode is an area that has received considerable investigation and depends strongly on the selected current operating range. Pin-type cathodes are normally used for steady-state operation into the 1,000 ampere range. A transition to a button-type geometry is required when current levels exceed about 1,000 amperes. For current levels higher than 4,000 amperes, a hollow cathode geometry is used. The most widely used hot cathode material is tungsten. Thoriated tungsten or barium-calcium-aluminate or lanthanum doped tungsten are also used; these cathodes are limited to the low-moderate pressure regime.

The anode, is subjected to very high heat fluxes, and consumes a large percentage of the total power in the plasma arc. Therefore, when used in a melting application, the material to be melted normally becomes the anode. In many of the plasma arc reactors used for melting, an isolated (electrically floating potential)nozzle assembly or so-called "constrictor" is used to aid in arc confinement and stabilization. Choice of gas flow patterns (e.g., coaxial, vortex, or combinations thereof) near the electrodes and within the plasma arc reactor used for melting and processing materials is important for providing good stability and improved lifetime.

Basic Types of Plasma Reactors

Different types of plasma arc reactors are used for materials processing. As shown in Table II, these include dc, ac, and rf as the primary power input. The most prevalent type is the constricted dc plasma arc reactor. Figure 2 is a simplified sketch of a typical thermal plasma torch configured in the nontransferred mode. In this mode, an arc is established between an axial water-cooled pin-type cathode and an annular water-cooled anode. A coaxial, vortex, or combined coaxial-vortex gas stream is introduced into the chamber; this flow moves the anode attachment into the constricted region and also provides stabilization. This lengthens the arc, raises the voltage, and results in higher temperatures due to constriction. Axial temperatures can range between 8,000-30,000 K with corresponding high radial temperature gradients. Typical power levels for plasma torches of this type are 20-75 kW; the thermal efficiencies normally range between 55-85%. One or more power supplies may be used to provide superimposed sources. In general the nontransferred mode is limited in lifetime due to erosion of the nozzle especially at high currents. Typical electrode lifetime with vortex augmentation is approximately 60-100 hours. With proper design and choice of operating parameters, contamination level in the plasma positive column can be held to about 5 ppm which is an acceptable level for most applications. Considerable development has occurred over the last ten years in the design and optimization of plasma torches for material spraying, welding, and cutting operations. The electrode and nozzle size and shape, gas mass flow rates, gas mixtures, associated flow patterns, etc. are all key (coupled) parameters used for optimizing a source for each particular application. Within the last five years, there has been about a three fold increase in the number of companies that market and use plasma torches of this type.

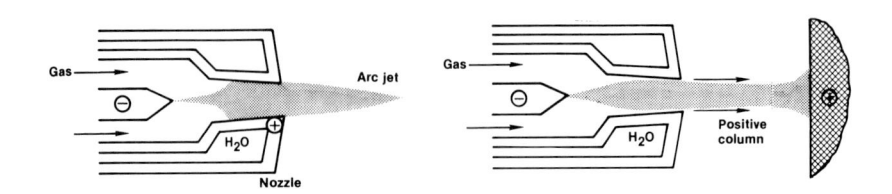

FIG. 2 DC Thermal Plasma Arc Jet (torch), Nontransferred Mode

FIG. 3 DC Thermal Plasma Arc, Transferred Mode

For applications where more power is required to be coupled into the workpiece (or metal to be melted), the d.c. transferred mode of operation is used as shown in Figure 3. In this case, the workpiece normally becomes the anode. This configuration permits much longer plasma arcs with higher currents achievable at significantly higher voltages. This results in more power into the plasma for the same current level as the nontransferred mode and a more efficient power deposition into the material since the anode attachment and corresponding high heat flux (see Fig. 1) is on the surface of the workpiece. The plasma discharge is also more stable in dc operation, relative to pulsed d.c. or a.c. The d.c. arc column does not have to cycle through zero current with the associated instabilities and significantly higher open circuit voltage requirements of a.c. operation. Ignition of the transferred mode of operation is normally accomplished by first starting a pilot discharge (using a high frequency (hf) starter system between cathode and nozzle) and then transferring the main discharge to the workpiece.

The a.c. type plasma arc reactors make up a separate category (normally for gas heater applications) and therefore only a few comments are included here. By using water-cooled coaxial copper electrodes together with external magnetic field augmentation for arc attachment spot movement, a wide variety of gases can be heated with this type device. Typical efficiencies are approximately 60%. Figure 4 is an example of an a.c. type reactor. In this case, three movable electrodes arranged about a central axis with a coaxial sheath gas are employed. Initially the electrodes are brought together in the plasma jet from an auxiliary d.c. torch system; this is followed by a three phase a.c. arc ignition. A major disadvantage of this type of system is the relative high electrode material losses. One variation to the conventional a.c. plasma reactor is the so-called centrifugal furnace (see Figure 4). As the material melts due to interaction with the plasma jet (or jets), it is contained as a stable layer on the inner wall of the furance due to centrifugal force. Attack of the furnace container is minimized by maintaining the required temperature gradient through the ceramic wall. Fusion of ceramics is the principal application of this type reactor. Additional variations to this centrifugal furnace may include using multiple plasma jets at each end of the furnace and transferring a long, high-voltage, high power a.c. (or d.c.) discharge between them. When a tilt mechanism is incorporated, the molten contents can be poured into molds or quick quenched to yield powder.

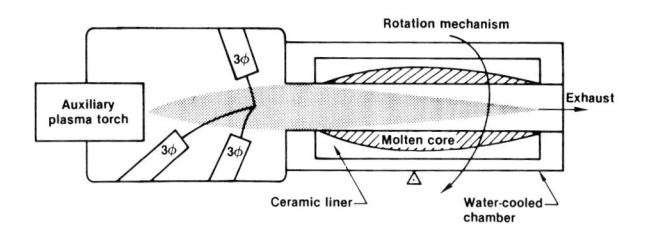

FIG. 4 Example of A.C. Thermal Plasma Reactor/Centrifugal Furnace

All the thermal plasma reactors described above use different forms of electrodes. Another class, primarily confined to laboratory scale, is the electrodeless rf plasma. Figure 5 is a sketch of the basic configuration. In this type reactor, the electrical power is coupled to the plasma via rf work coils (for inductively coupled) or rf plates (for capacitively coupled). Refer to Table I for typical operating range. In general, the rf plasma reactor is more expensive than other types, but it does possess the unique advantage of being able to heat all types of gases (including

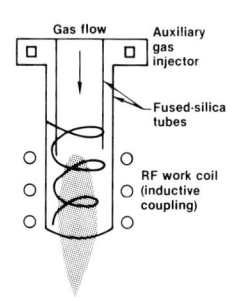

Table III. Major Advantages of Thermal Plasma Melting Techniques

- Steady-state uniform flow of very high temperature gas
- Complete control of atmosphere (inert or reactive)
- Minimum contamination and loss of volatile constituents
- Capability to process different forms of material
- Compact system with high throughput rates
- High thermal/electrical efficiency

FIG. 5 Example of R.F. Thermal Plasma Torch

highly reactive and corrosive types) to high temperatures with negligible contamination. A disadvantage of these devices is the magnetohydrodynamic problem of having a large fraction of the total gas feed not penetrating the main core of the plasma where the high temperature/reaction region exists.

In summary, from the standpoint of commercial applications the highest thermal efficiency is required, thus the selection of the transferred mode of operation with a d.c. source is favored (Fig. 3). When these d.c. type reactors are properly designed to incorporate nonconsumable electrodes, relatively long lifetime at high power levels is achievable with minimum contamination. The majority of all plasma reactors currently being used for melting and remelting operations operate in the transferred mode; therefore the following overview will emphasize this type.

Advantages

Several key advantages of employing thermal plasma reactors for melting and materials processing are shown in Table III. Use of a thermal plasma type heat source is characterized by a high concentration of energy and associated high gas temperatures well above those obtainable with chemical (combustion) flames or resistance heater systems. This results in both high throughput and melting rates. The flexibility afforded by operation under an inert environment minimizes contamination of the material to be melted or processed. Complete control of the atmosphere is achievable. Operation at elevated pressures with an inert gas minimizes the loss of high vapor pressure constituents and also eliminates the need for expensive vacuum pumping equipment as is required with vacuum arc remelt (VAR) and electron beam (EB) melting systems. With this type heat source, material melting rate and heat addition (superheat) to the molten pool are independent of the electrode feed rate as compared to VAR systems. By employing nonconsumable electrodes, negligible contamination is achievable. D.C. plasma reactors result in considerably reduced voltage fluctuation compared to competing systems. The option also exists for refining operations using slags and direct alloying as in the case of nitrogen injection. Remelting under a reactive gas environment permits in-situ chemical treatment. Other features include the ability to achieve dense ingots during primary melting that are free of shrinkage voids; in addition, the surface finish is good. Lastly, the thermal plasma reactor is a compact system that can process material in a variety of forms at high throughput rates and with relatively high electrical/thermal efficiency.

EXAMPLES OF PLASMA MELTING REACTORS

To date, only a few of the wide variety of plasma devices including the basic types described above have been developed for production-scale operation. This is especially true in the U.S. As of 1983, the U.S.S.R., GDR, and Japan have pioneered in establishing commercial scale plasma processing reactor systems for melting of materials. Normally those reactors and associated processes that have achieved commercialization are operated in a proprietary manner. In the following a brief description will be given of several selected examples of these reactors where information is available.

Furnaces

One of the first thermal plasma reactors used to melt steel was the Linde reactor [1], as shown in Figure 6. This reactor was introduced almost twenty years ago. The operating voltage under an inert argon environment was relatively low (less than 100 V); the d.c. plasma source was mounted centrally within the top of the reactor. The plasma arc operated in a stable mode due to a combination of low flow rates and a relatively short plasma positive column (apprx. 0.1 m). The bottom electrode was water-cooled copper and electrically connected to the molten pool of material to be processed. Over 100 hours operation was reported at current levels exceeding 1,000 amperes. Other features included a magnetic coil connected in series with the discharge for stirring the molten pool. The Linde results, including scale-up from 1 to 10 tons, indicated high quality steel could be produced relative to that obtained by vacuum melting technology. Rapid wear of the bottom electrode and high erosion of the refractory lining due to the abrasive action of the magnetically mixed molten material were identified as potential problem areas that terminated operation of this unit.

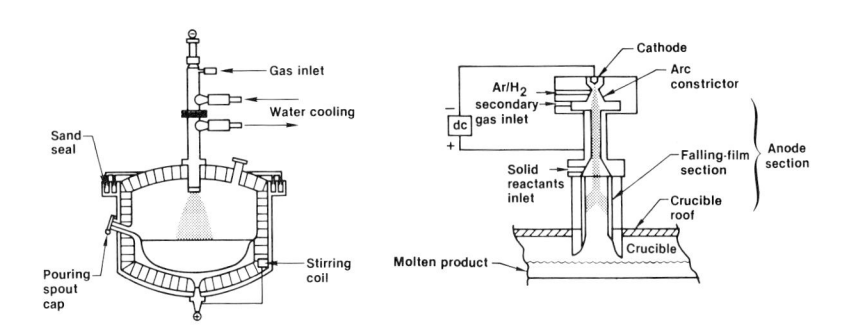

FIG. 6 Linde Plasmarc Furnace   FIG. 7 Bethlehem Falling Film Reactor

Bethlehem Steel Corporation also pioneered in demonstrating the advantages of thermal plasma reactor processing for commercial scale ore reduction. Bethlehem has a patent [2] for a falling film plasma reactor that uses a high power plasma arc (2,000 V, 500 A) to heat feedstock that melts down a vertically oriented hollow electrode. Being in the current carrying path of the plasma, the liquid film is subjected to ohmic heating together with convection and radiation heating. Figure 7 is a schematic of this type furnace. The plasma arc is stabilized by a vortex gas flow between a tungsten cathode and a cylindrical water-cooled copper anode. Iron oxide fines are pneumatically transported to the anode section; the ore fines melt on contact with the hot plasma gas and form a falling liquid film in the hollow anode section. Metallurgical reduction processes are reported to take place within the falling film. The main advantage of the process is claimed to be relatively long residence times obtainable thus enabling resultant high yields via improved heat transfer and direct contact of the material required for the reaction. In addition, the film serves to thermally insulate the anode.

The preliminary experiments on the direct plasma reduction of iron oxide were conducted at power levels of about 100 kW; favorable results led to a scale-up in power to the 1 MW range (2,000 V, 500 A). In the large power furnace, the lowest specific energy consumption was about 15% greater than the theoretical requirement. At the reference low specific energy consumption level, the electrical energy consumption of the falling film d.c. plasma reactor is approximately equal to the total energy equivalent input of conventional steelmaking. Full-scale commercial steelmaking would require plasma reactors of about 500 MW. Therefore, based strictly on energy requirements, the plasma reactor technique does not excel over conventional steelmaking. Bethlehem also investigated this type falling film plasma reactor for the reduction of vanadium ore concentrates into ferrovanadium and vanadium metal. Activity in this area has been reduced since ferrovanadium is readily and economically available on the open market. One field that does look promising for this type reactor is the so-called mini-steel-plant application.

Only a few other U.S. companies use thermal plasma reactors for melting on a commercial scale. These are Associated Minerals Consolidated (AMC), Plasma Materials Incorporated (PMI), and Plasma Energy Corporation (PEC). The AMC reactor is used in processing zircon via a carbon electrode scheme [3]. PMI employs a nitrogen d.c. plasma source to spheroidize particles of magnetic iron oxide used in photocopy machines with a magnetic developer system [4]. The PEC plasma pilot plant furnace is being used for scrap melting, electronic scrap processing, and cement production [5]. Typical power ranges of these systems is 150 to 300 kW.

A wide variety of reactor designs for the melting of metals have been reported in the foreign literature over the last ten years. The following discussion highlights some of the features and status.

Daido Steel Co. has been active over the last ten years in developing plasma arc remelting techniques applicable to steels, superalloys, and nonferrous metals. Their initial R&D started with several 50kW plasma arc furnaces and has progressed up to the MW range. Daido's R&D also includes plasma furnaces, plasma progressive casting (PPC), plasma skull casting, and oxide reduction using a hollow cathode plasma furnace. Emphasis has been placed on the plasma induction furnace (PIF) system [6]. This integrated thermal plasma technique provides both a protective inert argon atmosphere over the molten pool and the supplemental power required for keeping the thin layer of slag molten. A 0.5 T pilot plant and 2.0 T production unit are in operation and Daido has recently transitioned into a 5 Ton reactor. Figure 8 illustrates the basic reactor. For reference, the Daido 2 Ton PIF system operates at a total power of 1 MW (600 kw induction/400 kW plasma) and employs an additional 200 kW for induction stirring. Nominal melting capacity is 200 T/mo., melting time is about 3 hrs, and the measured power consumption is 1,000 to 1.3 x 1,000 kW hr/T. Table IV shows the comparison between the standard VIF and the PIF system. The major advantages include use of greater than 50% scrap, melt capability for wide range of alloys, good slag refining capability, high level of cleanliness, low melting costs, and high melting yield. The main disadvantage is the relatively high argon gas flow rates.

Plasma primary melting reactors using a plurality of sources projecting through the sidewalls has been commercialized in the GDR. The original reactors consisted of approximately 5 ton units with a single plasma source (transferred mode) located in the roof; maximum operating currents were 6,000 (at less than 500 V) amperes using a tungsten alloy disc shaped cathode.

FIG. 8 Daido Plasma Induction Furnace

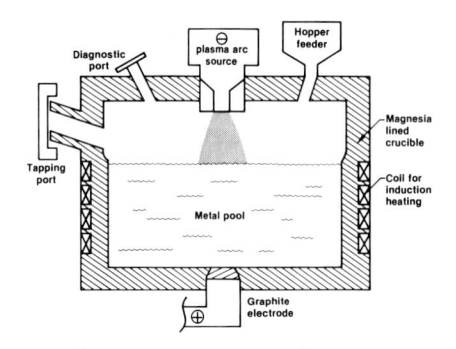

Table IV. Comparison of VIF and PIF Systems

Characteristic	VIF	PIF
Volatilization	High	Low
Mechanical property	Good	Good
Workability	Excellent	Good
Yield (melting)	98	98.9
Product yield (total)	75	85
Melting cost (index)	100	59
O	5-15 ppm	8-25 ppm
N	10-30 ppm	10-50 ppm
H	1 ppm	2-5 ppm
Desulfurization	10 percent	50-85 percent
Decarburization	90 percent	70 percent

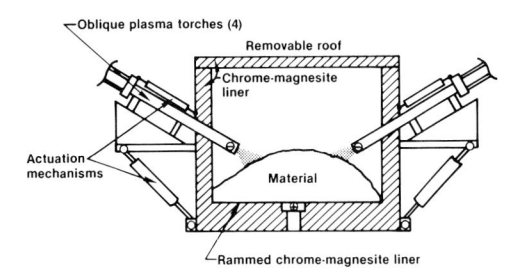

FIG. 9 GDR Plasma Furnace

Ten ton and 30 ton melting capability units (see Fig. 9) are in operation at the VEB Edelstahlwerks at Freital in GDR [7]. They are used in continuous operation for melting iron and nickel based alloys, stainless steels, and high alloy tool steels under an inert argon environment. The stainless steel melting is conducted both with and without nitrogen addition. The walls and roof are lined with bricks of chrome-magnesite; the bottom section is lined with rammed chrome-magnesite refractory. Due to the high heat loads, several sections of the reactor incorporate water-cooled assemblies to provide long lifetime with minimum erosion. The number (3 to 6) of high current (approx. 6,000 A) d.c. plasma sources used depends on the reactor size and processing requirement. The current level, column length, and separation distances are critical parameters that must be properly adjusted to obtain stable operation. As in many other plasma cathode electrode configurations, the degree of cooling applied to the tungsten alloy cathode is most critical in providing long lifetime. Additional details of the 10 and 30 Ton plasma reactors are given in Table V. Note that the 30 ton unit specific energy consumption is equivalent to updated electric arc furnaces. The efficiency increases with the size of the furnace and power input. Additional advantages over high power electric arc furnaces include reduced melting times; reduced melt contamination; very low residual oxygen and hydrogen levels; smoother power loading (no flicker or surges); high alloying species retention; reduced total iron losses (less than 20%); reduced carburization; potential for in-situ nitrogen gas phase alloying, and significantly reduced noise level. Related activity with this type of plasma melting reactor is also in progress in the U.S.S.R. [8].

Table V Comparison of Small and Large Scale GDR Plasma Furnaces

Characteristic	Type P15	Type P35
Design capacity (tons)	15 (nominal 10 T)	35 (30 T)
V (Volts)	700	700
Power (kW) (0.96 power factor)	10^4	20×10^3
Melt down rate (T/hr)	9	20
Specific energy consumption (kW hr/T)	600 nominal	500 nominal (equiv to conventional arc melting process)
Argon gas flow (m^3/hr)	18	45

Significant developments of large scale thermal plasma reactors and high power (megawatts) plasma sources is also evident at Acurex, Aerospatiale, ASEA, HULS, MINTEK, ONTARIO-HYDRO, SKF and Westinghouse. Details of some of these activities were reported at the 6th International

Symposium on Plasma Chemistry (Quebec, Canada July 1983) and elsewhere in M.R.S. proceedings under extractive metallurgy.

In addition to the falling film plasma reactor system, other plasma melting reactor schemes have been devised to enhance the residence times between the injected materials and the plasma source for extractive metallurgy applications. Tetronics uses a precessing plasma source technique (see Fig. 10) wherein the source is located vertically over a molten pool and the plasma positive column is moved in the transferred mode via precession of the source at about 30 rps.

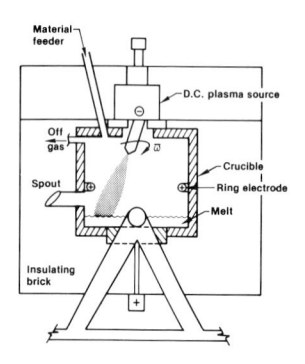

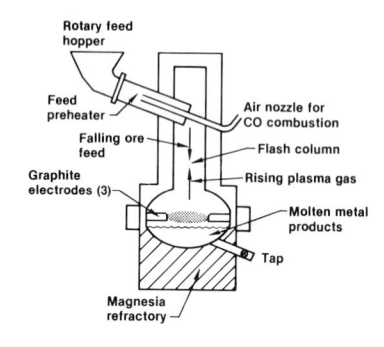

FIG. 10 Tetronics Plasma Reactor

FIG. 11 Extended Plasma Arc Flash Reactor

The anode attachment can be either to a circular water-cooled electrode located at the periphery of the chamber or directly to the surface of the molten material. This type furnace has been operated at power levels from 0.2 to 1.4 MW [9]. Applications include thermal treatment of sand (zircon) to facilitate leaching of silica from zirconia, concentration of low-grade ilmenite ores, spheroidization of nickel and iron powder, and removing copper from copper-bearing minerals.

The use of a closed-hearth type, extended carbon electrode plasma furnace for the reduction of many types of oxide ore fines has been developed in Canada [10]. Figure 11 illustrates the so-called "extended arc flash reactor". The hearth is formed within a magnesia refractory lining. Three hollow graphite electrodes at 120 deg spacing supply the 3-phase power; gas injection is through the hollow electrodes. Pulverized oxide materials are fed into the top of the flash column of the furnace; as the fines drop through the rising hot gas stream into the zone of the diffuse plasma, considerable preheating and prereduction occurs. Metal and slag, resulting from the reduction of the ores, collect in the hearth and are heated by radiation and convection from the plasma arcs. High carbon ferrochromium has been produced by reduction of low-grade chromite ores with a variety of reducing agents in this type of furnace. Metal recoveries between approximately 85 to 95% were achieved at energy requirements of about 13.2 kW hr/kg.

Cold Crucible and Other Casting Systems

Different types of consumable and nonconsumable electrode designs are in use. The consumable type systems are primarily employed in the melting and remelting of bulk weldable electrodes and in the steel industry. Consumable type systems are in commerical use in the U.S.S.R. for the production of nitrogen alloyed steels [8]. Figure 12 is a schematic of a design employing a hollow electrode that serves as a collimator/consumable nozzle to obtain a longer discharge with associated higher voltage levels. In contrast to typical VAR reactors with large shrinkage cavities, this design includes provision for an on-axis plasma source for hot topping.

Figure 13 is a schematic of another type of consumable electrode plasma melting reactor that is in use in the U.S.S.R. [8]. In this system, a plurality of plasma sources are located in the sidewalls and are used to continuously melt a large consumable electrode that is rotationally indexed downward toward the mold assembly. Use of a cold mold and controlled plasma heating assists in achieving the shallow melt profiles in this configuration that are conducive for ingots of good workability due to unidirectional grain solidification. The most recent advances have been made with cold crucible nonconsumable electrodes coupled with continuous ingot withdrawal.

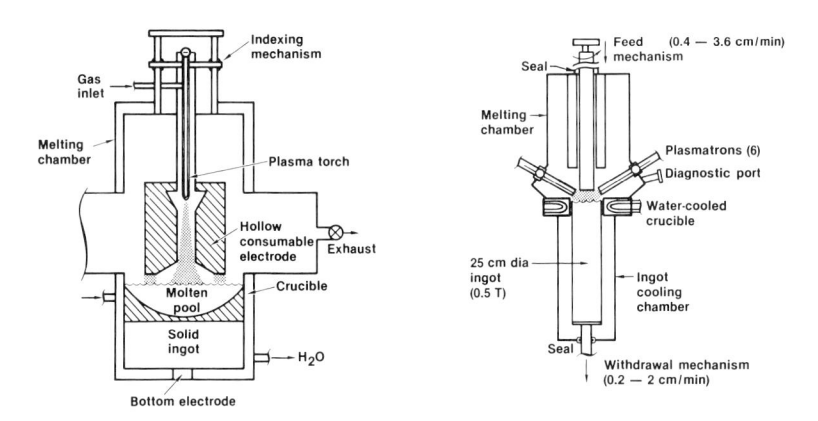

FIG. 12 Sketch of U.S.S.R. Production Scale Plasma Reactor for Hollow Electrode Melting

FIG. 13 Sketch of U.S.S.R. Production Scale Plasma Reactor for Continuous Ingot Withdrawal

The nonconsumable plasma melting reactor with cold crucible and continuous ingot withdrawal has been developed primarily for scrap (solids, turnings, powder) consolidation into ingot material and subsequently into final working billet form via several vacuum arc remelt (VAR) operations. Figure 14 illustrates the Daido Plasma Progressive Casting (PPC) reactor with a single thermal plasma source [6]. In this single source configuration, the power level is about 0.1 MW and a 12-cm-dia mold is used for continuous bottom withdrawal. Because of the strong affinity of titanium for oxygen/nitrogen, water-cooled copper molds with the plasma operating under an approximately 1 atm argon environment is employed. This system is used in the development of continuous ingot withdrawal technology using titanium sponge, scrap, commercially pure Ti and various alloy combinations thereof.

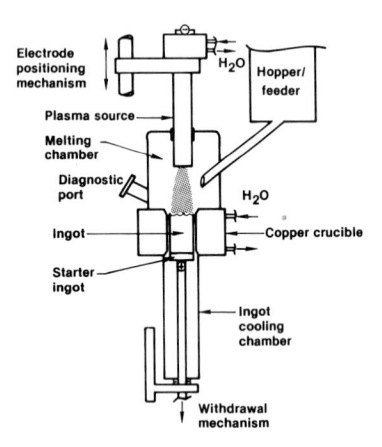

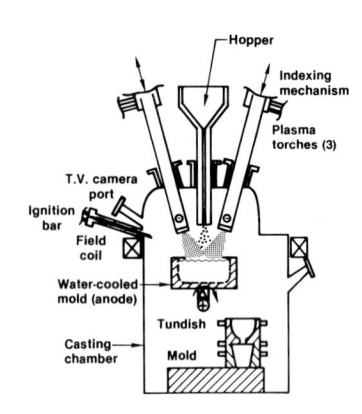

FIG. 14 Daido Plasma Progressive
Casting Reactor

FIG. 15 Daido Multiple Plasma
Source Reactor for Skull Casting

Figure 15 shows the basic plasma skull casting reactor under development at Daido over the last five years [6]. This reactor produces castings of reactive metals and alloys. Unique characteristics of this type reactor relative to vacuum arc or electron beam systems include: operation in a 1 atm inert (argon) environment with negligible vaporization losses of the alloying elements; negligible loss due to molten pool splash; wide variety of scrap forms can be used; and a significant degree of superheat to the molten metal is achievable thus leading to simplified production of thin wall castings. Recent emphasis is on production of reactive intermetallic compounds that have application for hydrogen storage. The exact geometry used, argon injector design, cathode type, degree of water cooling etc. are key parameters in providing a stable, uncontaminating, long-lifetime plasma source. Cathode tip lifetime is reported to be 300 to 1,000 hours.

Special Applications

The Daido Takakura works uses a modified thermal plasma system for reprocessing nuclear wastes. The objective is the volume reduction (approximate order of magnitude) and stabilization of the noncombustible solid waste generated at nuclear power plants. Figure 16 is a sketch of the basic thermal plasma melting furnace and solidification chambers [6]. The processing capacity of the reactor is about 200 T/yr using a total plasma power of 310 kW. In the solidification chamber three types of forms are being generated: (1) 0.3 m cubic blocks; (2) shot (0.1 to 0.7-cm-dia); and (3) direct solidification into a storage can. The thermal plasma characteristics of high temperature, long run time, stable operation at atmospheric pressure, and nonconsumability of the tungsten electrode has provided relatively easy and complete melting of all metals; inorganic materials, and organic materials. The overlying advantage of using the thermal plasma for this type of melting/consolidation process is the independent aspect of the primary heat source relative to other systems. A separate 80 kW plasma source is used in conjunction with a water-cooled hearth system to allow continuous pouring of the molten waste. The input nuclear waste material is contained within sealed 50 liter drums that enter a vertical feed chute through an airtight valve system. The maximum melting capacity is 4 T/day.

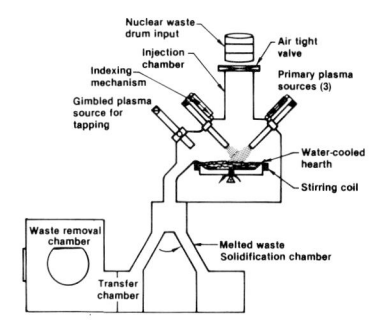

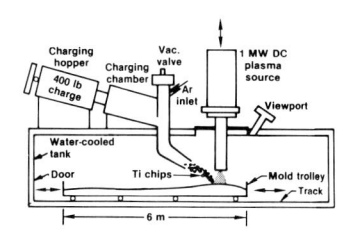

FIG. 16 Daido Plasma Reactor for
Nuclear Waste Processing

FIG. 17 Frankel Plasma Reactor for
Scrap Agglomeration

The Frankel Co. is the only facility in the U.S. that uses a thermal plasma reactor for a consolidation operation. Fig. 17 is a sketch of the reactor used for agglomerating (i.e., partially melting) Ti scrap, virgin sponge, or master melt alloys into porous bulk weldable electrodes (approximately 5 ton ingots) that are used in secondary VAR processing [12]. A single d.c. plasma source (approximately 1 MW) is vertically oriented within a large (2-m-dia) steel tank. Inside the tank is located a full length track upon which a mold trolley (water-cooled split halves -- 0.5-m-dia) traverses. The relatively low voltage plasma source (approximately 200 volts) operates in the transferred mode on argon gas with coaxial plus swirl flow for aerodynamic arc spot movement. A unit of this size can process about 0.5 T/hr. A batch loading procedure is used. After initial chamber pump down to vacuum conditions, the plasma melting operation takes place at slightly above 1 atm. Following electrode fabrication, the material is shipped to VAR melters for subsequent homogeneity and chemistry adjustment according to specifications and final product usage. This type plasma scrap alloy agglomeration process relies on proper scrap treatment and classification prior to the actual melting step. No provision is included for the removal of deleterious inclusions.

The USSR has had a continuous development effort in the area of thermal plasma reactors for the production of large monocrystals of refractory metals, directionally solidified eutectic alloys, and high melting point carbides. The primary effort in this area, in progress throughout the last seven years, is being conducted at the Baikov Institute [13, 11].

The basic operation of the system shown in Figure 18 begins with fusing the end of a seed cyrstal using the plasma source itself. The molten material is supplied by plasma melting of corresponding alloy metal (compacted or hot isostatic pressed) from small diameter feedstock. Continuous withdrawal of the ingot is conducted simultaneously with the solidification of the melted material on the top of the crystal. Relatively large diameter monocrystal ingots of tungsten have been produced possessing excellent physical properties, especially the property of ductility at normal temperatures. Monocrystal ingots of TiC, HfC, TaC, NbC have also been produced via the thermal plasma system. The distinct advantages of this thermal plasma aproach over systems such as electron beam and vacuum arc resides in its compact simple design coupled with its energy efficient, simplified mode of control and operation with a wide variety of gases. Commercial applications of this type system have been activated throughout the USSR.

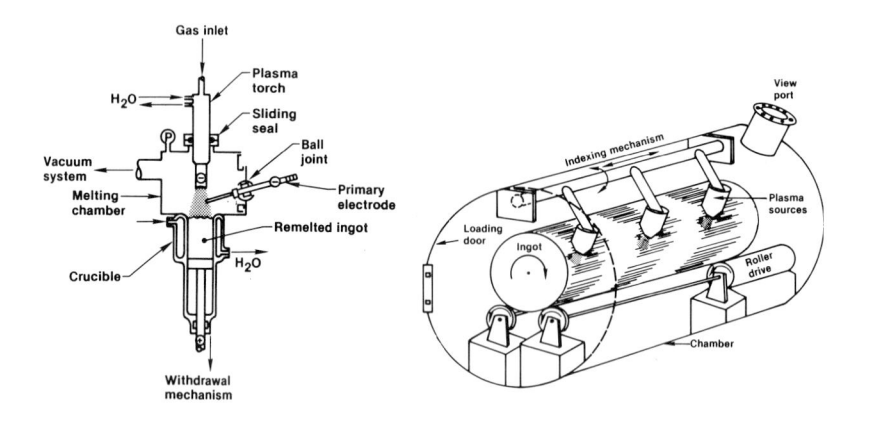

FIG. 18 USSR Plasma Reactor Used in Monocrystal Growing

FIG. 19 USSR Plasma System Used in Surface Treatment

Thermal plasma surface treatment, another special application that is similar to a laser glazing operation, consists of locally melting a relatively thin layer of material in a continuous operation. This operation is used to repair surface cracks, and other blemishes that have occurred in prior processing steps. The ability to improve the surface condition and concentricity to within designated specifications alleviates the need for use of expensive centerless grinders and other surface treatment equipment. Plasma surface treatment equipment development is currently being conducted in the USSR [8]. Figure 19 is a schematic of the type of equipment presently being used for surface treatment of steels, specialty steels, and titanium. A chamber large enough to contain the entire billet is operated under an inert atmosphere environment. The billet is supported on a series of mechanically actuated rollers that allow controlled rotation of the billet while being exposed to a series of thermal plasma sources.

Concluding Remarks

Thermal plasma melting and processing is an interdisciplinary technology in transition. It has become established in several fields of commercial application as described in this overview. New prospects for large scale implementation of plasma melting technology by the domestic mining, metallurgical, specialty metals, iron and steel and chemical industries appear promising based on recent state-of-the-art advances. Overall progress in the last several years has been encouraging, however, gaps still exist in our fundamental understanding of thermal plasmas and their interaction with the wide variety of materials to be processed. The importance of this area is being recognized by many industrial groups (including Hydro-Quebec, SKF, UTC, Westinghouse, et al.) where research is currently underway to better define new applications of thermal plasma technology and provide the demonstration hardware required to establish both technical and economic viability. The many inherent advantages that the plasma reactor offers should result in the commercialization of new melting and materials processing schemes in the near future.

REFERENCES

1. G. Magnola, The Plasmarc Furnace, Can. Min. and Metall. Bulletin, 57, 57-62 (1964).
2. D. R. MacRae, Plasma Processing in Extractive Metallurgy: The Falling Film Plasma Reactor, Amer. Inst. Chem. Engr, 75, 186, 25-30 (1979). Also U.S. Patent 3,661,764, (1972).
3. L. W. Scammon, U.S. Patent 3,661,764, (1972).
4. P. Wilks, Plasma Materials Inc., Manchester, N.H., Private communication June (1983).
5. S. L. Camacho, Plasma Melting Experience with the PEC Transferred Arc Torch, Int'l Ind. Seminar on Pilot Plant Experiences - Melting and Processing Technology, S.C., Oct. (1983).
6. K. Yamaguchi, et al, Daido's Plasma Technology, Daido Steel Co. Ltd., Machinery Division Tech. Rpt., Nagoya, Japan, June (1981).
7. W. Lugscheider, Large Scale New Melting Technology Developments, 1st Int'l Ind. Seminar on Pilot Plant Experiences -Melting and Processing Technology, Seabrook, S.C., Oct. 1980. Also 5th Int'l. Symp. on Plasma Chemistry, Edinburgh, Scotland, August (1981).
8. A. Bolotov, Institute of Energy, Alma-Ata, USSR. Private communication 6th Int'l Symp. on Plasma Chemistry, Quebec, Canada, July (1983).
9. J. R. Monk, Application of Plasma to Metallurgical Processes, Proceedings of 5th Int'l Symp. on Plasma Chemistry, Edinburgh, Scotland, August 1981. Also Canadian Patent 957133; (1974).
10. R. S. Segsworth and C. B. Alcock, Extended Arc Furnace and Process for Melting Particulate Charge. U.S. Patent 4,006,284, February (1977).
11. N. N. Rykalin, Plasma Processes in Metallurgy, Moscow Nauka 1973, Plasma Engineering in Metallurgy and Inorganic Material Technology, Pure and Appl. Chem. 48, 179-194 (1976). Private communications with N. N. Rykalin at 4th ISPC, Zurich, 1979, and A. Bolotov at 6th ISPC, Canada, (1983).
12. G. Herman, Frankel Co., Compton, California. Private communication January (1982).
13. Y. M. Savitsky, et al., Production of Monocrystals of High Melting Point Metals by Plasma Arc Heating, JPRS Trans. No. 61321 (1974).

PLASMAS IN EXTRACTIVE METALLURGY

W. H. GAUVIN AND H. K. CHOI
McGill University* and Hydro-Quebec Research Institute**,
*Montreal, Que., and **Varennes, Que., Canada.

ABSTRACT

It is becoming apparent that plasma technology will revolutionize certain of the conventional operations of extractive metallurgy. The use of plasma arcs in the iron and steel industry is already being demonstrated at powers up to 100 MW and these development will be briefly reviewed. In this connection, the advantages of plasma furnaces over electric furnaces are beginning to be well documented.

It is, however, in the area of the production of the refractory metals, such as molybdenum, zirconium, titanium, niobium, tungsten, tantalum, chromium and vanadium, that plasma technology offers unique opportunities as a result of advances in the design of transferred arc reactors. The work of the authors in this particular field will be described as well as other recent industrial developments.

PLASMA TECHNOLOGY IN THE IRON AND STEEL INDUSTRY

Following three decades of gestation, during which a considerable body of research data were nevertheless accumulated, there is every indication that successful applications of plasma technology to metallurgical operations are currently multiplying at an accelerated rate. To begin with, the reluctance of operators to switch from conventional fuels to electricity is slowly disappearing. The widespread use of electric furnaces in steelmaking based on scrap, in the production of highly-alloyed steels and of ferro-alloys, is ample proof that electrical energy can be used industrially and on an economic basis in many applications. More importantly, however, there is a growing realisation that, in spite of the very high degree of sophistication reached by electric furnaces (sizes up to 7 meters above the slag line, in capacities in excess of 100 tons per hour and powers up to 120 MW), they present many disadvantages which are inherent to their fundamental principles of operation, such as high cost of refractory replacement (16% of the total operating cost) and of electrodes (24%), while the cost of electrical energy is only about 37% [1]. The operation is a batch process. Each step of the process requires a different level of applied power to the electrodes and results in different power factors, refractory wear index and temperature level control. Because of the changing power demands, the shocks to the electric grid are very severe. Finally, electric furnaces are restricted as to the particle size and physical characteristics of the feed material. They certainly cannot accept materials in the liquid or gaseous states, which, as shall be seen later, may be the preferred state for the feed in certain applications.

Recent developments in plasma furnace operation have clearly shown that these furnaces have the capability of overcoming the shortcomings of the electric furnace. As far as metallurgical appllications are concerned, it is in the iron and steel industry that the major developments have taken place, and it is in this field that a direct comparison of the performance of the two types of furnace has become possible. Such a comparison was recently documented in an article by W. Lugscheider of Voestalpine, Austria [2]. This article is based on the operation of a 10-ton furnace first commissioned in Freital, in the German Democratic Republic (GDR, East Germany), followed by the commissioning of a 30 ton plasma furnace in the same plant in 1977.

The 10-ton plasma arc furnace has been in 3-shift operation since 1973 at the VEB Edelstahlwerke at Freital. This furnace was built by VEB Lokomotiv-und-Elektrotechnische Werke, "Hans Beimler" in Henningsdorf. It has a pivoting roof, a hearth, and a side wall which are constructed as individual segments and fitted together to form a vessel. Each segment when worn out can be easily replaced with minimum down-time of the melting system. The wall and roof of the furnace are lined with chrome-magnesia bricks to withstand surfaces temperatures of 1850°C. Sensors placed in the lining prevent overheating of the refractories by controlling the heat output of the plasma torches. The furnace hearth is prepared from rammed magnesite. The furnace is heated by means of three plasma torches of the transferred-arc type, each of 3-MW nominal power, externally mounted.

The return electrode of the transferred arc plasma system is located in the center of the hearth. It consists of a water-cooled copper block equipped with a temperature warning device which also shuts the delivery of the electric power to the system in the event the temperature is about to exceed a safe limit. The cathode tip is made out of tungsten, with lanthanum as electron emitter addition. The plasma gas is argon (8 Nm^3/h in each torch). The maximum adjustable current in the working range of 200 to 600 volts is 6 kA.

As previously mentioned, a 30-ton plasma furnace was commissioned in the Freital plant in 1977. This furnace had been constructed at the Electrothermal Equipment Manufacturing plant at Novosibirsk, USSR. The features of the 30-ton plasma furnace are very similar to those of the 10-ton unit except that a fourth torch was added. The operating voltage for the plasma torches in the 30-ton unit was from 150 to 660 V with current capability all the way up to 10 kA. However, the current density in the tungsten cathodes was not permitted to exceed 2 000 A/cm^2. A service life of at least 200 hours has been reported for the cathodes. The life of the anodes is of course much longer. The argon flow in the larger furnace is 60 m^3/h. The power factor in both furnaces is 0.96. Finally, the specific power consumption is 600-650 kWh/ton in the smaller furnace and 550-600 in the 30-ton furnace [3].

In the USSR, plasma furnaces of 100-ton capacity are now operating with six 3.5-MW plasma torches. Reportedly, these furnaces are being used for smelting ferroalloys and hot metal preparations using a combined charge of steel scrap and direct-reduced iron. The objective is to develop a continuous alloy steel production operation on a streamline basis from one concentrate to final mill product [4].

Based on the growing body of evidence available, it is justified to claim the following advantages for the plasma furnace [3, 4]:

(1) Higher melting efficiency compared to similar sizes of conventional electric arc furnace, without adjunct use of secondary refining.
(2) Ability to produce alloys with low carbon.
(3) Superior recovery of the alloy metals contained in the scrap and of those added in the melt.
(4) Ability to inject and alloy the melt with nitrogen through the gas

phase instead of through the use of expensive nitrogen-containing ferroalloys.
(5) Higher quality product compared to that from conventional electric arc or induction melting furnaces.
(6) Achieving low oxygen and hydrogen residuals in the melt.
(7) Reduction of iron loss to below 2 per cent.
(8) Suppression of noise to a level below 60 dB.
(9) Elimination of discontinuous shock loading of the electric power transmission mains.
(10) 30% lower capital investment compared to a modern Ultra High Power (UHP) electric arc furnace system.

Less well documented from an operational point of view, but nevertheless very promising are a number of recent developments, such as:

- The Plasmared Process (SKF, Sweden), to preheat and upgrade the reducing gas in iron reduction, at the 8-MW level.
- The Plasmasmelt Process (also SKF), for the direct smelting reduction of iron ore.
- The Plasmadust Process (also SKF), to recover metals such as zinc and lead, from steel mill oxide-containing dusts.
- The replacement of blast furnace tuyeres by plasma heaters (Westinghouse, USA).
- The oxidation of $TiCl_4$ to produce TiO_2 (Tioxide International, England).
- The decomposition of zircon to ZrO_2 and SiO_2 (Ionarc, USA).
- The recovery of strategic metals from the scrap generated by the aerospace industry (United Technologies, USA).

PLASMA FURNACE DESIGN FOR REFRACTORY METAL PRODUCTION

So-called refractory metals belong to Groups IVB, VB, and VIB of the Periodic Table. More specifically, this class of metals includes zirconium, titanium, niobium, vanadium, tantalum, chromium, molybdenum and tungsten. They exhibit high strengths and good corrosion resistance in a wide range of corrosive environments. Because of their high melting points and their free-energy-of-formation characteristics, they require a fairly elevated processing temperature. It is not surprising, therefore, that the possibility of using plasmas as a heat source would attract increasing attention, particularly in view of the fact that the plasmagen gas, in many cases, can also be used as a reactant. The reviews by Fauchais et al. [5, 6] contain details of the vast body of research carried out during the last 20 years.

Of the four major types of plasma generation - DC jet, RF torch, hollow tubular electrodes and transferred-arc plasmas - only the latter, generated between a non-consumable cathode and an anode consisting of the molten product, appears to be capable of meeting the exacting specifications demanded by the strategic metal market. None of the other three types can provide the uniformity and completeness (nearly 100%) of reaction required for most of these metals.

A very efficient design of transferred-arc reactor is the Hydro-Quebec-Noranda design shown in Figure 1. In this type of reactor, the powdered feed (which may consist of a very wide spectrum of size distributions from a few microns to several millimetres) is introduced tangentially in a high-velocity cold carrier gas and forcefully projected against the inner wall of a cylindrical "sleeve" where it is exposed to the intense radiation of the plasma column generated between the cathode shown in Figure 2 and the molten bath of product, which acts as the anode. (For powers over 1 MW the conical cathode tip shown in Figure 2 should be replaced by a cylindrical thoriated

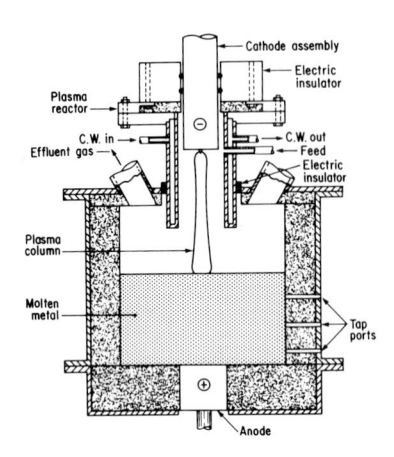

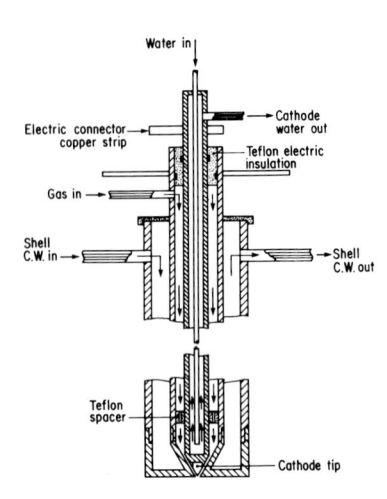

FIG. 1. Hydro-Québec-Noranda transferred arc reactor.

FIG. 2. Cathode design.

tungsten button embedded in copper). The feed begins to melt as soon as it hits the wall of this sleeve and flows down the inner wall of the latter in the form of a molten film, where the desired high-temperature reaction begins to occur. During this stage of the operation, all the radiant heat generated by the plasma column is absorbed by this falling film. At the lower edge of the sleeve, the molten stream agglomerates in the form of 200- to 500-µm drops, which fall into the molten bath below, continuing to absorb radiant heat in the process. Whatever additional residence time is required by the reaction is provided in the molten bath, before the product is tapped out. Depending on the arc length, from 50 to 60% of the power fed to the system is transferred to the anode bath as electron transfer effects.

Several small-scale (up to 125-kW) versions of this design have been demonstrated by Noranda and by McGill during the past three years, and a 1-MW reactor of this design is currently being tested by Davy McKee in their R & D division in Stockton-on-Tees, England. Finally, IREQ (the research arm of Hydro-Quebec) has just completed the construction of a research reactor designed to operate in the 50 to 250 kW range.

The Hydro-Quebec-Noranda design offers a number of advantages, which are not immediately obvious:

a) The molten film continues to absorb radiation as it falls into the molten bath, thus shielding the walls of the reactor and obviating cooling requirements.
b) The feed carrier gas, which is optically thin, picks up very little heat from the plasma by radiation and only a little by convection from the molten film on the sleeve; it thus helps to maintain the back of the sleeve and the roof of the reactor relatively cool, as it leaves through the effluent exit ports on the roof.
c) The cold plasmagen gas is used to cool the cathode assembly before passing over the cathode tip.
d) No cooling of the anode is required if (i) a large plate is used, of the appropriate surface area and (ii) a heel of molten metal is left at all times at the bottom of the reactor, when the latter is emptied.

e) Because of the efficient utilization of the electrical energy supplied to the system, no water cooling of the reactor is necessary. Judicious heat transfer design results in the formation of a skull of solidified metal against the reactor refractory, which protects the latter.
f) In larger installations, stirring of the anode molten bath should be effected by means of a magnetic field. This is not necessary for powers up to 1 to 1.5 MW.
g) Because of the high tangential velocity of the carrier gas for the feed, complete stabilization of the plasma column is effected, which permits the use of greater arc lengths and also avoids short-circuiting with the sleeve. This, in turn, allows a higher voltage to be applied, with a resulting smaller current, which is most desirable, particularly in view of the upward curve of the voltage-current (V-I) characteristics of this type of device. Very large currents are thus avoided, which considerably minimizes cathode erosion. Two additional factors contribute to increasing the voltage still further: firstly, a certain fraction of small particles may enter the plasma column in the sleeve region and thus decrease the electrical conductivity of the latter, and secondly the gases in vortex motion surrounding the plasma column undergo a sudden expansion as they emerge from the sleeve; this in turn causes an enlargement of the column diameter, an increase in the total energy radiated and a corresponding increase in voltage.

In the light of the above description, it is now possible to consider the operating characteristics of other transferred-arc systems. The reactor design developed by MacRae and colleagues in the Homer Laboratory in Bethlehem, U.S. undoubtedly constitutes a major pioneering effort (see Figure 3). Its distinguishing feature is the location of the anode, which takes the form of a vertical cylinder embedded in the wall on which the feed impinges to form a molten film. It should be noted that the molten bath in the bottom crucible is heated only by convection in contact with the fairly cool spent gases, and that the metallic anode is protected from the random arc roots only by the presence of a layer of molten material. Should the latter exhibit a break, catastrophic gouging of the bare wall might result. Good results have been obtained by MacRae on the production of iron (using CH_4 and H_2) and of ferrovanadium (using carbon).

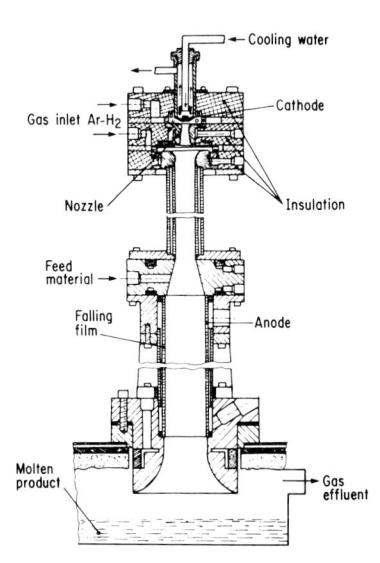

FIG. 3. Bethlehem Steel Co. reactor.

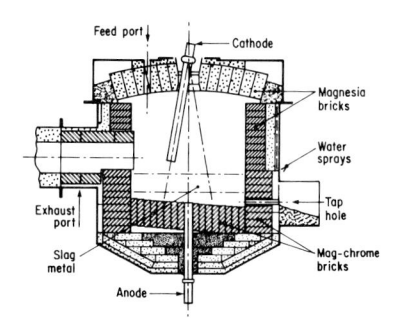

FIG. 4. Tetronics reactor.

The transferred-arc furnace developed by Tetronics Research and Development Co. Ltd. of Faringdon, England, is shown in Figure 4. It is characterized by a rotating cathode assembly making a small angle (9°) with the vertical which aids both stirring and anode energy distribution. This arrangement is called an Expanded Precessive Plasma Gun. Very little plasmagen gas is fed to the cathode and the feed material, which can range up to 1 or 2 cm in size, is simply dropped through the feed ports. Heat losses (to the cathode assembly and furnace walls) range from 20% to 30% for small furnaces and are down to about 15% for furnaces of 1.4 MW and higher (up to 3 MW). This furnace is simple in design and apparently quite rugged. One of its most recent applications was for the production of ferrochromium at a power of 0.55 MW, conducted at Faringdon in collaboration with Mintek (Council for Mineral Technology) of Randburg, South Africa [7]. This work clearly indicated the technical feasibility of producing a high-carbon (5% to 6%) ferrochrome analyzing 55-56% Cr out of a South African ore containing 44.6% chromite (Cr_2O_3), in the presence of fluxes consisting of quartz and lime, and with coal (54.3% fixed carbon, 33.4% volatile matter) as reductants, with good thermal efficiencies.

Based on this work, Curr et al. [8] of Mintek have very recently described an experimental 100-kVA furnace, shown in Figure 5, with a carbon lining on the roof, obviously designed to remedy a weakness in the roof construction of the previous furnace. Characteristic of this furnace is the hollow graphite cathode, 50 mm in diameter with a 10-mm bore, through the centre of which a mixture of argon and nitrogen was passed as the plasmagen gas at rates of 5 to 50 L/min. The feed, ranging in size from 0.2 to 6 mm, was fed under gravity at rates of 5 to 70 kg/h through the single feed port, and not through the central bore of the cathode. Based on this experimental work, Middleburg Steel and Alloys recently announced that they have ordered a 20-MW plasma furnace from ASEA, Sweden, to be commissioned in October 1984, specifically to expand their ferrochrome production and thus help South Africa to boost its share of world ferrochrome production from 42% in 1980 to around 65% in the near future.

In closing this description of plasma-generating devices, mention should be made of another type which, although not a transferred arc, may

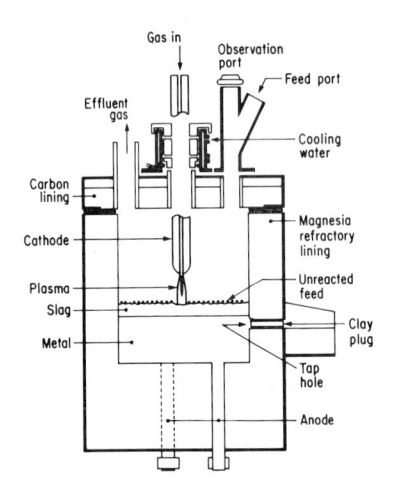

FIG. 5. Mintek reactor.

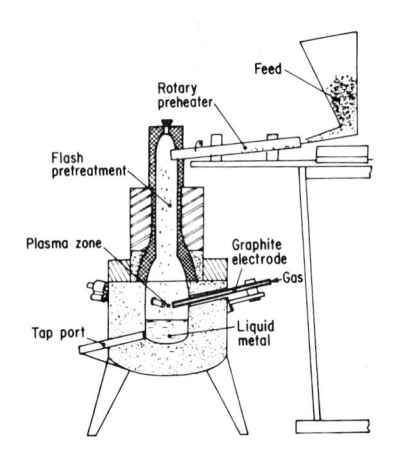

FIG. 6. University of Toronto reactor.

find applications in the production of ferrochromes. Developed at the University of Toronto [9] and schematically shown in Figure 6, this device operates on 3-phase AC current, with the arc struck between three hollow graphite electrodes, through the bore of which the plasmagen gas (Ar, Ar plus H_2, or CO) is fed. The feed falls by gravity through a chimney where it is pre-treated in counter-current contact with the effluent gases before passing through the arc and finally falling into the bottom crucible. This design has many advantageous features, although the fact that the graphite electrodes are consumable may be detrimental for certain applications since they may create unwanted carbon contamination of the product. It also requires that a feed consisting of fine particles be pelletized or agglomerated prior to treatment in this furnace.

RECENT PROCESS DEVELOPMENTS

Ferrochrome

It is probable that the production of this alloy is currently receiving the greatest amount of attention on the part of industry. Because of the availability of good ore grade (typically 44% to 46% Cr_2O_3) South Africa holds a dominant position in this very important component of the steel industry. By addition of suitable fluxes (quartz or SiO_2, and lime) and coal as the reducing agent, which may be either fed with the chromite, or dumped into the molten bath, a high-carbon (5% to 6% ferrochrome) is readily obtained. The metallurgical reactions occurring are fairly complex, but it is probable that the controlling rate mechanism is the solution of the iron and chromium oxides in the slag (governed by the oxygen activity) which are then reduced by the carbon in the coal to yield the metal and, also, the metal carbide. The process is heavily endothermic. Depending on the composition of the ore, the amount of fluxes and excess carbon must be carefully determined so as to minimize the losses of Cr_2O_3 in the slag (which should not exceed 2.5%). For example, for a Cr/Fe ratio of 1.69 in the ore, an approximate feed composition might be 60 kg of 44% chromite, 12 kg of SiO_2, 3 kg of lime and 25 kg of coal, all on a dry basis. Optimum temperatures appear to be 1900 K for the slag and about 1825 K for the liquid metal. The product would analyze about 56% Cr, 34% Fe and 5.5% C [7].

The production of ferrochromes by plasma techniques illustrates most decisively the many advantages of the latter over the conventional submerged-arc electric smelting furnace, the most important of which are:

a) no flicker on the power grid
b) little noise or gaseous pollution and few other environmental problems
c) no need for costly sizing, screening and agglomeration of the feed
d) no problems due to electrical conductivity of the carbonaceous reducing agent and the slag, since the plasma furnace operation is not affected by an uncontrollable rate of descent of the feed into the hot zone, as in an electric furnace
e) much lower installed cost of power supply. Costs for the plasma reactor are $165 per installed kilowatt ($125 for supply of power, transformer, SCR, power factor correction, and $40 for delivery to plasma reactor). The cost for a conventional electric furnace is $360 per installed kilowatt ($60 for the power supply and $300 for delivery to the furnace). These costs are in US dollars
f) no need for very large vessels to accommodate a large volume of unreacted burden
g) no need for expensive lumpy metallurgical coke; any carbonaceous reducing agent can be used, providing it will not contribute to undesirable impurities in the product

h) trivial cost of replacement (including down-time) compared to the cost of cathode consumable electrodes used in arc furnaces, despite the fact that the cathodes used in transferred-arc furnaces have a limited life (about 200 to 300 h at present).

Owing to the high endothermicity of the reduction reaction (it takes almost exactly as much energy to bring the raw feed to 1800-1900 K as it does to reduce it at that temperature) the operation is highly energy-intensive (about 3.2 kWh per kilogram of chromite charged).

In conclusion, it would appear that the possibility of producing high-carbon ferrochrome by the plasma route from medium-grade ores available in Canada and in the USA, is well worth considering. From a technical point of view, it illustrates the use of carbothermy to reduce an oxide, in preference to aluminum, silicon or magnesium as a reductant.

Zirconium

In contrast with chromium, little development work has been done on the plasma treatment of ZrO_2 to produce the metal. Is should be mentioned that the plasma decomposition of zircon sand ($ZrSiO_4$) to yield ZrO_2 has been carried out commercially by Ionarc in Bow, New Hampshire, for a number of years.

Starting with zircon, the production of zirconium sponge is complicated by the necessity to remove the hafnium (which is invariably associated with zirconium in the ore) down to 50 ppm in the sponge, to meet the very severe criteria imposed by the nuclear industry. This results in an extremely complex and ponderous sequence of over 60 steps in the overall conventional extraction process, as practised by Wah Chang, Western Zirconium or Pechiney, Ugine, Kuhlmann.

Figure 7 represents the last stage in the commercial process, often called the Kroll process. It is the most expensive stage in the whole process (about 35% of the total operating cost of US$16 a kilogram to produce the sponge) owing to the cost of magnesium (about a kilogram is required per kilogram of sponge) and of argon and helium, in addition to the high labor cost. Finally, this is a batch process and the final stages of sponge shearing, sorting, crushing, blending, etc. are particularly labor-intensive.

Replacement of the conventional Kroll process by a plasma process has been under study at McGill University in Montreal and at the Industrial Materials Research Institute (IMRI) in Boucherville, Que., for the past three years. Figure 8 is the flow sheet of the proposed plasma process. The technical feasibility of the process has been demonstrated [10, 11], but efforts are now being devoted to optimizing the recovery of the metal in the molten zirconium anode. Although complete capture is not required (unreduced $ZrCl_4$ or lesser chlorides can be recycled to the torch) a minimum of 30% collection is desirable. From an economic point of view, the plasma process, at about 10 kWh/kg of Zr, offers considerable advantages: no magnesium is required, the process is continuous and, most important, a dense metal is produced (not a zirconium sponge) which could be continuously cast.

Little incentive to pursue the development of a plasma route to nuclear zirconium has been evidenced, however, owing to the small size of the market. On the other hand, the production of zirconium metal containing the naturally-occurring hafnium would appear to present a real opportunity for a new industry in view of the low cost and abundance of zircon and the remarkable high strength and corrosion resistance of the metal, which are not affected by the presence of hafnium. With this in mind, the McGill Plasma

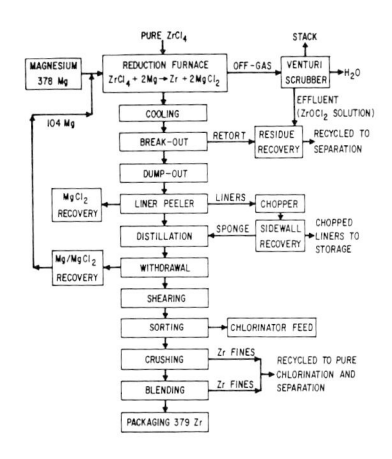

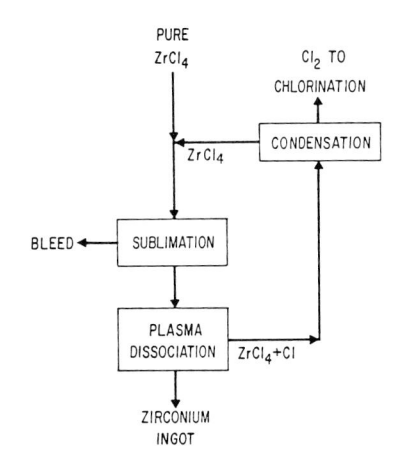

FIG. 7. Kroll process for last stage of zirconium sponge production.

FIG. 8. Plasma production of zirconium.

Group investigated the production of $ZrCl_4$ by the chlorination of ZrO_2 in the presence of a chlorine plasma [12, 13]. Chlorination occurred both with and without carbon addition but was faster in the presence of carbon, at temperatures higher than 1700 K. Without the necessity to remove the hafnium, the production of dense zirconium metal at a cost of $4 to $5 per kilogram appears possible.

From a technical viewpoint, it should be pointed out that efforts to reduce ZrO_2 directly to the metal have not been successful. It can now be generalized that to produce a metal from its oxide, the best route is to convert the oxide to the halide (preferably, the chloride) which, for most refractory metals, has low melting (or sublimation) and boiling points, and is easily prepared.

Titanium

Commercially, titanium metal is produced by the Kroll process (reduction of $TiCl_4$ with magnesium) or by the Hunter process (reduction of $TiCl_4$ with sodium). Both processes suffer from the complexities discussed in the case of the Kroll porcess for zirconium production. For the past four or five years, the Electricity Council Research Centre in Chester, England, has investigated the single-stage gas-phase reduction of $TiCl_4$ with sodium in a hydrogen plasma at temperatures of up to 2300 K. As recently described [14], the process has run up against a considerable setback owing to the difficulty of controlling the addition of the feed materials.

The carbothermic reduction of TiO_2 in the presence of an argon or argon-plus-hydrogen plasma has been attempted at temperatures up to 3200 K. Although yields of 94% Ti were obtained, the product was invariably contaminated with titanium carbide.

In a joint McGill-IMRI project, the production of titanium metal, in a manner identical to that previously mentioned for zirconium, has been initiated. Here again, the problem is to optimize the yield of titanium captured in the molten anode. In addition to the production of metal ingots by continuous or semi-continuous casting of the anode metal, there is considerable interest in the production of secondary products, such as carbides and nitrides, and of other alloys, principally in the form of powders for plasma spray coating.

As produced by the Kroll Process, titanium takes the form of a so-called sponge metal and must there-fore be melted in order to be cast in slabs, ingots, etc. Bhat [3] has given a good description of the major recent developments in the melting of titanium as well as those of speciall steels and alloys. Daido, in Japan, has pioneered the concept of combining plasma transferred-arc heating with induction heating, as shown in Figure 9. For 2-tonne furnaces several 400-kW torches are used, and the frequency level for the induction coils is 150 Hz. In addition to its remarkable flexibility, this furnace is capable of producing resistance alloys (Kovar, Permallov) and super-alloys with quality levels equal to that of vacuum induction furnaces, more-over at a much lower cost and with better conservation of the alloying elements.

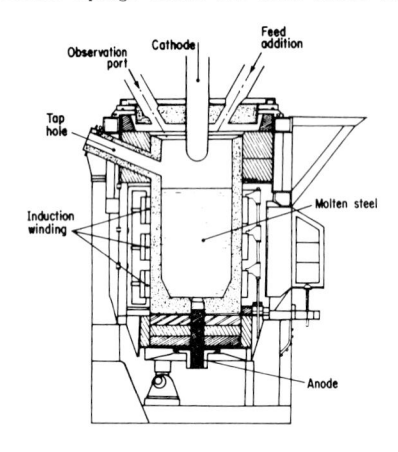

FIG. 9. Daido combined induction and plasma furnace.

Finally, mention should be made of the plasma beam torch [3] which is a hollow cathode transferring an arc to a molten bath which is contained in a continuous-casting mould. The feed material in a coarsely comminuted form is fed directly to the latter. The operation is carried out in a vacuum and very small amounts of high-purity argon are fed down the bore of the hollow cathode to create the plasma beam. Typical of this installation is the Ulvac (Tokyo) casting unit capable of casting ingots of commercial purity titanium and titanium alloys (such as the important 6Al - 4V - 90Ti) at a total power of 2.4 MW (six 400-kW plasma guns), operating at 10^{-3} to 10^{-1} Torr. The power source is rated at 34 000 A DC and only 90 V, and the power consumption is 9 000 kWh per short ton of alloy. It is interesting to note that the cost of titanium scrap to be remelted is $15 000 US per tonne.

Molybdenum

Ferromolybdenum and molybdic oxide are two of the most important additives for the production of specialty and HSLA steels. The conventional method of extraction from molybdenite, MoS_2 (the natural ore), begins with the roasting of the molybdenite concentrate (about 55% Mo and 39% S) to produce the oxide MoO_3 and, unfortunately, sulfur dioxide, which must be scrubbed out. This is followed by a stage of purification and, finally, by the reduction of the oxide with aluminum and ferrosilicon in what may be termed archaic conditions. It is not surprising, therefore, that decomposition of MoS_2 by the plasma route to produce the metal and gaseous sulfur which can be subsequently condensed to solid sulfur has attracted attention. The technical feasibility of this approach was proven by Munz [15] in 1975, who found that the rate of the decomposition reaction was controlled by the rate of heat transfer, at temperatures above 1883 K, which is the melting point of MoS_2. Although conceptually simple, this reaction presented an unexpected difficulty of considerable magnitude, namely the necessity of achieving 99.9% decomposition efficiency, to meet the maximum sulfur content of 0.15% in the product demanded by the steel industry. For the next five years, attempts were made to achieve this degree of sulfur elimination in eight different kinds of plasma reactor by in-flight particle contacting with a plasma tail flame, but without success. It is only when the transferred-arc reactor shown in Figure 1 was developed that a product containing 0.085% sulfur was finally obtained. It is interesting to note that considerable elimination of the more volatile impurities occurred while the sulfur was being driven off: magnesium, sodium and potassium were completely eliminated while 84% of the lead impurity, 85% of the antimony, 66% of the bismuth, 94% of the copper and 83% of the phosphorous were distilled off; the molybdenum remained in the liquid state.

Following an extensive development program, the process appeared so promising from an industrial point of view that a detailed engineering assessment was carried out [16]. With regard to the capital cost, the plasma furnace, ancillary power supply and instrumentation accounted for only 8% of the total cost. On the operating side, 4% covered the cost of nitrogen (as plasmagen gas), 18% for raw materials, 43% for labor, and only 11% for electrical power while 24% accounted for the loss of molybdenum in the effluent gas in the pilot plant available at that time (a deficiency which could be largely reduced).

Plasma thermal decomposition offers one of the most powerful techniques to extract metals from their naturally-occurring compounds. Certain compounds, however, appear to be unusually resistant to this approach. Thus silver selenide remains as an undissociated gas, even when exposed to a temperature of 5000 K. Similarly, although FeS_2, pyrite, rapidly loses one atom of sulfur, the remaining FeS appears to be impervious to thermal decomposition as a gas up to about the same temperature.

Vanadium

Very little difficulty was experienced in treating vanadium pentoxide in the IREQ-Noranda reactor with iron and carbon to produce a ferrovanadium containing 79.4% V, 18% Fe and 0.6%, which met the specifications of the 80% ferrovanadium required by the steel industry.

Tantalum and Niobium

Plasma production of the metals or of their compounds for these two elements is still in the laboratory stage. Recently, however, the carbothermic reduction of tantalite, Ta_2O_5, at a temperature of 3400 K, to yield 99.9% Ta with 50 ppm of oxygen and 400 to 500 ppm of carbon was reported [17].

The production of ferroniobium is currently being investigated at McGill University, following a promising preliminary study. The conventional process involves aluminum as the reductant and is carried out under vacuum; it is both complex and expensive.

CONCLUSIONS

The production of refractory metals and their alloys or secondary products is an area which has not received much attention,, in spite of the fact that the conventional extraction methods are complex and expensive. For the first time, plasma technology offers techniques of production which are simple and appear to be economically attractive. The plants would be small but efficient and highly productive. Required inventories and corresponding working capital, would be low. They would require little labor, but would be highly energy intensive. Continuous in operation, they would exhibit the unique ability of instantaneous start-up or shut-down, typical of plasma operations. Being totally enclosed and almost noiseless in operation, little environmental problems can be anticipated.

The field is admittedly new, but is based on a technology which is already in a fairly advanced state of development. Many opportunities exist for exciting new areas of research. For example, plasma technology offers an almost ideal method of producing the so-called glassy (or amorphous) metals. Equally challenging research opportunities should be kept in mind in the area of highly specific metal powders for plasma spray coating, which can be prepared by atomization of the molten metallic solutions of the appropriate composition, prepared by the transferred-arc technique previously described. It is hoped that this short review will stimulate still other areas of research in this promising field.

REFERENCES

1. Electric-Arc Steelmaking A Review , Metals and Materials, 27-40 (April, 1981).
2. W. Lugscheider, "Utilisation du Four à Plasma en Aciérie Electrique", J. du Four Electrique, No. 10, 29-33 (December 1981).
3. G.K. Bhat, "Plasma Technology and Its Industrial Applications", A survey report prepared for EPRI, Pittsburgh, PA (December 1981).
4. G.K. Bhat, Private Communication, following his visit to the USSR Electrotherm Plant, Novosibirsk, USSR.
5. J. Aubreton, B. Pateyron, P. Fauchais, "Les Fours à Plasma de Traitement Métallurgique", Rev. Int. Hautes Tempér. Réfract. 18, p. 293-319, (1981).
6. P. Fauchais, "Les réacteurs et Fours à Plasma", J. Four Electrique, No. 7, p. 9-16 (August-September 1982); No. 9, p. 7-12 (November 1982); No. 10, p. 29-35 (December 1982); No. 1, p. 28-30 (January-February 1983).
7. N.A. Barcza, T.R. Curr, W.D. Winship and C.P. Heanley, "The Production of Ferrochromiun in a Transferred-Arc Plasma Furnace", 39th Electric Furnace Conference, p. 243-260 (December 1981).
8. T.R. Curr, N.A. Barcza, K.U. Maske and J.F. Mooney, "The Design and Operation of Transferred-Arc Plasma Systems for Pyrometallurgical Applications", Proceedings of the Sixth International Symposium on Plasma Chemistry, Montreal, p. 175-180 (July 1983).
9. C.A. Pickles, S.S. Wang, A. McLean, C.B. Alcock and R.S. Segworth, Trans. ISIJ, 18, p. 369-377 (1978).
10. A. Kyriacou, "Characteristics of a Thermal Plasma Containing Zirconium Tetrachloride", M. Eng. Thesis (McGill University) (March 1982).
11. P. Spiliotopoulos, "Characteristics of Zirconium Tetrachloride Thermal Plasmas", M. Eng. Thesis (McGill university) (August 1983).
12. O. Biceroglu and W.H. Gauvin, "Chlorination Kinetics of Zirconium in an R.F. Plasma Flame", AIChE Journal, 26, No. 5, p. 734-743 (September 1980).
13. O. Biceroglu and W.H. Gauvin, "The Chlorination Kinetics of Zirconium Dioxide in the Presence of Carbon", Can. J. of Chem. Eng., 58, p. 357-

366, (June 1980).
14. K.A. Bunting, "Sodium and Titanium Tetrachloride Feed System for a Plasma Route to Titanium", Proceedings of the 6th International Symposium on Plasma Chemistry, Montreal, p. 193-198 (July 1983).
15. R.J. Munz and W.H. Gauvin, "The Decomposition Kinetics of Molybdenite in an Argon Plasma", AIChE Journal, 21, No. 6, p. 1132-1162 (1975).
16. W.H. Gauvin, G.R. Kubanek and G.A. Irons, "The Plasma Production of Ferromolybdenum - Process Development and Economics", J. Metals, p. 42-50 (January 1981).
17. K. Taniuchi and K. Mimura, "Arc-Plasma Reduction of Tantalum Oxide with Carbon", Proceeding of the Australia/Japan Extractive Metallurgy Symposium, Sydney, Australia (July 1980).

RAPID SOLIDIFICATION BY PLASMA DEPOSITION

D. APELIAN
Professor and Head, Department of Materials Engineering,
Drexel University,
Philadelphia, PA 19104

ABSTRACT

Rapid solidification processes (RSP) have been reviewed. The structural refinement one may obtain via RSP and the consolidation methods which need to be utilized have also been reviewed. The merits of low pressure plasma deposition (LPPD) or rapid solidification by plasma deposition (RSPD) as a means of combining both the atomization and consolidation steps of RSP is highlighted. The applications and challenges of RSPD are discussed.

INTRODUCTION - RAPID SOLIDIFICATION

The interest in the production of rapidly solidified materials and their subsequent consolidation into near-net shaped components can be traced to segregation of the alloying elements which arises during conventional solidification processing. Engineering alloys freeze over a range of temperature and liquid concentrations; as a consequence, during the freezing process the alloying elements segregate out. In the mushy zone which is what exists over this temperature range during the freezing process, the solidified fraction is in the form of many dendritic crystals and coexists with the liquid phase. This liquid phase in the mushy zone can be thought of as the interdendritic fluid. Segregation in castings and ingots is divided into two categories: macrosegregation (long range segregation) and microsegregation (short range segregation). Macrosegregation occurs over distances approaching the dimensions of the casting whereas microsegregation occurs across the dendritic scale.

The source of macrosegregation in castings which evolves during the solidification process is due to localized trapping of the interdendritic fluid. Thus macrosegregation during processing can to a great extent be controlled by manipulating the fluid flow and heat flow conditions of the interdendritic fluid such as in Electroslag Remelting (ESR), Vacuum Arc Remelting (VAR), and Vacuum Double Electrode Remelting (VADER) processes. On the other hand, macrosegregation can be alleviated by sintering together (into the final shape of interest) small rapidly solidified alloy powders or particulates (splats) that are essentially identical in average composition.

Microsegregation, the second type of segregation, manifests itself as a concentration gradient across the dendrite arms. The regions between the dendrite arms are usually rich in solute elements and contain equilibrium and/or non-equilibrium second phases and microporosity. Moreover, it has been found that the spacing between dendrite arms is related to the cooling rate of the casting - high cooling rates giving rise to extremely fine spacings and thus castings with a very fine structure, and slow cooling rates giving rise to a coarse structure and large dendrite arm spacings[1]. This is illustrated in Figure 1. The significance of the effect of the cooling rate on the dendritic spacing is that with high cooling rates one can produce "microcastings" which can be heat treated to yield a homogeneous product. The diffusion distance of the solute element(s) across the segregation spacing is essentially decreased due to rapid solidification and thus

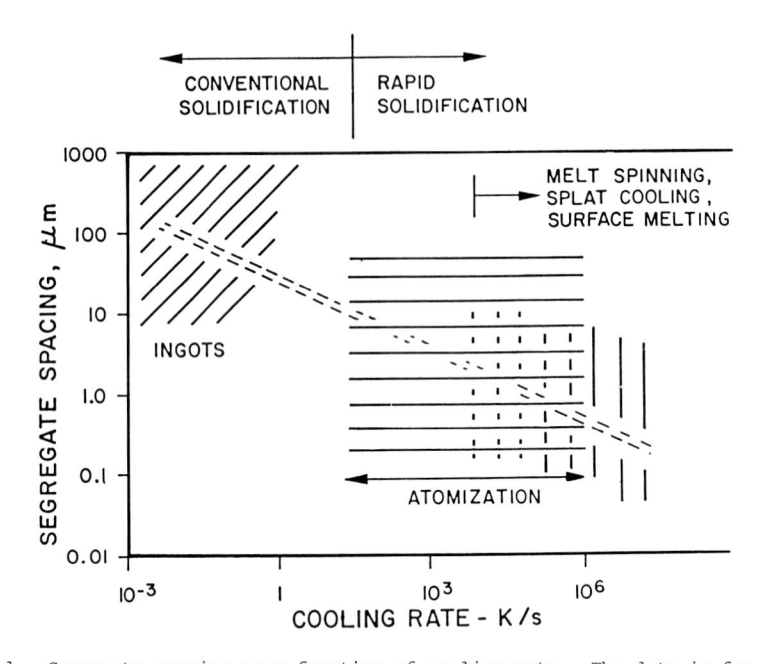

FIG. 1. Segregate spacing as a function of cooling rate. The data is for aluminum alloys; however a similar relationship exists for ferrous and non-ferrous alloys [1].

microsegregation can be "removed" by heat treating.

There are other microstructural achievements of rapid solidification processing (RSP) in addition to obtaining a fine dendrite arm spacing. Mehrabian [2] has reviewed these achievements; in brief however, they are: (i) the attainment of microcrystalline structures having grain sizes in the order of 1/8 to ¼ µm, (ii) production of undercooled structures such that non-equilibrium phases may be produced in the as-cast component, and (iii) production of amorphous rapidly solidified structures.

RAPID SOLIDIFICATION AND CONSOLIDATION PROCESSES

Several rapid solidification processes are shown in Figure 2. The atomization processes (gas and centrifugal) produce fine powder particulates or microcastings which subsequently need to be consolidated [1], whereas the melt spinning and the self-quenching processes directly give rise to a rapidly solidified thin-layer casting and thus consolidation is not required.

The cooling rates which can be achieved through some of these rapid solidification processes for Al, Fe and Ni melts are given in Table I [1]:

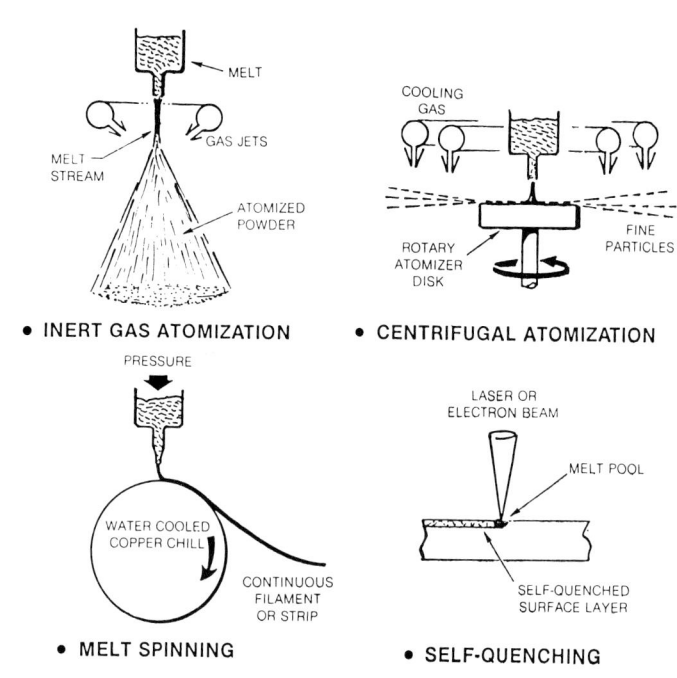

FIG. 2. Rapid solidification processing methods [1].

TABLE I. Cooling Rate Limitations in RSP [1].

Process	Heat Transfer Coefficient, $\frac{W}{m^2 \cdot K}$	Spacing Limit, μm	Maximum Cooling Rate, K/s
Atomization	$\sim 10^5$	~ 10	$10^4 - 10^7$
Melt Spinning	$\sim 10^5$	~ 25	10^6
Self Quenching	$\to \infty$	~ 10	10^8

It must be pointed out that the lower cooling rate limit of RSP is $\sim 10^3$ K/s.

Consolidation methods of RSP particulates and powders may be carried out by either hot isostatic pressing or hot extrusion, see Figure 3. If the compact has superplastic properties then the extruded billet can be isothermally forged to the near-final shape. Much research has been concerned with attempts to identify an optimum technique of consolidating powders to produce high-density, high-performance components. A paradox exists since the product of most RSP is in powder or flake form which requires further processing to produce useful shapes. Unfortunately, subsequent processing steps such as consolidation and sintering do significantly alter the desirable as-quenched structure. Sintering and other thermal treatments coarsen the structure... thus defeating the purpose of the RSP route and the production of the initial fine microstructure.

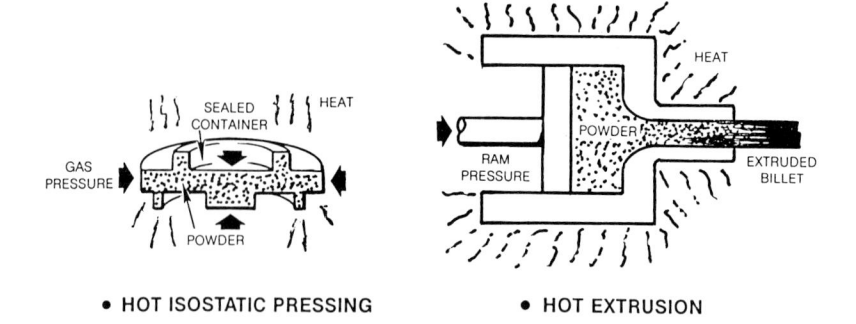

FIG. 3. Consolidation methods of RSP powder or splat particulates [1].

RAPID SOLIDIFICATION BY PLASMA DEPOSITION (RSPD)

Rapid solidification by plasma deposition circumvents this paradox because it combines melting, quenching, and consolidation in one single operation. Plasma arc metallizing is a powder-consolidation technique that has been used over the past 25 years to consolidate structures. The process involves the injection of powder particles into a high-temperature plasma gas stream created by heating an inert gas using an electric arc in a confined water-cooled nozzle. The powder particles injected into the plasma jet (at temperatures of $\sim 10^4$ K and higher) are melted, and subsequently resolidify when they impinge on a substrate. The cooling rates achieved are typically 10^5-10^6 Ks^{-1} and resulting grain sizes are 0.25-0.5 µm [3].

Plasma spraying in air typically results in deposits which exhibit low densities owing to oxidation of the deposited material. The problem of oxidation can be reduced by the use of an inert atmosphere or a shielded plasma jet; however, the achievable deposit densities are still unacceptable for high-performance structural components. Conventional plasma arc sprayed deposits have inherent defects, such as unmelted particles and oxide inclusions, which preclude their use as high-performance structural parts. Co/Cr/Al/Y deposites made by various plasma spraying processes is shown in Figure 4. It is quite evident that by spraying in a reduced pressure environment a product is obtained which has (metallurgically speaking) structural integrity.

Recent developments in the field of plasma arc spraying, namely the introduction of low-pressure plasma deposition (LPPD), have resulted in a renewed interest in the capabilities of plasma processing. In conventional processes the deposition is carried out at atmospheric pressure, whereas in LPPD the process is carried out in an evacuated chamber, thus permitting higher pressure ratios which yield much higher gas velocities, in the range Mach 2-3. The advantages of LPPD over conventional plasma spraying may be summarized as:
(a) higher particle velocities which create >98% theoretical density deposits,
(b) broad spray patterns which produce large deposit areas, and
(c) transferred arc heating of the substrate which improves deposit density.

These characteristics mean that LPPD has the potential of becoming a viable method of consolidating powders for high-performance applications. Moreover,

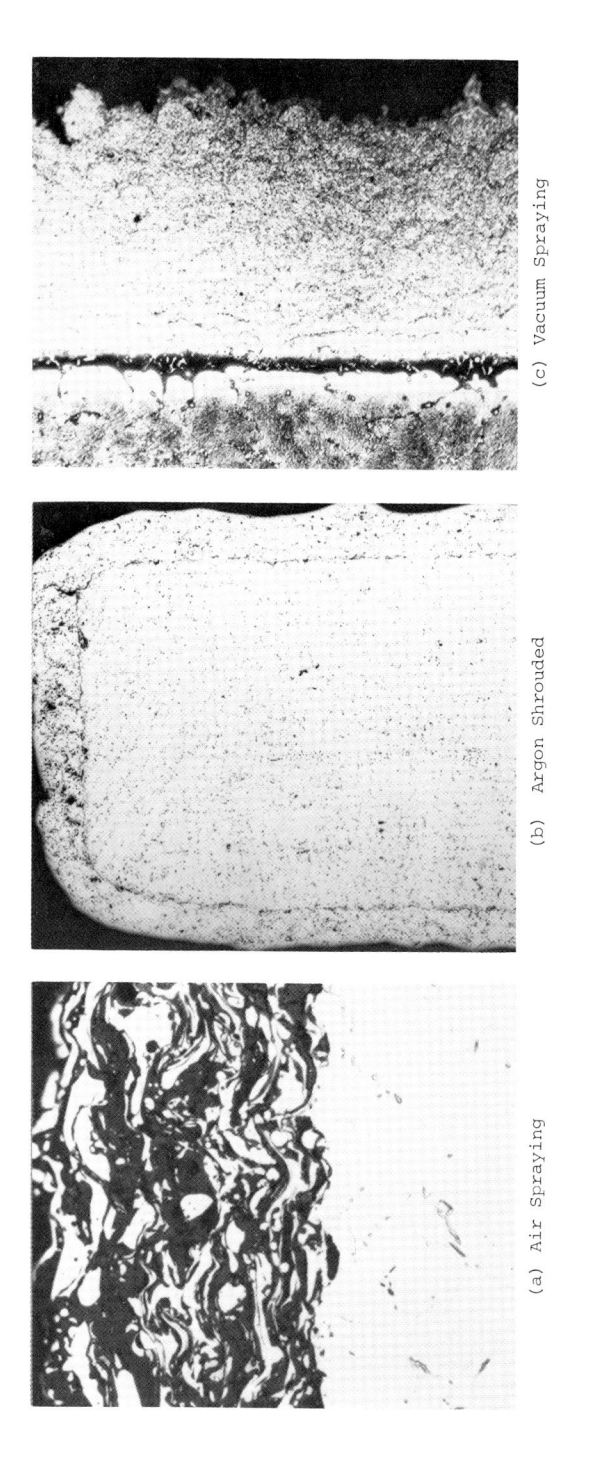

FIG. 4. Co/Cr/Al/Y deposits made by various plasma spraying processes.

(a) Air Spraying (b) Argon Shrouded (c) Vacuum Spraying

LPPD can be regulated automatically to make controlled deposits on complex geometries at reasonably high deposition rates (up to 50 kgh^{-1}). In sum, LPPD offers unique manufacturing flexibility not attainable by alternative RSP processes.

The structure of deposits made via LPPD are shown in Figure 5. By this new technology large deposits as well as high deposition rates can be achieved; dense deposits (>98%) in the as-sprayed condition can be produced; cleaner bondlines exist and the presence of oxide inclusions is essentially non-existent.

The LPPD process is quite complex in that there exists a multitude of operational variables. Table II is a listing of the various parameters which should be accounted for and controlled to produce high quality rapidly solidified deposits.

The production of dense, high-strength deposits requires that:

(i) a large fraction of the injected powder particles be heated to a molten state before they impinge on the substrate or the previously deposited particles

(ii) besides being in a molten state, the particles should also have sufficient velocity to be able to spread out and flow into the irregularities of the previously deposited layer

(iii) a strong interparticle or particle-substrate bond should be formed.

Thus a knowledge of the particle-plasma (melting) and droplet-substrate (solidification) interactions during plasma spraying is vital for a fundamental understanding of the process as well as for process optimization. Melting and solidification (including spreading and adhesion) phenomena during plasma arc spraying have been critically reviewed [4]. Unfortunately most of the literature deals with conventional plasma spraying at relatively low jet velocities and ambient pressure spraying conditions. Several papers presented in this conference (those by Wei and Paliwal) point out the significantly different conditions which are applicable during LPPD.

OBSERVATIONS AND IMPLICATIONS

As shown in Table II, plasma deposition is a complex phenomenon which requires the control of many variables in order to produce high-quality deposits. Many of these variables are not well understood and require additional attention. Among these are included a better understanding of melting and solidification during plasma spraying. These two phenomena are highly interrelated and the extent of plasma-particle interaction has a dominant effect on particle-substrate interaction. For example, the momentum, viscosity, temperature, thermal conductivity, and size of the impinging particles (droplets) onto the substrate greatly affects the solidification process at this interface.

It must be remembered that the entire plasma deposition process occurs in a highly dynamic environment. The particles being sprayed are not of the same size and hence follow different trajectories through the plasma jet. In applying any of the governing heat-transfer equations, one must take into account the changes in the physical properties of the gas and the applicable drag and heat-transfer coefficients because of the spatial variation in the plasma gas velocity and temperature. A review of the literature [4] shows that there is no general agreement with respect to the applicable drag and heat-transfer coefficients for plasma-spraying conditions. Further work is

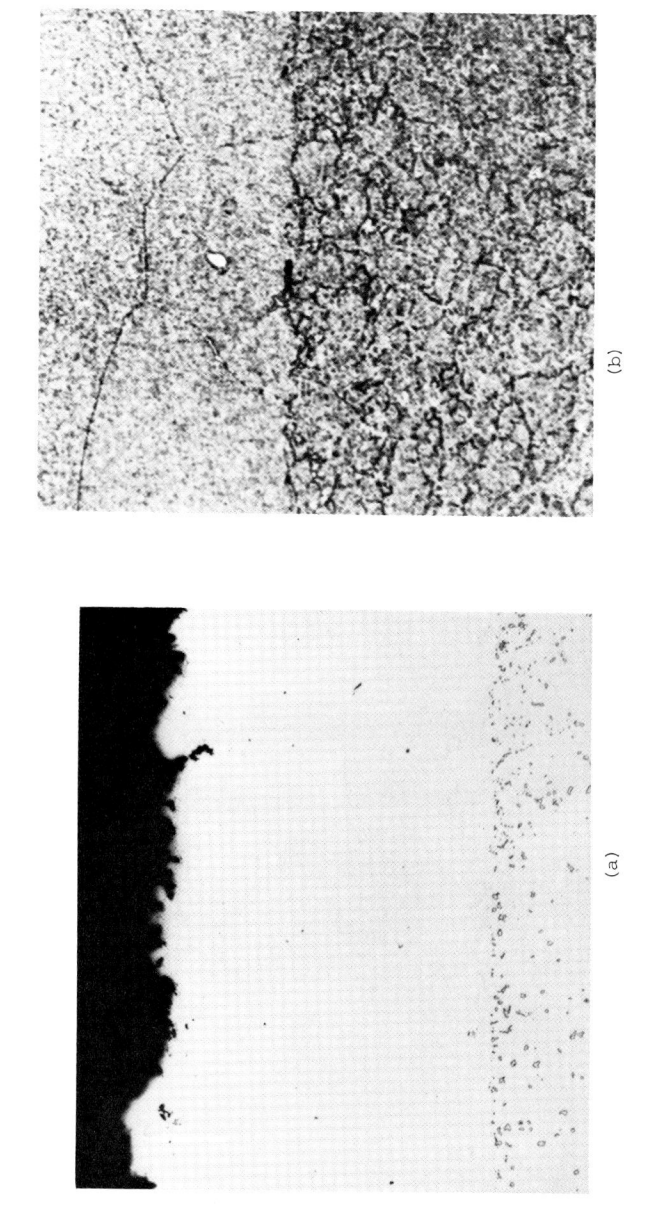

FIG. 5. Vacuum plasma sprayed deposits of (a) Co-29Cr-6A-1Y coating, and (b) nickel base superalloy, IN-738.

TABLE II. Operational Variables in LPPD.

Power and Energy	Spraying Material	Substrate	Powder Feeder	Spraying Conditions
Voltage	Size of the powder particles	Physical and chemical properties of the substrate (thermal linear expansion coefficient, etc.)	Nature and mass flow rate of the carrier gas of the particles	Distance between the nozzle exit and the substrate
Current intensity	Particle shape		Powder mass flow rate	Relative movement between the substrate and the torch
Nature of the plasma gas	Particle diameter			
Flow rate of the gas	Particle size distribution	Surface roughness, oxidation and cleanliness of the substrate	Location and inclination of the powder injector	Ambient atmosphere or controlled atmosphere
Nature and design of the electrodes of the plasma torch	Physical and chemical properties of the powder (melting point, thermal conductivity, etc.)	Substrate temperature during spraying process	Number of powder injectors	Method cooling of the substrate
			Powder injection velocity	

needed. Most of the investigators have used relationships developed (via dimensional analysis and empirical correlations) for relatively large particles in gas streams of low temperatures and velocities. It is only recently that Sayegh and Gauvin [5,6] have analyzed the problem of heat transfer to a sphere, taking into account the changes in gas properties in the vicinity of the sphere owing to the steep temperature gradient existing there. Moreover, the available experimental data are mainly for stationary particles of relatively large size (2-6mm) and no attempts have yet been made to show that the correlations obtained are valid for the smaller particles used in plasma spraying (typically 40 µm). In a review paper, Waldie [7] has stipulated that for smaller particle sizes (and in the limit as the particle size approaches the mean free path of the plasma gas) noncontinuum effects become significant and the Nusselt number would be lower than that calculated by any theory based on continuum mechanics. That this is indeed the case has been shown for atmospheric plasma spraying by Chen and co-workers [8]. When plasma spraying under a reduced pressure environment these noncontinuum effects will be even more dominant and cause significant lowering of the momentum and heat transfer to the powder particles [9]. Experimentally measured particle velocities during LPPD are reported to be much lower than those expected from calculations based on continuum mechanics [10].

Considering droplet-substrate interactions, droplet shape immediately after impact has been shown to influence both cooling rate and metallurgical bonding quality [4]. Additional studies are clearly required better to define the range of droplet shapes which can occur during plasma deposition and to determine whether preferred shapes can consistently be controlled. Substrate temperatures will also affect droplet shape and bonding, but their effects have not yet been systematically evaluated. One also needs to study the effect of the relative gun-to-substrate motion for its effect on substrate temperature and the relative uniformity of the deposit which can be achieved. The allowable impact angles for droplets impinging upon the substrate need to be evaluated if parts of complex shape are to be manufactured.

Off-design operation of the plasms gun will affect particle melting, solidification, and deposition problems. This occurs in a real sense during actual processing, owing to anode wear by arc erosion and, to a lesser extent, by fluctuations in power supplies and associated control equipment. The challenge for the success of plasma deposition as a fabrication method for high-performance components may well include the need for a low level of process sensitivity to off-design conditions. A better understanding of the importance of these effects and how they can be controlled is required before high-performance structural shapes can consistently be produced by plasma deposition.

The technical challenges are clear, as are the rewards for meeting them. Plasma deposition, especially in low-pressure environments, is one of the few processes by which fully dense, near-net-shape structures can be produced, such that both rapid solidification and consolidation are combined in one step.

REFERENCES

1. R. Mehrabian, B.H. Kear, and M. Cohen (eds.): Rapid Solidification Processing - Principles and Technologies II; 1980, Baton Rouge, LA, Claitor's Publishing Division.

2. R. Mehrabian: "Rapid Solidification", Intern. Metal Reviews; 1982, Vol. 27, No. 4.

3. M.R. Jackson, J.R. Rairden, J.S. Smith, and R.W. Smith: **J. Met.**, 1981, 33, (11), 23.

4. Apelian et al.: "Melting and Solidification in Plasma Spray Deposition"; **Intern. Metals Reviews**; 1983; Vol. 28, No. 5.

5. N.N. Sayegh and W.H. Gauvin: **A.I.Ch.E. Journal**, 1979, Vol. 25, No. 3, p. 522 .

6. N.N. Sayegh and W.H. Gauvin: **A.I.Ch.E. Journal**, 1979, Vol. 25, No. 6, p. 1057

7. B. Waldie: **Chemical Engineer**; May 1971, p. 188.

8. X. Chen, Y.C. Lee, and E. Pfender: Proc. 6th Int. Symp. on 'Plasma Chemistry', International Union of Pure and Applied Chemistry, Montreal, Quebec, 1983, p. 51.

9. D. Wei, D. Apelian, M. Paliwal, and S.M. Correa: "Melting of Powder Particles in a Low Pressure Plasma Jet", Materials Research Society Symposium on Plasma Processing and Synthesis of Materials, November 1983, Boston.

10. G. Frind, C.P. Goody, and L.E. Prescott: Proc. 6th Int. Symp. on 'Plasma Chemistry', International Union of Pure and Applied Chemistry, Montreal, Quebec, 1983, p. 110.

RESEARCH NEEDS IN ARC TECHNOLOGY

J. V. R. Heberlein
Westinghouse R&D Center
1310 Beulah Road
Pittsburgh, PA 15235

ABSTRACT

The state of the art and the research needs of selected applications of arc technology are described. The selected applications include plasma torch design, plasma synthesis of materials, plasma chemical synthesis and metallurgical processing. The research areas where results can have the broadest impact on the technology are electrode effects, plasma fluid dynamics and non-equilibrium effects.

INTRODUCTION

The arc plasma is generated by a current of at least 1A passing through a gas. The gas is heated to temperatures between 5000K and 25,000K, and the ionization of this gas at these high temperatures allows the current to pass through it.

The atmosphere in which the arcing takes place can be controlled - any mixture of gases can be used as arcing medium. The special characteristics of the electric arcs have led to its use in a number of applications. For example:

(a) Circuit Breakers.

The electrical conductivity of the arc plasma can be controlled by the choice of gas and by controlling the heat transfer from it to its surroundings. This characteristic is used in circuit breaker applications where the arcing space is cooled by a variety of means to promote arc extinction at current zero in ac circuits and changing of the gap to a dielectric.

(b) Discharge Lamps.

The high radiation intensity of the arc has long been utilized in discharge lamps. Metal vapor arcs provide the most efficient light sources (other than monochromatic light), which is an important consideration in view of the fact that approximately 25% of the electricity in the U.S. is used for lighting.

(c) Process Gas Heating.

The high energy density of the arc heated plasma is a particular advantage in plasma processing. It increases the efficiency of process gas heating, process reactor volumes can be reduced, and the high temperature gas can be used advantageously for metal and refractory processing.

(d) Plasma Processing.

The power input into the process can be controlled directly and independently of the process gas by adjusting the arc current. This leads to simplified and improved process control.

A considerable amount of research has been performed which has enhanced our understanding of electric arcs, and some of the results of this research have been translated into improvements of practical devices or processes. However, this transfer to practical applications is very often impossible because in many experiments, the choice of the particular parameters was guided by the desire to facilitate the study, and because we do not know quantitatively how a change of some parameters will influence the results. For example, electrode erosion results obtained with an argon arc moving over a copper electrode can hardly help us predict erosion for an arc in oxygen. The difference in thermal conductivities between high temperature argon and oxygen, as well as surface reactions in case of the oxygen arc will influence the current and heat flux densities, which, in turn, will influence arc root motion and electrode material evaporation, both of which will again change the heat flux density. These cumulative effects make a general description of many arcing phenomena extremely difficult, and usually one is forced to simulate the conditions of a specific application very closely if one wants to obtain results that would be relevant to this application.

In the following sections I will describe the state of the art in selected plasma applications and give my opinion on the research needed to advance the state of the art. The selected applications are chosen to provide minimum overlap with the other presentations in this session and emphasize arc technology in materials processing.

No economic evaluation is given of the selected application, although the improvement of the process economics is naturally a necessary consideration in defining the research needs. A good summary on the commercial potential of plasma technology in materials processing is given in a report by M. G. Down [1].

PLASMA GENERATORS

The considerable advance in plasma generator technology over the past twenty years is demonstrated by the fact that the duration of high power process experiments has increased from a few seconds to several days or even weeks. Long lived plasma torches are commercially available ranging in power from 100W to 6 MW for industrial applications and specialty torches with up to 50 MW for laboratory experiments. In the following I want to describe three different plasma heater designs and derive research needs from the current limitations.

(a) Wall Stabilized Arc Heater.

In 1956 Maecker [2] has described how an arc can be forced to have a cylindrical shape and a specific length by confining it inside a cascade of water-cooled copper rings. This arrangement is shown schematically in Figure 1. The individual rings have to be insulated from each other to avoid a current path through the copper. The electrodes are either the outside rings or are separated from the stack, e.g., in form of a tungsten rod cathode and a separate ring anode. The process gas can be introduced at either end or between the constrictor discs, and frequently has a swirl component to stabilize the electrode attachment. The torch is operated on dc, and typically an auxiliary arc between one electrode and an adjacent disc is required to break down the gap between anode and cathode and establish the main arc.

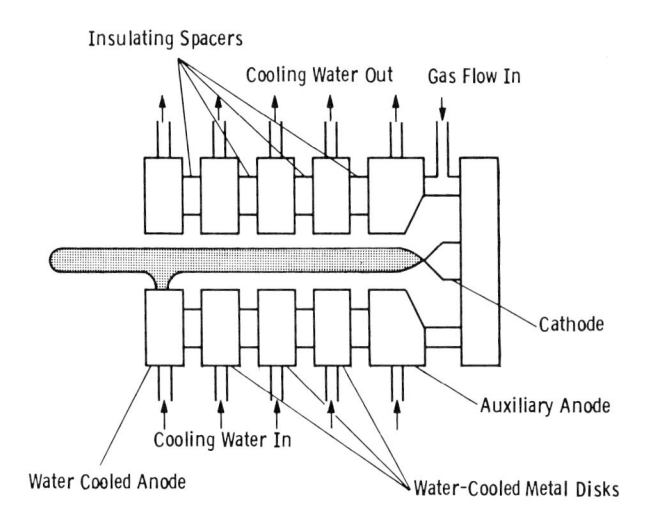

Fig. 1 — Schematic of wall-stabilized plasma heater

Since the arc length can be adjusted independently of gas flow rate, one can obtain high arc voltages, high power levels and high gas enthalpies at moderate currents [3]. On the other side, one has a complex cooling system, and the arc heating efficiency drops when high enthalpy levels are achieved. Plasma torches based on this design principle have been used in re-entry simulation tests at power levels up to 60 MW for generation of high enthalpy gas flows at high pressures [4].

Two-dimensional models furnish correlations between arc current, arc voltage, channel diameter and length and gas enthalpy distributions for several gases (air, argon, H_2/He and H_2/CO mixtures). Such a model can be used for the design of a torch for a specific application [3], however, modeling of the radiative energy transport is made difficult by the incompleteness of radiation property data for some process gases.

(b) Vortex Stabilized Arc Heater.

When the arc heated gas is rotating inside a cylindrical tube, the heavier cold gas will be forced to move away from the axis along the tube wall while the lighter, arc heated gas will be confined to the center of the swirl. A typical arrangement is shown in Figure 2. The two electrodes

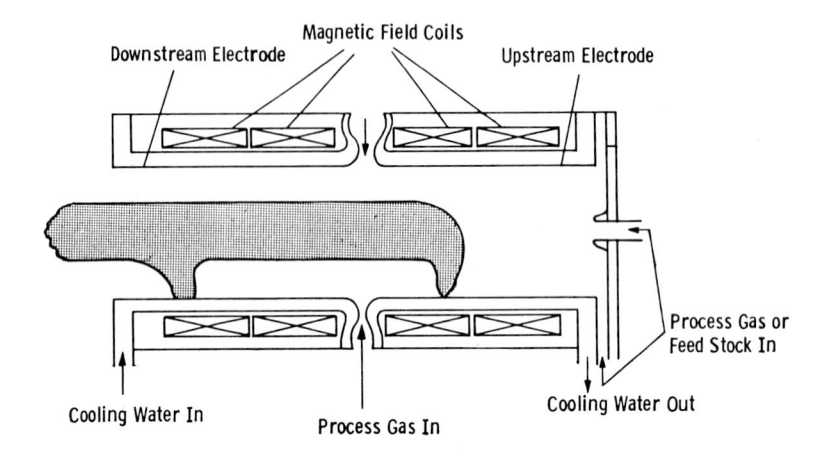

Fig. 2 — Schematic of vortex stabilized plasma heater

consist of watercooled copper cylinders being axially aligned. The process gas is introduced through the gap between the electrodes with a tangential velocity component. Part of the gas can be introduced at one end of the torch, and the arc heated gas leaves the torch at the other end. The stabilizing swirl is generated by either one of two means or a combination thereof: by the tangential velocity of the gas entering the torch, or by a magnetic field generated by an array of coils surrounding the electrodes and forcing the arc roots to rotate. The arc length depends on arc current, gas type, gas flow rate, electrode channel diameter and swirl rotational velocity and adjusts automatically such that the arc voltage is a minimum for the experimental conditions. This type of torch can be operated with a.c. or d.c., and high efficiencies (80-95%) can be obtained at power levels in excess of 5 MW. Its simple construction has demonstrated long operational life in industrial applications [5,6]. Detailed modeling of the plasma in this configuration has not been accomplished. It requires a three-dimensional model, description of turbulent transport, and the forces due to the self-magnetic field have to be considered in the description of the arc root motion [7]. Scaling of this type of torch for a specific application is typically done according to empirical relations.

It should be noted that the described configurations of the wall-stabilized and of the swirl stabilized torches are extreme cases, and many torch designs employ elements of both stabilization mechanisms [3,4,6].

(c) Transferred Arc Plasma Heater.

In this type of plasma heater the arc is transferred from a plasma torch to an external electrode. The plasma torch then serves typically as one "non-consumable electrode" (usually the cathode). The other electrode is typically a cylindrical ring or a pool of molten metal (see Figure 3). This arrangement is somewhat inbetween a plasma gas heater and a conventional arc furnace, combining some of the advantages of both. It is particularly useful for metal heating and melting because the anode attachment at the metal pool provides very high heat flux densities. Also metal fines injected into the reactor are heated by the intense arc radiation in addition to the hot gas heating while they fall. High power levels are achieved by using high arc currents (up to 10 kA per torch are reported) [8]. Multiple torches have been used in one reactor and one metal pool to increase the power loading up to 50 MW. However, difficulties are experienced at these high power levels with the arc control, e.g., coalescing of multiple arcs occurs, or arc attachment to the walls due to stray arcing [9].

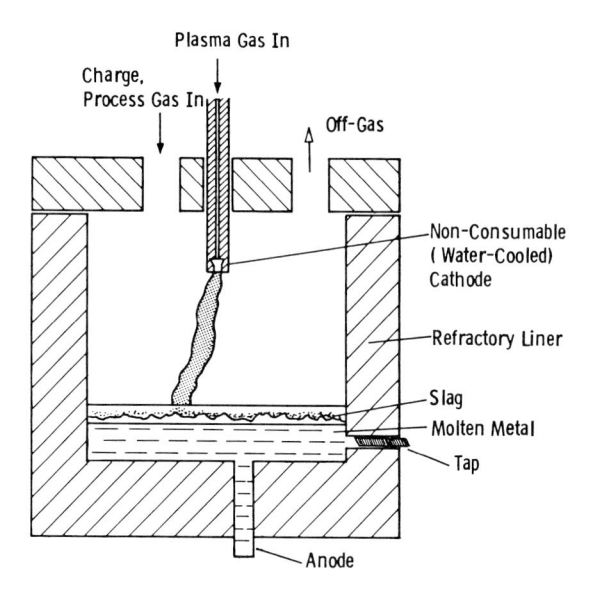

Fig. 3 — Schematic of transferred arc reactor

(d) Research Needs for Plasma Generators.

There is no doubt that modern plasma generators work well for a variety of applications, even though an understanding of the details of the arc heating process in commercial torches is incomplete. A parallel approach of model development and experimental verification is suggested. A three-dimensional model of arcs in turbulent flow would be required, in particular the interaction of the arc with cold gas injection must be better understood. Such a model will provide a firm base to the design of a plasma torch or furnace of a specific size and performance characteristic as dictated by the application. Diagnostics on practical plasma generators, in particular, flow and temperature distributions, are needed to assist in the model development. Care should be taken that adaptation of the plasma generator to the diagnostics, is minimized. Rather, the diagnostics should be chosen such that the actual plasma heater device is least modified, to minimize the disturbance of the phenomena to be investigated.

The interaction between the arc roots and the electrode surface is even less understood. Although considerable work has been done on describing electrode erosion rates for specific experimental conditions, [10,11,12] the number of parameters influencing this erosion is so large that it is in general not possible to predict erosion rates under different experimental conditions. The heat flux density at the electrode surface depends on the arcing gas and gas pressure, the arc root velocity, the arc current, the arc current density, the electrode material and the surface condition. The surface condition, in turn, is influenced by the heat flux density, the arc root velocity and the arcing medium. For example, when the arc passes over the electrode surface it may not only melt or vaporize the surface, but it may also cause a variety of chemical reactions such as metal oxidation or metal oxide reduction. The change of the surface condition, in turn, will influence the arc motion as well as current density and heat flux density at the surface.

A better understanding of these phenomena will lead to longer electrode life of plasma torches and, consequently, improved process productivity, and improved control over the melting/slag formation process in plasma furnaces. It is particularly important when the plasma generator is scaled to higher power by using higher arc currents, because both, electrode erosion effects and magnetic instabilities increase strongly with current.

PLASMA SYNTHESIS OF MATERIALS

Metal compounds or ceramics can be synthesized in a plasma by injecting the metal in some form into the plasma where it will react with the plasma gas or with another chemical which is also injected into the plasma. Two examples of such reactions are [13,14]

(1) $2 \; NbCl_5(v) + 5 \; H_2 + N_2 \rightarrow \delta\text{-}NbN(c) + 10 \; HCl(v)$

(2) $SiCl_4 \; (v) + CH_4 \rightarrow SiC(c) + 4 \; HCl(v)$

The principal advantages of the plasma process are:

(i) High temperature phases of compounds can be obtained by fast quenching (> 10^6 K/s) to room temperature. Examples are as in reaction (1) above the superconducting niobium nitride [13], or the diamond like tungsten carbide [15]. Even amorphous powders of metals or compounds have been obtained.

(ii) A high purity product can be obtained with fewer processing steps by collecting the product at a temperature where the by-product is in the vapor phase. The reaction (2) above is an example for such a process.

(iii) The process lends itself to continuous operation, whereas many corresponding conventional processes are batch processes.

For refractory compounds, the product generally is generated in the form of ultrafine powders [16], i.e., with particle sizes between 50 Å to 500 Å diameter, depending on the experimental conditions, and with a narrow size distribution. The small particles form when during the quenching a high supersaturation of the vapor exists leading to the formation of a large number of nucleation sites. Larger particles have been obtained [16] but only with a large percentage of the particles still within the size range of a few hundred Angstrom diameter. The small size of these particles makes the handling during further processing difficult, and process development for obtaining larger size particles without losing the narrow size distribution is needed to realize the far reaching opportunities of the plasma synthesis process.

Closer to practical applications is the generation of high purity reactive metals in a plasma reaction. An example is the generation of high purity silicon in a plasma process [17,18]:

$$SiCl_4(v) + 4\ Na(v) \rightarrow Si(\ell) + 4\ NaCl(v)$$

The liquid silicon condenses on a skull wall and is then collected in a crucible while the by-product NaCl remains in the vapor phase. Work is being pursued on a similar process for the generation of titanium [19,20]. It should be noted that this collection mechanism has been modeled, and the model experimentally verified [18].

Research Needs for Materials Synthesis

The major need for the processes resulting in ultra-fine particles of compounds is the improvement of the control over particle size. This requires better control over the reaction product residence time in a certain temperature region, over the vapor pressures in this region, and over the temperature gradient in the quench zone. It is necessary to evaluate what the influence is of each of these parameters on increasing the particle size. In addition, formation of alloys of metals with dissimilar boiling points (e.g., Ti-Al) from a vapor phase reaction in a single step process could be of interest. And there is always the possibility that a compound will be discovered with a very attractive combination of properties.

For the processes where the product is collected in liquid form, the major need is the operation of larger scale reactors for sufficiently long time periods to allow a thorough evaluation of the potential of the process under consideration.

CHEMICAL PLASMA SYNTHESIS

The driving forces for introducing plasma processes in the chemical industry are:

(a) reduction of processing steps can lead to smaller installations and lower capital requirement;

(b) the high energy density of the plasma leads to improved energy utilization;

(c) since the arc discharge can heat any gas, one has complete control over the atmosphere in the reactor, which may simplify the reactor design;

(d) substitution of electric power for oil or gas may reduce the dependence on imported fuels, which is particularly advantageous when inexpensive hydroelectric power is available.

We will discuss the state of the art in chemical plasma synthesis and the corresponding research needs on hand of three examples.

Example 1: Acetylene Production from Hydrocarbons or Coal.

Acetylene is an alternative starting material to ethylene in the manufacture of synthetic compounds such as PVC. It is the most stable hydrocarbon compound above 1600K, and is usually produced by pyrolysis and subsequent quenching. In the conventional process, oil or gas is used as reactant as well as energy source. A commercial plasma process has been in operation since 1940 in Huels, Germany, based on natural gas as reactant. AVCO and Chemische Werke Huels, Germany, both have been developing plasma processes based on powdered coal injection as reactant with 1 MW and 0.5 MW reactors, respectively [21,22]. Research has focussed on the injection of coal powders into the arcing region and designing the reactor such that optimum residence times are achieved. This optimization is needed for maximizing the yield. Although thermal history and flight path of the injected particles have been calculated, the optimization of the reactor design has been largely empirical. Further improvement may be gained from a realistic three dimensional model. However, little has been reported on studies of possible yield enhancement using non-equilibrium or catalytic effects.

Example 2: Nitrogen Fixation for Fertilizer Production.

NO_2 can be produced by heating a mixture of oxygen and nitrogen with a plasma heater and then quenching the mixture. Such a plasma generation process has the advantage of fewer processing steps compared with the natural gas/ammonia processing which can be characterized by:

$$\text{natural gas} \xrightarrow[\text{reform}]{+ \text{ steam}} H_2 \xrightarrow[\text{catalyst}]{+ \text{ air}} NH_3 \xrightarrow[\text{catalyst}]{+ O_2} NO, NO_2 \xrightarrow{+H_2O} HNO_3$$

compared to the plasma process:

$$\text{electricity} \xrightarrow{\text{air}} \xrightarrow{\text{plasma}} NO \xrightarrow[\text{quench}]{O_2} NO_2 \xrightarrow{H_2O} HNO_3$$

The process economics depends on the process yield and specific energy requirement, and on the local natural gas/electricity cost ratio. There has been a considerable laboratory effort with a 30 kW reactor concentrated on increasing the plasma process yield and reducing the specific energy requirement [23,24]. Sophisticated diagnostics are being used to characterize the reaction region, in conjunction with modeling of the reaction kinetics using rate equations. The results are being used in optimizing the reactor design and operation, i.e., to determine the plasma gas heating rates, the location of the reactant injection and quenching. Figure 4 shows schematically how the reactor can be optimized [24]. Experimental yields have been found that exceed by far those predicted by equilibrium thermodynamics, 11% vs. 6%, an effect explained by non-equilibrium in the reaction region, i.e., use is made of the different kinetic rates for the different reactions. Use of WO_3 catalysts can increase the yield even further. It appears that the needed research is being performed, and the research devoted to this application may serve as a model for other applications of plasma technology. Of course, what remains to be done, once an economical yield and specific energy requirement have been achieved, is the scaling to industrial size.

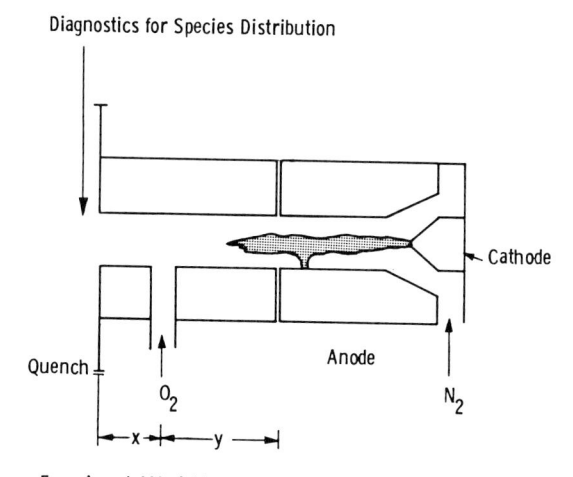

Experimental Variables Influencing Yield:
  Arc Current, Power
  N_2, O_2 Mass Flow Rates
  Distance y of Oxygen Injection from Anode
  Distance x of Final Quench from O_2 Injection

Fig. 4 — Illustration of research for yield optimization

Example 3:   Destruction of Toxic Wastes.

Although this application is a decomposition of an unwanted compound rather than the synthesis of a wanted one, it should be mentioned here because it has become increasingly important, and the research needs are essentially the same. The advantages of the pyrolysis by a plasma process are the very high temperatures in the reaction zone and the possiblity to choose any plasma gas. The wastes decompose at the high temperatures, and the plasma gas and the quench rate are chosen to eliminate their reformation.

Research Needs in Chemical Synthesis/Decomposition

The extreme temperature gradients result in very rapid heating and quench rates, coupled with deviations from equilibrium. A model of the reaction kinetics is needed for optimization of the reactor. For most processes, the reaction rate coefficients are only partially known, and theoretical or experimental determination of these constants is needed. Diagnostics has to be developed to measure these rate coefficients under realistic conditions. The effects of non-equilibrium, e.g., high energy electrons or metastable atoms or molecules, on the process is very little known, and an understanding of such effects may further enhance the potential for plasma processing. Similarly, it would be beneficial to have a better understanding of the catalytic action of solid surfaces in contact with the gas.

METALLURGICAL PLASMA PROCESSING

There are many applications for plasma processing in the metallurgical industries, and the commercial potential for plasma technology is vast. From an arc technology point of view, the applications can be divided into two major groups: (1) a plasma torch, usually of the swirl stabilized type, heats the process gas which then reacts with the material to be processed; (2) the material to be processed is heated largely by direct interaction with the arc in a transferred arc type reactor. In the first case, there is a considerable overlap with the research requirements in plasma torch technology, and many of the research needs for the applications in the second group have been mentioned in the section on transferred arc plasma generators. I will, therefore, on the basis of a few examples highlight the problems that need to be addressed and are receiving attention.

Blast Furnace Firing

It has been predicted theoretically that the coke consumption of existing blast furnaces can be reduced by up to 75%, and the productivity increased by up to 120%, if arc heated reducing gas is injected together with the blast air and fuel through the tuyere [26]. Preliminary tests with a small experimental blast furnace appear to support the theoretical predictions. A variation of this approach is the generation of the reducing gas in the plasma by injecting pulverized coal into the plasma heated air stream. Another variation is the arc heating of the blast air [6]; this would reduce the amount of coke needed in the furnace for heating, and it would not depend on the availability of reducing gas. The requirements for the plasma torch are extreme with regard to reliability, ruggedness, stability of operation and length of electrode life.

Research Needs: Again, the major need are test data from operation on a realistic scale, and tests at CRM at Seraing, Belgium, are proceeding to provide these [36]. On a smaller scale, research on the factors influencing electrode erosion in oxidizing or reducing arcing media may lead to increased electrode life and increased operation times between plasma torch shut-downs. There is less certainty about the operation with pulverized coal injection. In order to achieve the desired effect, intimate mixing of the coal powder with the plasma is required, and a minimum of carbon deposition on the wall should take place. There are a variety of models describing the interaction of single particles with a laminar plasma flow [27-30], and larger scale experiments are needed to determine to what degree these models help the design of a coal injection system into a turbulent plasma flow, and more larger scale experiments such as reported by Meyer et al [31] are needed to determine to what degree these models help the design of a system injecting coal into a turbulent plasma flow.

The process in which arc heated gases provide heat in a ferro - alloy smelting furnace [32] - or in a furnace reducing steel mill baghouse dust or other waste oxides is from the plasma point of view very similar to the blast furnace firing process, and the research requirements are essentially the same.

Transferred Arc Reactors for Ferro-Alloy Production or for Remelting

A large number of different transferred arc reactors have been designed to replace the traditional submerged arc furnace. Reactors have been operated on a pilot plant scale for generation of ferro-vanadium [33], ferro-chrome [34] and manganese [9], and of pure or alloyed molybdenum [35]. The major advantages of the plasma reactor compared to the submerged arc furnace are: (1) better process control because the power can be adjusted independently of the charge, (2) inexpensive coal fines can be used as a reductant instead of high quality coke, (3) non-consumable electrodes are used instead of expensive carbon electrodes, and (4) the ore can be introduced in the form of fines. The ore injected into the arcing region, is heated by the arc, but the melting and the reduction is thought to occur mainly in the slag covering the molten metal pool. The processes, however, have yet to be scaled to commercial size.

Research Needs: The already mentioned problems of arc control have to be solved, i.e., arc attachment on the reactor walls has to be avoided without sacrificing arc motion over the slag resulting in a more uniformly distributed heat input. Modelling of the reactor including arc instabilities and self-magnetic fields, as well as radiative transport would help the designer. The physics and chemistry of the arc anode attachment would lead to a better understanding of the slag processes. Finally, when current levels of 10 kA are used, it is crucial that the cathode emission process is optimized to assure a practical cathode life. Research is focussed on obtaining a mixture of materials assuring a low work function for sufficient electron release at a temperature where the erosion rate is sufficiently low.

The same considerations apply for plasma melting furnaces. However, for remelting of high value scrap, such as Ti or spray coating alloys, economical furnaces may be obtained without scaling to large sizes, and the present pilot scale reactors may offer already the commercial process.

SUMMARY OF RESEARCH NEEDS

As mentioned previously, there are several application areas of arc technology where larger scale experiments are needed to identify possible additional research needs. Several metallurgical applications fall into this category, like plasma torch melting, blast furnace firing, etc. In another group of applications, the research needs are clearly identified, and research is progressing to eliminate them. Examples are above all plasma spraying and chemical synthesis of nitric oxide. Finally, there are the applications where a breakthrough is needed to achieve full realization of the potential for arc technology. Particle size control for plasma materials synthesis is one example in this group.

In Table I, the major areas requiring attention of researchers are listed, together with the specific application areas which would benefit from the research, and the benefits themselves. Electrode effects are listed first not because research in this area will have the largest impact on any one application, but because it will influence the largest number of applications. Plasma-solid particle interactions have been studied in detail by several groups, however, detailed descriptions of heterogeneous reactions and simultaneous injection of a large number of particles with a certain velocity distribution are certainly incomplete.

The largest impact on any single application is probably to be made in the areas of plasma chemical synthesis or decomposition, because non-equilibrium or catalytic effects produced either by high electron temperatures or by surface enhanced reactions have been studied comparatively little and are very little understood.

Concluding, I would like to emphasize two points: firstly, the complexity of plasma processes requires intimate knowledge of both, gas physics and materials science, and interdisciplinary research is necessary for meaningful results.

Secondly, as pointed out in the introduction, a choice of parameters close to the operating parameters of an actual application is needed to assure that the results are valid for this application. Considering the large number of present and potential applications, we can be certain that there will be objectives for research for many years to come.

ACKNOWLEDGMENTS

Discussions with Drs. P. G. Slade, T. N. Meyer and M. G. Down were very helpful in the preparation of this paper, and I am indebted to them for their review of the manuscript. I gratefully acknowledge the speedy preparation of the manuscript by Ms. T. McElhaney, and of the final paper by the Word Processing Center.

TABLE I. Summary of Research Needs

Research area/specific tasks	Applications	Benefits
Electrode effects - arc root motion - heat flux, current density - surface chemistry - high electron emission materials	Plasma generators Plasma reactors Circuit breakers Lamp	Longer life, less product contamination New generation of breakers, low wattage lamps, longer life.
Plasma fluid dynamic and heat transfer - radiation properties - temperature and velocity distribution, models and experiments	Plasma generators Plasma reactors Circuit breakers Plasma spraying	Scaling laws, design criteria, increased process yield New generation of breakers Improved spray coatings
Plasma - solid particle interaction - heat and momentum transfer and measurements - heterogeneous reactions/ particle growth - injection of large number of particles, trajectory distribution	Plasma synthesis Plasma spraying Metallurgical plasma processing	Particle size control Improved spray coatings Inexpensive reductant by coal injection
Gas physics/Reaction kinetics - rate equation models - rate coefficients - non-equilibrium (multi-temperature) models - catalytic effects	Chemical plasma synthesis	Improved yield, specific energy requirement

REFERENCES

1. M. G. Down: "Plasma Processing for Materials Production," EPRI Report EM-2771, December 1982.

2. M. Maecker: "Ein zylindrischer Bogen für hohe Leistungen," Z. Naturforschung 11a, 457, 1956.

3. G. N. Liu, F. Y. Chu, C. J. Simpson: "Design and Laboratory Utilization of a Thermal Plasma Generator," Proc. 6th Intl. Symp. Plasma Chemistry, Montreal, Canada, July 1983, p. 156.

4. W. Winovich and W. C. A. Carlson: "The 60 MW Shuttle Interaction Heating Facility," ISA Transactions 19/2, p. 75, 1980.

5. M. G. Fey and F. J. Harvey: "The Role of Plasma Heating Devices in the Electric Economy," Metals Engineering Quarterly May 1976, p. 27.

6. M. G. Fey, T. N. Meyer, W. H. Reed, W. O. Philbrook: "Thermal Plasma Systems for Industrial Processes," Proc. Industrial Opportunities for Plasma Technology, Toronto, October 1982.

7. S. A. Wutzke: Ph.D. Thesis, University of Minnesota, June 1967.

8. G. Scharf: "Operational Experiences in the Plasma Steelplant at Freital," 1st European Electric Steel Congress, Aachen, Germany, September 1983.

9. T. R. Curr, et al: "The Design and Operation of Transferred Arc Plasma Systems for Pyrometallurgical Applications," Proc. 6th Intl. Symp. Plasma Chemistry, Montreal, July 1983, p. 175.

10. M. G. Fey and J. McDonald: "Electrode Erosion in Electric Arc Heaters," A.I.Ch.E. Plasma Chemical Processing Symposium, August 1976.

11. A. E. Guile: "Arc Electrode Phenomena," IEE Rev. 118, p. 1131, 1971.

12. A. E. Guile and B. Juttner: "Basic Erosion Processes of Oxidized and Clean Metal Cathodes by Electric Arcs," IEEE Transactions Plasma Science, PS 8/3, p. 259, 1980.

13. P. Ronsheim, et al: "Thermal Plasma Synthesis of Transition Metal Nitrides and Alloys," Plasma Chemistry and Plasma Processing.

14. G. Perugini: "Arc Plasma Reactions for Special Ceramics," Proc. 3rd Intl. Symp. Plasma Chem., Limoges, France, 1977.

15. P. C. Kong, et al.: "Synthesis of β-WC_{1-x} in an Atmospheric-Pressure, Thermal Plasma," Plasma Chemistry and Plasma Processing, 3/1, p. 115, 1983.

16. P. Ronsheim, E. Pfender, L. E. Toth: "Characteristics of Particle Growth in a Thermal Plasma Jet," 5th Intl. Symp. Plasma Chemistry, Edinburgh, U.K., 1981, p. 844.

17. M. G. Fey, T. N. Meyer, W. H. Reed: "An Electric Arc Heater Process to Produce Solar Grade Silicon," Proc. 4th Intl. Symp. Plasma Chemistry, Zurich, Switzerland, August 1979, p. 708.

18. J. V. R. Heberlein, J. F. Lowry, T. N. Meyer, D. F. Ciliberti: "The Reduction of Tetrachlorosilane by Sodium at High Temperatures in a Laboratory Scale Experiment," Proc. 4th Intl. Symp. Plasma Chemistry, Zurich, Switzerland, August 1979, p. 716.

19. M. G. Down: "Titanium Production by a Plasma Process," Wright Aeronautical Materials Laboratory, Technical Report AFWAL-TR-82-4018, May 1982.

20. K. A. Bunting: "Sodium and Titanium Tetrachloride Feed System for a Plasma Route to Titanium," Proc. 6th Intl. Symp. Plasma Chemistry, Montreal, July 1983, p. 193.

21. A. J. Patrick and R. E. Gannon: "A 1 MW Prototype Arc Reactor for Processing Coal to Chemicals," Proc. Symp. Industrial Opportunities for Plasma Technology, Toronto, Canada, October 1982.

22. R. Muller and C. Peuckert: "Recent Developments for the Production of Acetylene from Coal by the Huels Arc Process," Proc. 6th Intl. Symp. Plasma Chemistry, Montreal, July 1983, p. 270.

23. J. F. Coudert, et al: "Applications of the Plasma Jet Generator to the Nitrogen Oxide Synthesis," Proc. 3rd Intl. Symp. Plasma Chemistry, Limoges, France, 1977.

24. J. M. Baronnet, et al: "Nitrogen Oxides Synthesis in a DC Plasma Jet," Proc. 4th Intl. Symp. Plasma Chemistry, Zurich, Switzerland, August 1979, p. 349.

25. T. G. Barton: "Problem Waste Disposal by Plasma Heating," Int. Recycling Congress, Berlin, Germany, 1979, Vol. I, p. 733.

26. N. Ponghis, R. Vidal, A. Poos: "Operation of a Blast Furnace with Superhot Reducing Gas," Met. Reports CRM, Vol. $\underline{56}$, p. 9, 1980.

27. A. J. Harvey and T. N. Meyer: "A Model of Liquid Metal Droplet Vaporization in Arc Heated Gas Streams," Metallurgical Transactions B, $\underline{9B}$, p. 615, 1978.

28. A. Vardelle, M. Vardelle, P. Fauchais: "Influence of Velocity and Surface Temperature of Alumina Particles on the Properties of Plasma Sprayed Coatings," Plasma Chemistry and Plasma Processing, $\underline{2/3}$, p. 255, 1982.

29. Xi Chen and E. Pfender: "Heat Transfer to a Single Particle Exposed to a Thermal Plasma," Plasma Chemistry and Plasma Processing, $\underline{2/2}$, p. 185, 1982.

30. D. Apelian, et al: "Melting and Solidification in Plasma Spray Deposition - A Phenomenological Review," Intl. Metal. Reviews.

31. T. N. Meyer, et al: "Metal Powder Vaporization Using a Plasma Torch," Proc. 6th Intl. Symp. Plasma Chemistry, Montreal, Canada, July 1983, p. 369.

32. S. Santen: "Plasma Smelting," Proc. 6th Intl. Symp. Plasma Chemistry, Montreal, Canada, July 1983, p. 174.

33. D. R. MacRae, et al: "Ferrovanadium Production by Plasma Carbothermic Reduction of Vanadium Oxide," Proc. 34th Electric Furnace Conf., St. Louis, MO, Dec. 1976.

34. N. A. Barcza, et al: "The Production of Ferro-Chromium in a Transferred-Arc Plasma Furnace," Proc. 39th Electric Furnace Conf., Houston, December 1981.

35. W. H. Gauvin, G. R. Kubanek, G. A. Irons: "The Plasma Production of Ferromolybdenum-Process Development and Economics,' J. Metals, January 1981, p. 42.

36. N. Ponghis, R. Vidal, A. Poos: "PIROGAS - a new process allowing diversification of energy sources for blast furnaces," High Temperature Technology $\underline{1/5}$, p. 275, 1983.

THE PRODUCTION OF METASTABLE METALLIC PARTICLES DIRECTLY FROM THE MINERAL
CONCENTRATE BY IN-FLIGHT PLASMA REDUCTION

J.J. MOORE,* K.J. REID,* AND J.M. SIVERTSEN**
*Mineral Resources Research Center, University of Minnesota,
56 East River Road, Minneapolis, MN 55455; **Chemical
Engineering & Materials Science, University of Minnesota,
421 Washington Avenue S.E., Minneapolis, MN 55455

ABSTRACT

Combining the reduction of the mineral oxide to a liquid
metal and its rapid solidification provides an energetically
favorable route to provide metastable metallic particles or
powders. Such an option is available with the Sustained
Shockwave Plasma (SSP) reactor in which the mineral oxide is
reduced using a carbon-based reductant within the plasma
medium and subsequently rapidly solidified. This paper
examines the degree of metastability of these metal particles
using optical, electron and Auger microscopy and discusses the
potential of this processing route.

INTRODUCTION

There has been considerable research interest recently in the production and characterization of metastable metallic alloys, as indicated by two recently held symposia sponsored by the Metallurgical Society of AIME [1,2]. The nonequilibrium state of these metallic products has resulted, in many instances, in new and often unique microstructures and consequently significant changes in electrical, magnetic, corrosion and mechanical properties.

However, in most cases, the metastable condition is produced by melting an alloy or group of metals of the required chemistry followed by rapid solidification to effect the nonequilibrium, metastable condition, the metals and alloys having been previously processed through conventional extraction and refining routes. Considerable savings in energy and processing costs could, therefore, be achieved if metallic particles could be produced in the required metastable state directly from the mineral concentrate. The application of plasma reduction smelting [3] of the required blend of minerals makes this latter proposition possible. The sustained shockwave plasma (SSP) reactor [4], unlike other plasma reactors, utilizes the unique properties and highly reactive nature of the plasma medium itself to effect the required reduction to metallic form. If this is followed by rapid solidification, a high degree of metastability may be achieved in the reduced metallic product. The details of the SSP reactor have been described elsewhere [5] and in another paper [6] within this section and will not be given again here. In brief outline, the SSP reactor is a low volume, high throughput reactor in which a carbothermic reduction of oxide minerals takes place in less than 100 ms due to the highly reactive ionic, atomic and molecular reactant species present within the plasma medium. In this case the carbon-based reductant used was lignite char. The very small reaction or residence time in the plasma necessitates a fine particulate reactant feed.

EXPERIMENTAL PROCEDURE

Commercially available taconite and chromite concentrates were used in these initial experiments aimed at determining the metastability of the metallics produced by in-flight reduction of minerals within the SSP reactor. The chemical composition and sizing of these concentrates are given in Tables I, II and III, and the composition of the lignite char reductant is given in Table IV.

TABLE I. Chemical Analysis of Taconite Concentrate

Total Fe, Fe_T (%)	66.81
Fe^{3+} (%)	44.49
Fe^{2+} (%)	22.16
FeO (%)	28.5
Fe_2O_3 (%)	63.6
Fe_3O_4 (%)	92.1
Metallic Fe (%)	0.16
SiO_2 (%)	6.23
S (%)	0.004
C (%)	0.37

TABLE II. Chemical Analysis of Chromite Concentrate

SiO_2 (%)	6.6
Al_2O_3 (%)	15.49
FeO (%)	21.3
Fe_2O_3 (%)	1.6
MgO (%)	15.25
Cr_2O_3 (%)	38.95
Na_2O (%)	0.20
K_2O (%)	0.01
TiO_2 (%)	0.4
CaO (%)	0.51
S (%)	0.021
P (%)	0.011
Cr/Fe	1.51

TABLE III. Screen Analysis of As-Received Taconite and Chromite Concentrates

Mesh Size	Microns	Taconite, Wt %	Chromite, Wt %
48	300	0.00	0.00
-48 +65	-300 +210	0.00	25.44
-65 +100	-210 +149	0.00	29.46
-100 +150	-149 +105	0.77	18.54
-150 +200	-105 +74	2.72	9.79
-200 +270	-74 +53	7.88	5.59
-270 +325	-53 +44	11.34	2.74
-325	-44	77.29	8.44
Total		100.00	100.00

These two concentrates were chosen since they would result in Fe-C and the more complex Fe-Cr-C microstructures, both of which have been well documented in the literature and from which the degree of metastability may be relatively readily determined metallographically.

The feed into the SSP reactor was based on a 200% stoichiometric carbon level for the reactions:

$$Fe_3O_4 + 4C \rightleftarrows 3Fe + 4CO$$

$$3(FeO \cdot Cr_2O_3) + 17C \rightleftarrows Fe_3C + 2Cr_3C_2 + 12CO$$

The SSP reactor conditions are given elsewhere in the proceedings of this symposium [6]. The reduction products were allowed to cool by falling from the plasma medium in a free fall chamber into a collecting vessel. Thus the quenching medium is essentially one of mixture of argon (the plasma gas) and air.

TABLE IV. Chemical Analysis of Lignite Char Reductant

Fixed Carbon, %	61.28
Sulfur, %	2.06
Volatiles, %	17.59
Ash, %	15.47
Fe^{3+} %	3.22
Fe^{2+} %	1.61
SiO_2 %	3.56
Al_2O_3 %	1.48
MgO %	1.19
CaO %	1.05

EXPERIMENTAL RESULTS AND DISCUSSION

The extent of metastability of these metallics was determined using optical, scanning electron microscopy (SEM) and scanning Auger microscopy while the point analysis of the individual phases was determined qualitatively using an energy dispersive x-ray (EDX) facility on the SEM, and semi-quantitatively on the SAM using the internal sensitivity factors. Typical photomicrographs of the metallics produced by this SSP-treatment are shown in Figure 1 for the reduction of taconite (Fe-C) and in Figure 2 for the reduction of chromite (Fe-Cr-C). Typical elemental point analyses of the metallic particles as determined with the internal standards in the SAM are given in Table V.

TABLE V. SAM Point Analysis of Metallic Particles

Specimen	Phase	Fe (Wt %)	C (Wt %)	Cr (Wt %)
Metallics produced from plasma-processed taconite				
Hypereutectoid Steel	Metallic	98.2	1.8	
Hypoeutectic Iron	Metallic	96.2	3.8	
	Ledeburite { Fe_3C		5.1	
	Pearlite		1.8	
Metallics produced from plasma-processed chromite		55.2	12.8	32.0

This metallographic examination and elemental analysis revealed that all metallics had very high levels of carbon present. The reduction of taconite resulted in compositions ranging from hypereutectoid steel with a

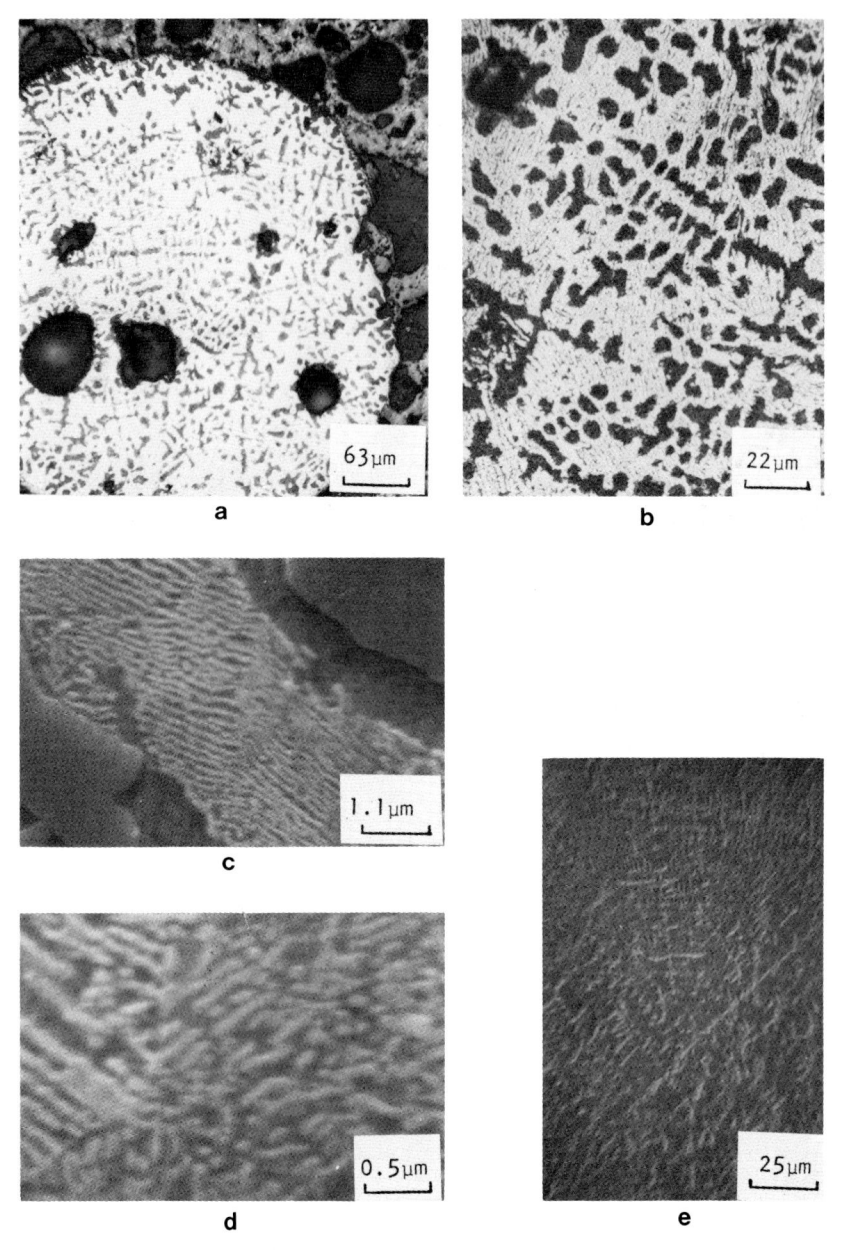

FIGURE 1. Photomicrographs of the Fe-C Metallics Produced from the Plasma-Reduced Taconite. a) General view of Fe-C metallic etched in nital surrounded by FeO-SiO$_2$ slag; b) High magnification of metallic structure in a) showing pearlite (dark phase) surrounded by ledeburite in which the γ has transformed to pearlite; c) Lamellar pearlite; d) Signs of break-up of lamellar pearlite; e) Dendritic structure (dendritic etch).

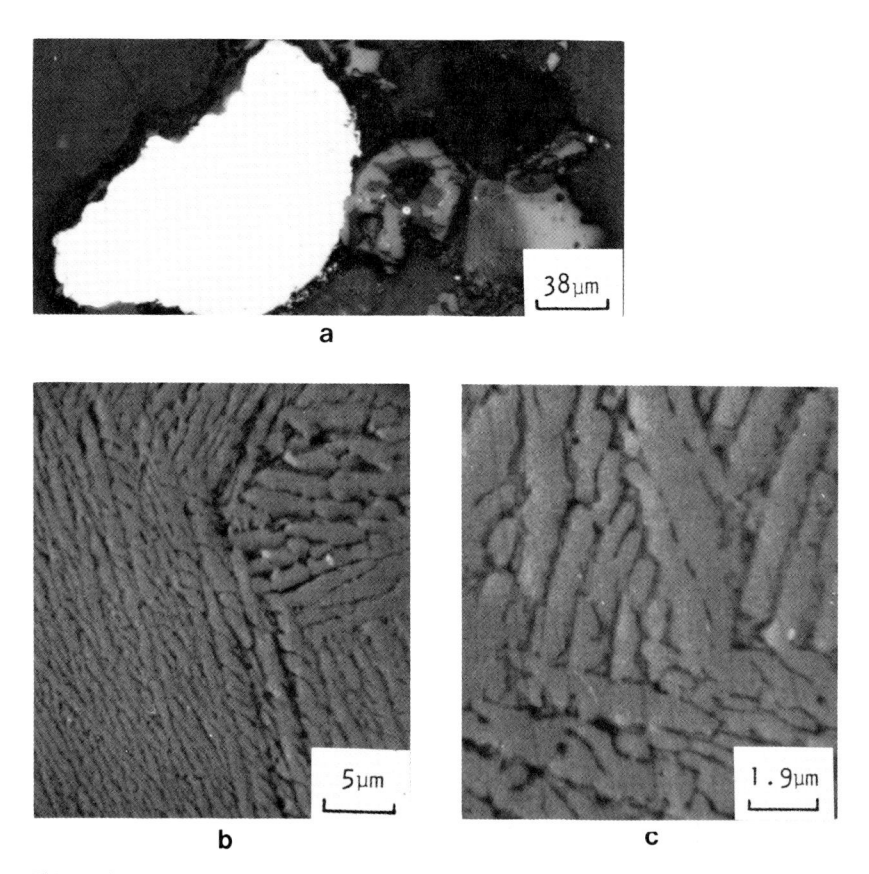

FIGURE 2. Photomicrographs of the Fe-Cr-C Metallics Produced from Plasma Reduced Chromite. a) General View of Fe-Cr-C Metallic (Unetched) Surrounded by Slag and Partially Reduced Chromite; b) Primary FeCr Carbides in Fe-Cr-C Metallics; c) Higher Magnification of b).

carbon level of 1.8% to hypoeutectic iron with a carbon level of 3.8% carbon. In the latter composition the 3.8% C-Fe liquid produces austenite dendrites on freezing with the remaining liquid subsequently transforming through the ledeburite (γ + Fe$_3$C) eutectic reaction. The γ present as dendrites and in the ledeburite eutectic subsequently transforms to pearlite (α + Fe$_3$C) on cooling through the eutectoid reaction. It is well established that secondary dendrite arm spacing, λ_2, is controlled by the rate of solidification which in turn will determine the degree of metastability. The relationship between secondary dendrite arm spacing, λ_2, and the average cooling rate, ε, has been empirically found to be $\lambda_2 = a\varepsilon^{-n}$ where a and n are constants largely dependent upon composition, solidification and heat transfer conditions. Therefore, secondary dendrite arms, λ_2, of the primary austenite dendrites in these metallic particles were measured and determined to be approximately 2.5 µm.

Quoted values for the constants a and n for a 0.9% C steel are 70, 0.42, respectively [7]. Substituting these values into the above relationship indicates an average cooling rate of 2.8×10^3 K s^{-1} for the metallics produced directly from the taconite concentrate within the plasma reactor. This rate of solidification is three to four orders of magnitude higher than in conventional casting processes.

The interlamellar spacing of the pearlite in Fe-C alloys can also be used to determine the degree of supercooling of the $\gamma \rightarrow$ pearlite eutectoid reaction [8] as represented in Figure 3. This parameter was determined using the SEM and found to be of the order of 0.1 μm. From examination of Figure 3 this value for the interlamellar spacing would indicate an eutectoid transformation temperature of 550-600°C, suggesting approximately 200°C of supercooling. Such supercooling would correspond with the nose of the Fe-C transformation diagram which would provide a microstructure which is transitory between a very fine lamellar pearlite and a modified pearlite in which the lamellae are broken up. This is consistent with the microstructures obtained as shown in Figure 1(d).

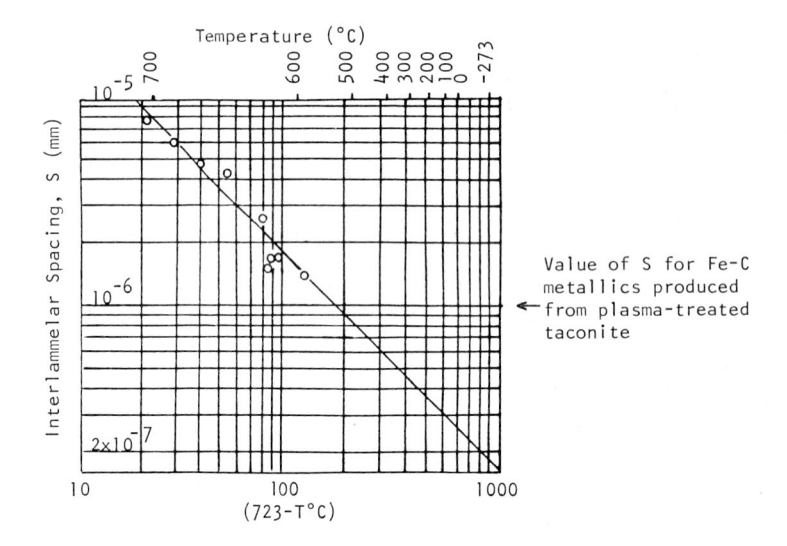

CIRCLES ARE EXPERIMENTALLY DETERMINED VALUES, THE LINE IS THE RELATIONSHIP FOR S BASED ON:

$$S = \frac{4\sigma V_m}{\Delta H_m} \cdot \frac{T_o}{T_o - T}$$

where: σ = specific surface energy for $\gamma \rightarrow$ pearlite reaction
V_m = molar volume of the material
ΔH_m = heat of transformation
T_o = equilibrium eutectoid transformation temperature
$T_o - T$ = degree of supercooling

FIGURE 3. Variation of Interlamellar Spacing with Temperature, Plotted According to Method of Mehl.

However, it could be expected that a cooling rate of 10^3 K s^{-1} should produce a much more modified structure than is evident in the photomicrographs in Figure 1. This may be explained in that although a high cooling rate of 10^3 K s^{-1} is present during quenching from a high superheat condition in the plasma medium, e.g., 3000 K through the liquidus. The subsequent cooling rate through the eutectoid transformation is then considerably decreased due to the surrounding slag or partially reduced mineral providing very poor heat conduction rates during the solid transformations in the metallics. This situation is further exacerbated by the latent heat of transformation that would need to be removed during the eutectoid transformation. Thus the cooling rate for the solid transformations may be several orders of magnitude lower than that through the liquidus-solidus range.

The reduction of chromite produced various compositions of high carbon, chromium irons in which the primary solidifying phase was a Fe-Cr carbide. X-ray analysis of the metallic particle shown in Figure 2 indicated that the carbide present was of the $(FeCr)_3C$ type. The overall chemistry of this particle as determined by SAM analysis (Table V) suggested that this was a hypereutectic alloy cast iron with a Cr:C ratio of 2.5. Examination of the liquidus diagram for the Fe-Cr-C system as produced by Jackson [9] (Figure 4) also indicates that the primary carbide should be one of $(FeCr)_3C$ for a Cr:C ratio of 2.5.

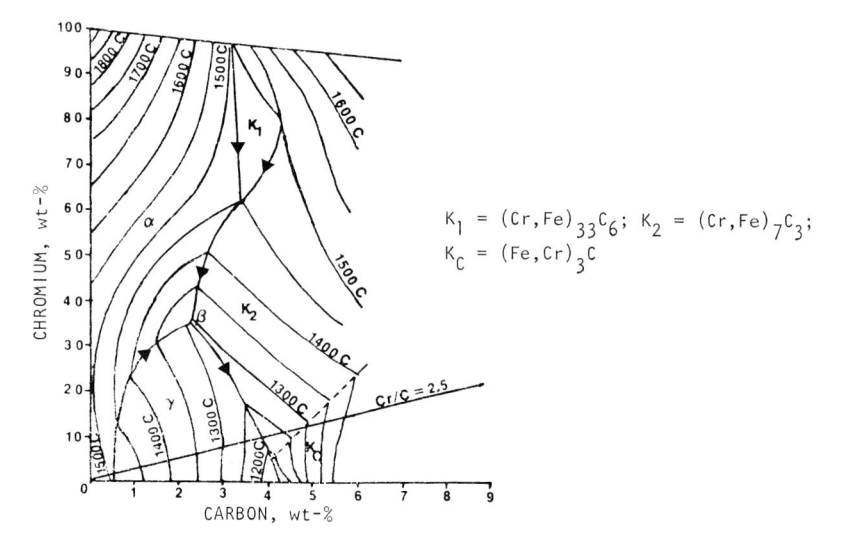

FIGURE 4. Liquidus Surface for the Fe-Cr-C System [9].

The typical microstructure of this metallic particle is shown in Figure 2. There is some indication that the primary $(FeCr)_3C$ carbide is present in a modified dendritic form as evidenced by side branches being observed at the higher magnification in Figure 2(c). Such morphology suggests that a high degree of supercooling occurred ahead of the solid-liquid interface. Measurement of the secondary dendrite arms of this structure gave a value of λ_2 = 2.8μm again indicating an average cooling rate of 10^3 to 10^4 K s^{-1}, using appropriate values [10] for constants a and n for high alloyed Fe-C metallics. The unidirectional primary dendrite arms with an average spacing of 0.7μm also supports the point that rapid solidification of the metallics followed the in-flight reduction of the chromite mineral within this plasma environment.

The very high carbon levels present in these Fe-C and Fe-Cr-C metallic particles is evidence of the rapid diffusion of the carbon reductant into the momentarily liquid state of the metallic particles as it reacted and passed through the plasma. SAM analysis of the hypoeutectic Fe-C metallic produced from taconite suggested a large excess (compared with equilibrium) of C in the pearlitic eutectoid structure transformed from the original austenite dendrite while a C deficiency (compared with equilibrium) existed in the cementite transformed from the eutectic (ledeburite) transformation. Even allowing for the inaccuracy that may be present on using the internal sensitivity factors, this trend demands a more detailed examination of such nonequilibrium carbon levels in these metallics as produced with this plasma reduction process.

CONCLUSIONS AND FUTURE WORK

In-flight reduction of taconite and chromite has been achieved with a solid carbon-based reductant using a small 60 kW sustained shockwave plasma (SSP) research reactor.

Examination of the metallic products produced from the plasma-processed taconite revealed fine, high carbon Fe-C metallics, i.e., hypereutectoid steel or hypoeutectic iron while the plasma-processed chromite produced fine high carbon Fe-Cr metallics with a predominance of primary carbides in a dendritic morphology.

The degree of metastability produced by this in-flight plasma reduction technique was determined from measurements of the secondary dendrite arm of the primary austenite and carbide in these Fe-C and Fe-Cr-C metallic particles which indicated an average cooling rate of between 10^3 to 10^4 K s^{-1} while measurement of the fine interlamellar spacing in the Fe-C metallics indicated a supercooling of approximately 200°C for the eutectoid temperature. At the same time, point elemental analysis determined by the scanning Auger microprobe using the internal sensitivity factors suggested that the pearlite in the Fe-C metallics and the primary carbides in the Fe-Cr-C metallics contained excessive (i.e., very much greater than equilibrium) amounts of carbon.

These states of high metastability were achieved in these metallics by quenching in a mixture of argon and air in the freefall chamber of the SSP reactor. It is, therefore, reasonable to expect much higher levels of metastability if liquid nitrogen were to be used as the quenchant. Liquid nitrogen could be sprayed onto the products at the top of the free fall chamber as they exit the plasma medium. It is also feasible to add glass forming elements, e.g., B, P, to the feed mixture in order to achieve greater metastability or even amorphous metallic product. In this respect these elements could be added in the form of phosphates or borates which would subsequently be reduced to P and B and diffuse into the liquid metallic particles within the plasma environment. This would greatly aid the economics of production of amorphous metal powder since the separate extraction of boron and phosphorus would be avoided.

The application of plasma technology to metals extraction and processing is in its infancy but there is no doubt that considerable potential exists in this field of in-flight reduction of minerals to achieve, in one operation, a high value, metastable metallic product.

ACKNOWLEDGMENTS

The authors are grateful to Mr. Pat Ryan who conducted the SAM and SEM analyses.

Joint financial support for this research program was provided by the Mineral Institutes Grant Title III of OSM and the Minnesota Legislative Commission for Mineral Resources.

REFERENCES

1. E.S. Machlin, T.J. Rowland, Eds., "Synthesis and Properties of Metastable Phases," pub. TMS-AIME, Warrendale, PA (1980).
2. B.J. Berkowitz, R.O. Scattergood, Eds., "Chemistry and Physics of Rapidly Solidified Materials," pub. TMS-AIME, Warrendale, PA (1982).
3. K.U. Maske, J.J. Moore, "The Application of Plasmas to High Temperature Reduction Metallurgy," High Temperature Technology (August 1982).
4. J.K. Tylko to Plasma Holdings, N.V., UK Patent No. 1,390,351-3.
5. J.J. Moore, J.K. Tylko, K.J. Reid, "Reduction of Lean Chromite Ore in a New Type Plasma Reactor," Proceedings Extractive Metallurgy of Refractory Metals, TMS-AIME, Chicago, 377-417 (1981).
6. K.J. Reid, J.J. Moore, J.K. Tylko, "In-Flight Metal Extraction in a Novel Plasma Reactor," to be published in Proceedings of Symposium on 'Plasma Processing and Synthesis of Materials' by Elsevier Science Publishing Co., Inc., MRS Annual Meeting, Boston (November 1983).
7. A. Suzuki, T. Suzuki, Y. Nagaoka, Y. Iawata, Nippon Kingaku Gakkai Shuho, 32 (1968).
8. M. Hillert, "Some Thermodynamic Aspects on Phase Transformation in Ferrous Alloys," Proceedings of the Darken Conference on Physical Chemistry in Metallurgy, pub. by U.S. Steel, 445-462 (1976).
9. R.S. Jackson, "The Austenite Liquidus Surface and Constitutional Diagram for the Fe-Cr-C Metastable System," Journal of the Iron and Steel Institute, 208, 163/167 (1970).
10. H. Jones, "Some Principles of Solidification at High Cooling Rates, Rapid Solidification Processing," edited by R. Mehrabian, B.H. Kear, M. Cohen, published by Claitor Publishing Div., Baton Rouge, LA, 28-45, (1978).

INTERACTION OF COAL PARTICLES INJECTED INTO ARGON AND HYDROGEN PLASMAS

K. LITTLEWOOD
Department of Chemical Engineering and Fuel Technology, The University, St. George's Square, Sheffield, S1 3JD, United Kingdom

ABSTRACT

Rapid heating of coal to temperatures in excess of 1500 K yields lower hydrocarbons which, if conditions are correct, will consist mainly of acetylene. Such conditions involve reaction times of 0.1 to 10 ms, together with rapid rates of quenching. Appropriate conditions have been achieved on a laboratory scale using a d.c.-arc plasma jet.

INTRODUCTION

During the past two decades there has been a number of investigations into the high-temperature reactions of coal [1-8]. Although the temperatures and residence-times to which the coal was subjected differ widely, there is a similarity in the nature of the products formed. Little or no tar is produced and at the high temperatures employed acetylene formation is thermodynamically favoured and appreciable quantities are found in the pyrolysis products.

Work here at Sheffield on the high-temperature decomposition of coal began in 1962. Early experiments [9,10] were aimed at comparing the yields of volatiles from the high-temperature pyrolysis of coal with those obtained during the standard volatile matter test [11]. The most convenient device available then for achieving temperatures above 2273 K was a small arc-image furnace. Results showed that the volatiles yield was significantly greater than that of the standard test and that about 9% by weight of the coal carbon had been converted to acetylene.

To improve the acetylene yield by increasing the heating rate and shortening the residence-time, subsequent pyrolyses were carried out in an argon plasma jet [12]. Complete devolatilisation of the coal was not achieved, but a maximum of 24% by weight of the coal carbon was converted to acetylene. Cold-model studies revealed that the penetration of the coal particles, which were injected radially into the plasma, was a critical factor in the devolatilisation process.

In a subsequent investigation [13], the coal was fed axially into the plasma through holes drilled in a hollow cathode holder. The maximum conversion of coal carbon to acetylene was 36% by weight. It was concluded that, due to the unfavourably low H/C atomic ratio of coal, this conversion represented the upper limit when pyrolysing coal in inert atmospheres. On the basis of equilibrium considerations, it was decided to use argon-hydrogen mixtures as arc gas in an effort to enhance the yield of acetylene. Unfortunately, stable operation was possible over a very limited range of experimental conditions because of the unfavourable geometry of the plasma head. However, one run was carried out successfully using a plasma containing 10% by volume of hydrogen, 74% by weight of the coal carbon being converted to acetylene.

The investigations up to this stage have been fully described in the literature [14,15].

The objective of the most recent investigation was to pyrolyse coals in plasmas generated with argon-hydrogen mixtures and hydrogen alone as arc gas in order to ascertain the upper limit of conversion to acetylene. For this purpose, a plasma jet capable of operating on such gases was designed and built.

PLASMA JET APPARATUS

The final design of the plasma head was a result of extensive trials with an experimental plasma jet. The latter was built such that its geometry could be easily altered to attain optimum conditions when operating on argon-hydrogen mixtures and hydrogen alone. Details of the design together with the ancillaries have been reported elsewhere [16,17].

EXPERIMENTAL

Coal, crushed to −200 mesh in a hammer mill, was fed into the carrier-gas stream by means of a screw feeder served from a continuously stirred hopper. The coal feed rate into the hollow cathode holder was monitored by a photo-electric cell and the output was displayed on a chart recorder. On emerging from the plasma jet, the coal-plasma stream was passed to a hot reaction chamber and the resulting pyrolysis products were subsequently quenched in a water-cooled chamber. Gas samples were withdrawn by means of a water-cooled probe, filtered and collected in sample bottles. Residual gas exited via an exhaust duct.

RESULTS AND DISCUSSION

Pyrolysis of four different coals under identical conditions revealed that there was no correlation between acetylene yield and volatile matter content.

Pyrolysis of the coals in the plasma jet gave rise to quantities of gaseous products in excess of those given by the standard volatile test. No liquid products or carbon dioxide were detected. With argon as arc gas, the major products were acetylene, hydrogen and carbon monoxide, with traces of methane and ethylene. Argon-hydrogen and hydrogen plasmas yielded acetylene, methane, ethylene and carbon monoxide, with traces of ethane and hydrogen cyanide.

Experiments using total gas flowrates in the range 90 - 140 ℓ/min revealed that maximum yields of acetylene were obtained with an arc-gas composition of approximately 50:50 argon-hydrogen (see Fig. 1.). The yields of methane and ethylene, however, increased with increasing hydrogen content of the arc gas.

At a lower total gas flowrate of 45 ℓ/min using arc-gas compositions containing up to a maximum of 58% by volume of hydrogen, maximum acetylene yields were obtained with a 54:46 argon-hydrogen mixture at an optimum power input of 15 kW (see Table 3 of reference 16).

The initial increase in acetylene yield with increasing hydrogen content can be attributed to several contributory factors. Firstly, the presence of hydrogen will have an inhibitory effect on the decomposition of the product acetylene by causing the equilibrium

$$C_2H_2 \rightleftharpoons 2C + H_2$$

Figure 1.

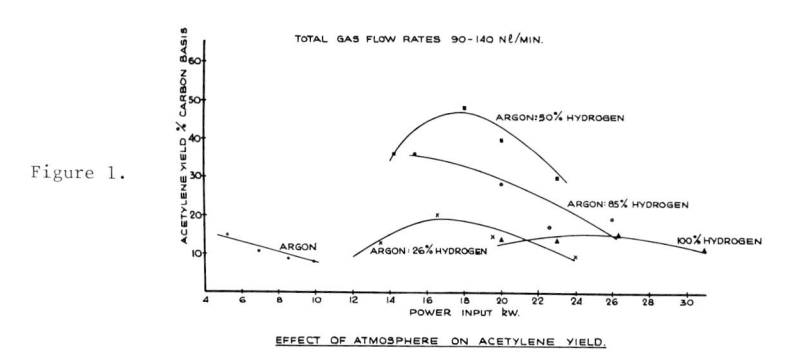

EFFECT OF ATMOSPHERE ON ACETYLENE YIELD.

to move to the left. Secondly, any hydrogen present in the plasma as atoms or in an excited state can act as a chemical intermediate to enhance the acetylene yield by combining with any carbon-containing radicals formed at the higher temperatures. Finally, confirmation that hydrogen from the arc gas was involved in the plasma reactions was afforded by hydrogen balances which revealed that the product hydrocarbons contained more hydrogen than was present in the original coal.

Decreases in the acetylene yield accompanying increases in the arc-gas composition above a 50:50 mixture can be attributed to the fall in the mean bulk plasma temperature that is associated with increased hydrogen content. Furthermore, lower temperatures favour methane formation and this is reflected in the fact that higher methane yields were obtained with plasmas operating on higher hydrogen concentrations.

With tangential entry of the arc gas, the acetylene yield decreased from 73% at 25 ℓ/min to 60% by weight of coal carbon at 55 ℓ/min. Although the highest acetylene yields were obtained at a total gas flowrate of 25 ℓ/min (Fig. 2), continued operation at this flowrate resulted in noticeable electrode damage. Consequently, gas flowrates were maintained between 40 and 45 ℓ/min. Acetylene yields also increased with increasing ratio of tangential/axial arc gas.

Figure 2.

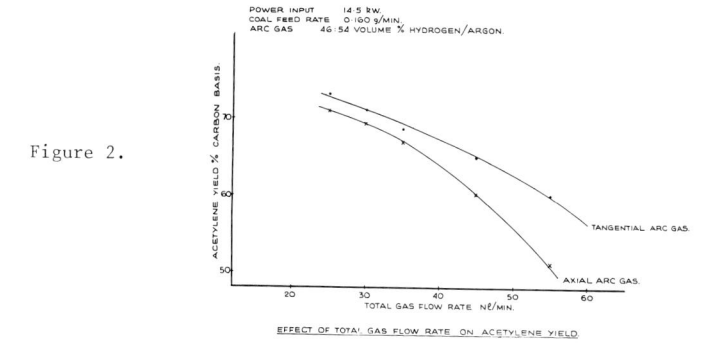

EFFECT OF TOTAL GAS FLOW RATE ON ACETYLENE YIELD.

The coal feed rate was varied over the range 0.045 to 0.270 g/min. Above 0.270 g/min, the plasma became unstable and led to marked non-reproducibility in the experimental results. At coal feed rates less than

0.045 g/min, the coal feeder did not operate uniformly. The results of these runs are presented in Fig. 3.

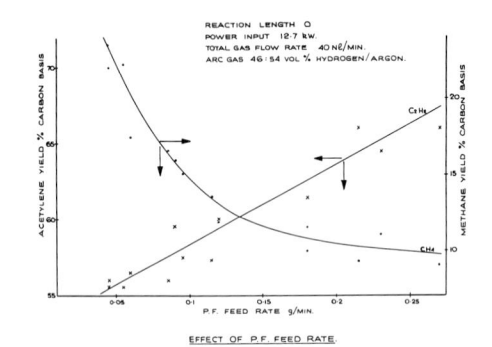

Figure 3.

EFFECT OF P.F. FEED RATE.

To increase the residence time of the coal particles in the hot zone, extensive sections, fabricated from alumina cement, were fitted below the anode nozzle. However, maximum acetylene yields were obtained when the pyrolysis products were quenched immediately.

A water-cooled sampling probe, capable of initial quenching rates of 1.4 MK/s, was used to withdraw the pyrolysis products from the quench chamber. The optimum probe position, equivalent to a residence time of about 1 ms, coincided with the maximum acetylene yield of 70.5% by weight of coal carbon. The effect of residence time is illustrated in Fig. 4.

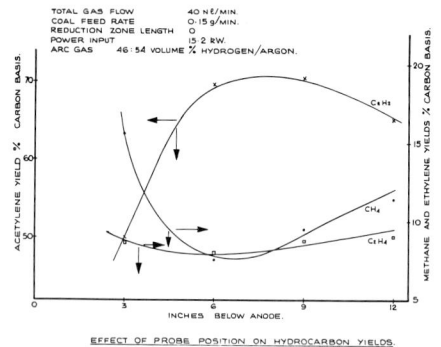

Figure 4.

EFFECT OF PROBE POSITION ON HYDROCARBON YIELDS.

The results of the pyrolysis experiments revealed that when the yield of acetylene was maximised, that of the methane was minimised. The converse was also true. Methane is considered to be an important precursor to acetylene [18]. To obtain information on the possible mechanism of acetylene from coal, methane was pyrolysed in a 54.46 argon-hydrogen plasma under conditions similar to those used for coal. The conversion of methane to acetylene attained a maximum of 53% by volume; the conversion decreased with increasing residence-time. The effect is depicted in Fig. 5. A methane route to acetylene does not seem to explain satisfactorily the much higher yields obtained in the pyrolysis of coal, but the similarities between Figs. 4 and 5 would suggest that methane does play an important role in the formation of acetylene from coal.

Figure 5.

METHANE PYROLYSIS IN ARGON/HYDROGEN PLASMA JET

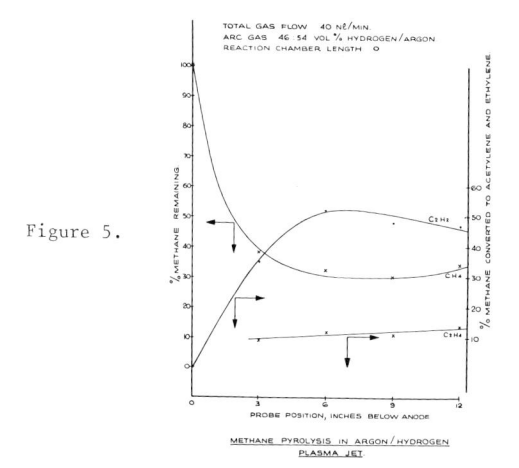

The decomposition of acetylene was investigated by injecting metered quantities into the plasma. Approximately 60% by volume of the acetylene was decomposed in the argon plasma compared with 25% in the 54:46 argon-hydrogen plasma. In the latter, the acetylene was converted mainly to methane, ethane and ethylene. In the former case, however, a hydrogen balance revealed that the acetylene had decomposed principally into carbon and hydrogen. As anticipated, it would appear that the presence of hydrogen retards the decomposition of the product acetylene.

In conclusion, it may be said that acetylene and/or methane can be produced directly in considerable quantities by pyrolysing coal in hydrogen-enriched plasma. Unfortunately, the specific energy requirement (kWh/kg product) is at least two orders of magnitude too high for the process to be commercially viable. This is because the coal throughput is far too low. Consequently, current investigations at Sheffield are aimed at increasing the coal throughout while maintaining the same degree of conversion.

REFERENCES

1. K. Littlewood & I.A. McGrath, "The Use of an Arc-Image Furnace for the Formation of Acetylene Directly from Coal - Some Preliminary Results", 5th Int. Conf. on Coal Science, Cheltenham (1963).
2. E. Rau & R.T. Eddinger, Fuel, 43, p. 246 (1964).
3. A.G. Sharkey, S.L. Schultz & R.A. Friedel, Nature, 202, p. 988 (1964).
4. E. Rau & L. Seglin, Fuel, 43, p. 147 (1964).
5. C.D. Hawk, M.D. Schlesinger & R.N. Hiteshue, U.S. Bureau of Mines Rep. RI/6264 (1963).
6. R.L. Bond, W.R. Ladner & G.I.T. McConnell, Fuel, 45, p. 381 (1966).
7. A.H. James, R. Nicholson & K. Littlewood, Gordon Int. Conf. on Coal Science, Tilton, New Hampshire, United States (1969).
8. S.C. Chakravartty, D. Dutta & A. Lahiri, Fuel, 55, p. 43 (1976).
9. A.H. James, Research Project Report, Sheffield University (1962).
10. S.K. Deyasi, M.Sc. Thesis, Sheffield University (1966).
11. "Proximate Analysis of Coal", British Standard 1016, Part 3 (1975).
12. A.H. James, Ph.D. Thesis, University of Sheffield (1968).
13. R. Nicholson, Ph.D. Thesis, University of Sheffield (1970).

14. K. Littlewood, Proc. of Symp. on Chemicals and Oil from Coal, Paper 59, p. 517, Central Fuel Research Institute, Dhanbad, Bihar, India (1969).
15. R. Nicholson & K. Littlewood, Nature, 236, p. 397 (1972).
16. M.J. Garratt & K. Littlewood, Proc. of Int. Symp. on Coal Science and Technology for the Eighties, Paper 14, p. 147, Central Fuel Research Institute, Dhanbad, Bihar, India (1979).
17. M.J. Garratt & K. Littlewood, "The Direct Production of Acetylene by the Pyrolysis of Coal in Plasma Jets", Paper presented at the 4th Int. Symp. on Plasma Chemistry, Zurich (Aug/Sept. 1979).
18. P.W. Gent, Ph.D. Thesis, Sheffield University (1972).

DESIGN AND USE OF AN EFFICIENT PLASMA JET REACTOR FOR HIGH
TEMPERATURE GAS/SOLID REACTIONS

F. W. GIACOBBE* AND D. W. SCHMERLING
*Cardox Corporation - Division of Liquid Air Corporation
5230 South East Avenue
Countryside, IL 60525

ABSTRACT

A unique and efficient plasma jet reactor has been developed and used to study the high temperature production of carbon monoxide from a reaction between powdered carbon and a pure carbon dioxide plasma. The plasma jet reactor was designed to allow the injection of powdered carbon above the arc discharge region rather than into the plasma flame below the arc discharge region. High yields of carbon monoxide, produced at relatively high efficiencies, were a direct result of this technique. The plasma jet was also designed to enable rapid changing and testing of various anode inserts.

Average yields of carbon monoxide in the product gases were as high as 80-87% in selected experimental trials. Carbon monoxide was produced at rates exceeding 15,000 l/hr (at STP) with a power expenditure of 52 Kw.

INTRODUCTION

In a recent article [1], a plasma jet reactor employed to produce carbon monoxide from powdered carbon and carbon dioxide was described. The objectives of that study were to determine the feasibility of and the economics related to the production of carbon monoxide in a plasma jet reactor. In this paper, a modified version of a plasma jet reactor used during an extension of the same study, has been described. One objective of this study was to determine the economics related to the production of carbon monoxide in a larger scale plasma jet reactor. An additional objective of this study was to determine the effect of plasma jet cathode diameter and anode design upon system performance and efficiency.

DESIGN FEATURES

General Construction

A diagram illustrating all of the essential features of the modified plasma jet reactor may be seen in Figure 1. The diagram also includes a detailed view of an automatic electrode feeding device employed during operation of the plasma jet reactor. Except for the automatic electrode feeder, this plasma jet reactor was operated in the same way as the plasma jet reactor described in reference 1.

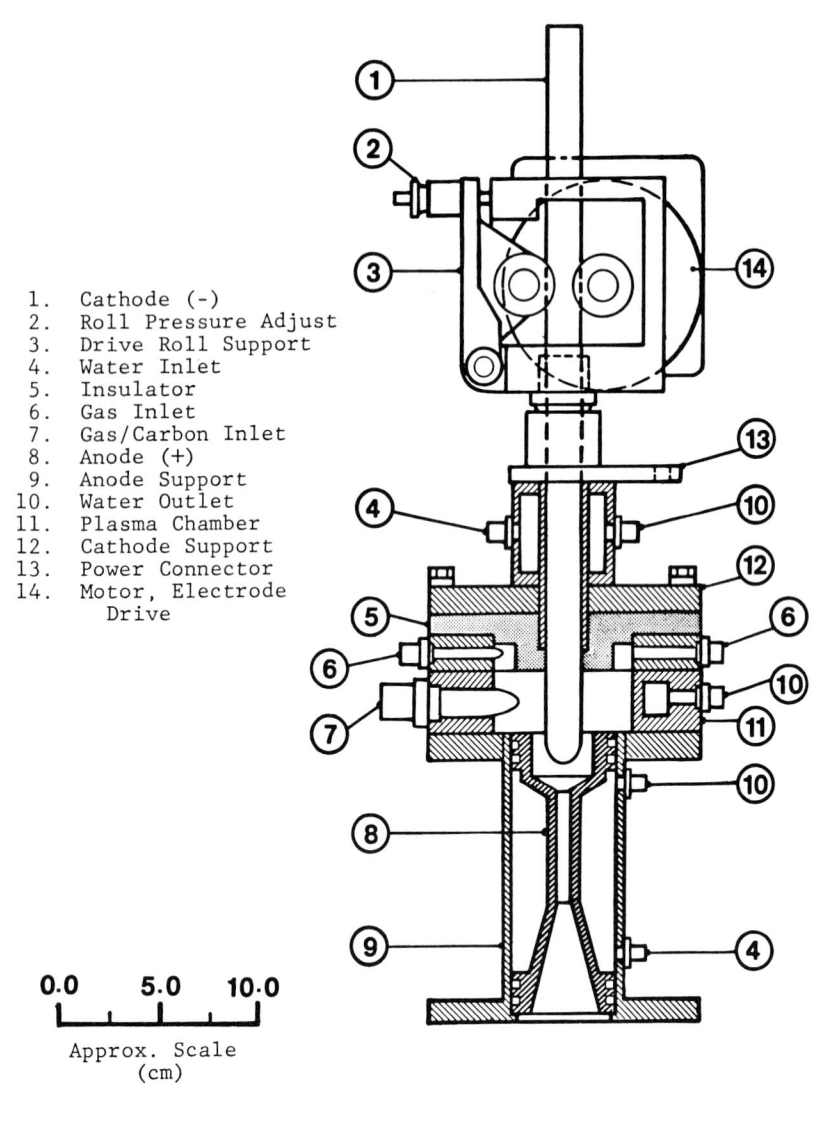

1. Cathode (-)
2. Roll Pressure Adjust
3. Drive Roll Support
4. Water Inlet
5. Insulator
6. Gas Inlet
7. Gas/Carbon Inlet
8. Anode (+)
9. Anode Support
10. Water Outlet
11. Plasma Chamber
12. Cathode Support
13. Power Connector
14. Motor, Electrode Drive

FIG. 1. Plasma Jet Reactor

Electrode Feeding Device

The automatic electrode feeding device shown in Figure 1 was a modified electrode wire feeder normally used in conjunction with a standard MIG welding torch. The main modification consisted of machining the wire feeding rollers so that they could efficiently drive the carbon electrode into the plasma jet reactor.

Carbon Electrodes

Relatively large consumable carbon electrodes (1.27 cm or 1.90 cm o.d. X 43.2 cm long) were fed into the plasma jet reactor by the electrode feeder. The electrodes (made by Arcair - Air Products and Chemicals - Lancaster, Ohio) were made for arc air gouging of weld metal. They were originally electrochemically coated with a thin layer of copper which was stripped off prior to their use. After initiation of the plasma arc in pure carbon dioxide, the carbon electrode was fed automatically into the plasma jet reactor in order to maintain a relatively constant arc voltage. An electronic circuit and control relay was used to detect small changes in the plasma arc voltage. An increase in arc voltage, due to electrode burn off, activated the electrode feeding motor. The electrode feeding motor then advanced the electrode into the plasma jet reactor until the plasma arc voltage returned to the setpoint voltage. An electrode motor drive timing circuit was used to control the electrode feed rate into the plasma arc reactor.

Due to the large size of the carbon electrodes, they were not consumed at a very rapid rate. Normally, the plasma arc reactor consumed energy at the rate of approximately 50-60 Kw. At this rate of electrical energy consumption, the 1.9 cm o.d. carbon electrodes were consumed at a rate of approximately 5.1 cm/hr.

Plasma Anodes

The plasma jet reactor was designed to operate with an anode insert sealed in place with O-rings. This is not a unique design feature [2-4]. However, a variety of different interchangeable copper anode inserts were machined for testing within this plasma jet reactor. By this means, a systematic attempt to study a single reaction system under a reasonably wide range of reactor configurations was made. Very few studies such as this are reported in the extensive literature involving the effects of electrical discharges upon chemical reactions [5].

Figure 2 illustrates the specific design features of six of the seven anodes tested during the course of this work. The only anode not shown in Figure 2 (1/4 FL) had the same configuration as the 3/8 FL anode except that the diameter of the central section of the anode was 0.635 cm.

Gas and Gas/Solid Injection

A mixture of carbon dioxide gas and powdered carbon was injected tangentially into an annular chamber surrounding the

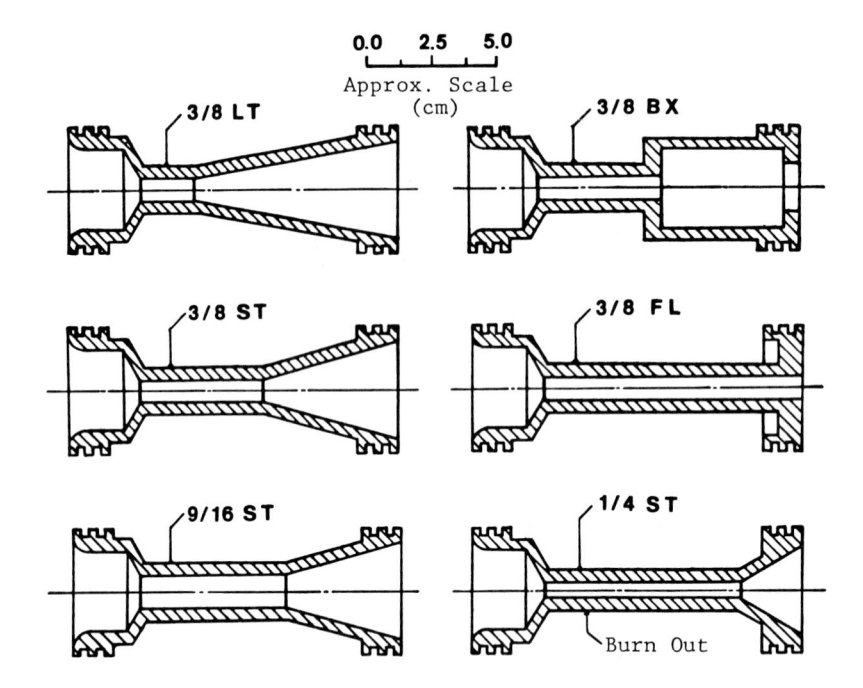

FIG. 2. Plasma Jet Anodes

carbon electrode. This annular chamber was located immediately above the arc discharge occurring between the carbon cathode tip and the uppermost section of the copper anode. An additional quantity of pure carbon dioxide gas was injected into another annulus located above the carbon dioxide/powdered carbon injection site. Injection of the stream of pure carbon dioxide was employed to inhibit short circuiting within the plasma jet reaction chamber. Before this modification of the plasma jet reactor, very erratic electrical behavior was exhibited during its operation. It was presumed that this erratic electrical behavior occurred due to short circuiting through a layer of internally deposited carbon. Internal carbon deposits were observed during periodic inspections made in conjunction with anode changes. Carbon deposition was not eliminated by the injection of pure carbon dioxide. However, a complete path of deposited carbon between the cathode and anode of the plasma jet reactor was eliminated. A U.S. patent [6] has been granted covering this special feature of the plasma jet reactor.

In order to optimize carbon monoxide yields using the plasma jet reactor, the injected carbon/carbon dioxide solid to gas ratio should have been maintained at the ideal ratio of 0.54 g/l. The ideal ratio of 0.54 grams of carbon per liter of carbon dioxide gas (at STP - 0°C and 1.0 atm) was obtained from the balanced chemical equation:

$$C(s) + CO_2(g) = 2CO(g)$$

Due to difficulties in controlling the powdered carbon feeding device, the ideal carbon/carbon dioxide solid to gas ratio was almost never attained in practice.

A few different types of industrial grade carbon blacks, made by the Cities Service Company - Columbian Division, were tested in the plasma jet reactor. The Raven 22 type seemed to work most effectively in the carbon feeder used in conjunction with this plasma jet reactor. This material has a mean particle size of 80 millimicrons, an apparent density of approximately 0.24 g/cm^3, and a BET surface area of about 17 m^2/g [7].

RESULTS AND DISCUSSION

A summary of experimental data, collected during some of the experimental trials carried out using the plasma jet reactor described herein, has been listed in Table I. This table contains a representative sample of averaged data from a few of the experimental trials actually carried out.

Although a substantial part of the data presented in Table I was collected during the use of two different electrodes and two anode types, data collected with other anode types was comparable. It should also be noted that etching of the anodes occurred, in all cases, after extended periods of use. In one trial run, the 1/4 ST anode was actually perforated by the plasma jet. The perforation occurred through the small diameter section of the anode, about 1.3 cm beneath the anode cup. The location of this perforation has been indicated in Figure 2.

The most significant, and probably most obvious, effect noted during the use of the various anodes tested was related to plasma jet head pressure. The plasma jet head pressures, associated with the data presented in Table I, ranged between 1.75 and 2.36 atm. Generally, the plasma jet head pressure rose and fell in inverse proportion to the smallest internal cross sectional area of the various anodes tested.

The tabulated data indicates that a plasma jet method can be used to produce carbon monoxide from carbon and carbon dioxide in relatively high yields. The data from the best run (trial 13, because the total carbon monoxide production volume to energy ratio was highest) indicates that carbon monoxide can be made at the rate of approximately 15,600 l/hr (about 260 l/min) with a power expenditure of 52 Kw. The carbon monoxide production volume/energy ratio which results from this data is about 300 l CO/Kw hr or 13.4 g moles CO/Kw hr. In the earlier study, described in reference 1, carbon monoxide was made at the rate of approximately 9,000 l/hr (about 150 l/min) with a power expenditure of 26 Kw. The carbon monoxide production volume/energy ratio which results from this data is 346 l CO/Kw hr or 15.5 g moles CO/Kw hr.

There are at least two possible factors responsible for the superior plasma jet synthesis of carbon monoxide described in reference 1. One of these factors is related to the average

TABLE I. Summary of operating conditions and results obtained during synthesis of carbon monoxide in a plasma jet reactor.

Trial*	Total CO_2 Input Flow Rate at STP (l/min)	Average CO Yield In Product Gases (% by vol)	Total CO Output Flow Rate at STP (l/min)	ELECTRICAL CHARACTERISTICS Current (amps)	Voltage (volts)	Power (kw)	Electrical to Chemical Conversion Efficiency (%)	C/CO_2 Ratio (g/l)	Total Run Time (min)
1	155	87	240	220	230	51	30	0.86	18
5	195	70	210	240	240	58	23	1.61	12
6	210	80	280	230	250	58	31	1.04	17
9	280	50	185	170	280	48	25	0.43	17
13	230	72	260	200	260	52	32	1.07	9
16	360	40	180	200	300	60	19	0.40	16
17	190	72	215	220	250	55	25	1.02	19
18	165	78	210	250	230	58	23	0.80	18

*Trials 1,5,6,9 - Anode Type 3/8 ST.
Trials 13,16 - Anode Type 3/8 LT.
Trial 17   - Anode Type 3/8 BX.
Trial 18   - Anode Type 9/16 ST.
Trials 1,9,13,16,18 - 1.90 cm diameter electrode.
Trials 5,6,17 - 1.27 cm diameter elctrode.

carbon/carbon dioxide solid to gas ratio actually delivered into the plasma jet reactors during their operation. The other factor is related to the size of the cathodes used in each of the plasma jet reactors studied.

During the most efficient trial run of the plasma jet reactor described in reference 1, the average carbon/carbon dioxide solid to gas ratio, actually delivered into the reactor, was about 0.45 g/l. During the most efficient trial run of the plasma jet reactor described in this paper, the average carbon/carbon dioxide solid to gas ratio actually delivered into the reactor was about 1.07 g/l. Therefore, much of the available energy was wasted heating the excess carbon.

A second factor which may have contributed to the superior performance of the plasma jet reactor described in reference 1 is related to cathode diameter. The smaller cathodes (0.794 cm dia.) used during that study may have resulted in the production of a higher plasma energy density than the plasma energy density produced with the largest cathodes (1.90 cm dia.) used during this study. The cross sectional areas of these cathodes are about 0.50 cm^2 and 2.84 cm^2. However, the power expended during the use of the larger cathode (from trial 13 of this study) was only twice as great as the power expended during the best trials with the smaller cathodes used during the study reported in reference 1. Additional data would have to be collected in order to verify these ideas and optimize the performance of the plasma jet system described in this paper.

Due to a restriction in the funding necessary to continue with this study, only a relatively small quantity of additional experimental data was collected. The additional data was not comprehensive enough to allow any definite conclusions concerning either the relative effectiveness of the various anodes tested or whether the system performance could have been improved. If additional experimental work on this system had been carried out, a major emphasis would have been placed upon optimizing the performance of the powdered carbon injection system used in conjunction with the plasma jet reactor. In addition, the influence of cathode diameter and anode design upon system performance would also have been studied in more detail.

CONCLUSION

A unique and relatively efficient plasma jet reactor has been designed, fabricated, and used to study the high temperature synthesis of carbon monoxide. The carbon monoxide was produced during a reaction between powdered carbon and a pure carbon dioxide plasma. The plasma jet reactor was designed to allow the injection of powdered carbon above the arc discharge region. High yields of carbon monoxide, produced at relatively high efficiencies were a direct result of this technique. The plasma jet reactor was also designed to enable rapid changing and testing of various anode inserts.

Average yields of carbon monoxide in the product gases were as high as 80-87% in selected experimental trials. Carbon monoxide was produced at rates exceeding 15,000 l/hr (at STP) with a power expenditure of 52 Kw. The data collected during this study may be used to estimate the material and energy costs related to the production of carbon monoxide from powdered carbon and carbon dioxide in a plasma jet reactor.

The plasma jet reactor design and some of the operating techniques reported in this paper may be applied to many other plasma jet studies involving high temperature gas/solid reactions. In particular, plasma jet reaction studies between powdered carbon (or powdered coal) and many plasma gases, in addition to carbon dioxide, can be studied with plasma jet reactors similar to the one described herein.

REFERENCES

1. F. W. Giacobbe and D. W. Schmerling, Plasma Chemistry and Plasma Processing, In Press.
2. C. S. Stokes in: Chemical Reactions in Electrical Discharges, R. F. Gould, ed. (American Chemical Society 1969) pp 390-405.
3. C. S. Stokes, J. A. Cahill, J. J. Correa, and A. V. Grosse in: Final Report "Plasma Jet Chemistry" Air Force Office of Scientific Research, Contract AFOSR-62-196 (Research Institute of Temple University 1964) p 7.
4. A. V. Grosse, H. W. Leutner, and C. S. Stokes in: First Annual Report "Plasma Jet Chemistry" Office of Naval Research Contract NONR 3085(02) (Research Institute of Temple University 1961) p 36.
5. J. D. Thornton in: Chemical Reactions in Electrical Discharges, R. F. Gould, ed. (American Chemical Society 1969) pp 372-389.
6. D. W. Schmerling and F. W. Giacobbe, U.S. Pat. No. 4,190,636.
7. Anon. in: Columbian Industrial Carbon Blacks Bulletin 10-75-7M (Cities Service Company - Columbian Division; Akron, Ohio) p 3.

HEAT TRANSFER ANALYSIS OF THE PLASMA SINTERING PROCESS

E. PFENDER AND Y.C. LEE*
*Heat Transfer Division, Department of Mechanical
Engineering, University of Minnesota, Minneapolis,
MN 55455

ABSTRACT

In the process of r.f. plasma sintering of Al_2O_3 at reduced pressure, a number of puzzling observations have been reported. They include higher heating rates and higher maximum temperature of a sample at increased propagation velocities of the sample through the plasma and a sudden temperature drop of approximately 700° C if the sample motion in the plasma is stopped. An analysis of the sintering process indicates that the change of the surface area and of the catalytic properties of the sample surface during sintering seem to be responsible for these observations. The change from an initially negatively charged to an essentially non-charged surface causes a drastic decrease of the heat transfer from the plasma to the sample. A simulation of the plasma sintering process based on a r.f. argon plasma provides a semi-quantitative explanation of all the observed phenomena.

1. INTRODUCTION

Plasma sintering represents an alternate approach for sintering of refractory materials, in particular of ceramics. A number of investigations [1-4] have demonstrated a much faster rate of sintering compared to conventional technology.
In some of this work anomalous heating has been reported [4,5] associated with the sintering of small alumina specimen in an argon plasma at pressures from 12 to 80 Torr. These observations will be briefly described in the following.
(a) As the sample moves through the plasma, it assumes surface temperatures up to and including the melting point. The surface temperatures recorded during this process increase with increasing translation velocity.
(b) The maximum surface temperature remains around 1500 K if the fired sample is again passed through the plasma under the same conditions.
(c) The surface temperature of the sample decays from its maximum value to approximately 1500 K whenever the motion of the sample in the plasma is stopped.

At a first glance, these observations seem to be contradictory. Therefore, a detailed analysis of the heat transfer process has been undertaken, based on a one-dimensional model.
In the first part of this paper, possible contributions of different conventional heat transfer mechanisms will be discussed and in the second part, the plasma heat transfer process will be analyzed. In the third part, the results of numerical simulations will be presented, followed by a number of conclusions.

2. POSSIBLE HEAT TRANSFER MECHANISMS

As the sample passes through the plasma region of a rf-discharge (Fig. 1), three zones of the sample may be distinguished:

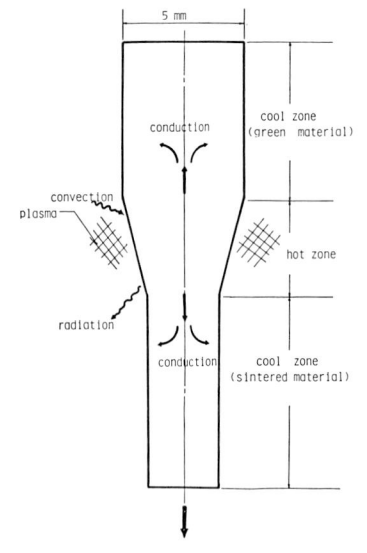

FIG. 1. Schematic of the plasma sintering process.

(a) A cold zone, representing the "green" material with a high porosity and low thermal and electrical conductivities. This zone will receive some heat from the hot zone of the sample which is partially released to the environment.
(b) A hot zone in which sintering takes place and in which the porosity as well as the thermal and electrical conductivities are changing. The sample receives heat from the plasma and some of this heat is transferred by conduction to the neighboring colder zones and some is released by radiation.
(c) A colder zone consisting of sintered material with low porosity and high thermal and electrical conductivities. This zone receives some heat from the hot zone which is partially released to the environment.

These considerations indicate that at least three heat transfer mechanisms are involved, i.e. conduction within the sample, convection/conduction to the sample in the plasma region, convection/conduction from the sample to the environment in the colder zones, and radiation from the sample to the environment. Radiative heating of the specimen from the plasma may be neglected.

The change of the properties of the sample during the sintering process is of particular importance for the following analysis. Besides enhanced density, thermal and electrical conductivities, the surface emissivity (ϵ) of the specimen will also increase during sintering. The specific heat of the sample which changes little during sintering, affects together with the density the rate of temperature changes, but not the values of steady state temperatures, i.e. the change of density and specific heat cannot explain the observed phenomena.

Initially, the low thermal conductivity of the "green"

sample contributes to the fast rise of its surface temperature. During the sintering process, the thermal conductivity increases by approximately an order of magnitude. Assuming that the heat flux from the plasma to the sample does not change during sintering, the increasing thermal conductivity cannot explain the magnitude of the observed temperature drop (from approximately 2,200 K to 1,500 K). Although the higher thermal conductivity will enhance heat conduction from the heated zone to the already sintered material, or in the case of a completely sintered rod, to the end faces of the rod, this axial heat conduction is by far too small to account for the observed phenomena.

A simplified analysis of the radiation losses leads to a similar conclusion. Without considering conduction, the steady state surface temperature follows from

$$\sigma_s \epsilon T^4 = h(T_\infty - T) \tag{1}$$

where σ_s is the Stefan-Boltzmann constant, T the surface temperature of the sample, h the heat transfer coefficient, and T_∞ the plasma temperature far away from the sample. If the increase of the emissivity during sintering is responsible for the observed temperature drop from $T_1 = 2,200$ K to $T_2 = 1,500$ K, then the ratio of the emissivities must be

$$\frac{\epsilon_2}{\epsilon_1} = \left(\frac{T_1}{T_2}\right)^4 \approx 4.6 \tag{2}$$

i.e. during sintering the surface emissivity should increase by a factor of 4.6. Typically, increases from 0.35 to approximately 0.4 are observed which is far below the required values. Therefore, radiation from the sample cannot be responsible for the observed phenomena.

These findings suggest that the heat flux to the sample must decrease during the sintering process. Based on the simplified energy balance of Eq. 1, heat transfer should decrease by a factor of 4 to 5 in order to be compatible with the observed phenomena. Such changes of the heat transfer are indeed possible as the following analysis will demonstrate.

3. THE PLASMA HEAT TRANSFER PROCESS

Heat transfer from a plasma to a neighboring wall depends strongly on the state of the plasma in the thermal boundary layer [6,7] which separates the free-stream plasma from the wall, and on the surface conditions of the wall (catalytic or non-catalytic). Fig. 2 shows qualitatively the behavior of of temperatures and charged particle densities in the boundary layer between a wall and a plasma produced from a monatomic gas. The wall is assumed to be electrically floating and the gas or plasma in the boundary layer is treated as a continuum.

In the case of an equilibrium boundary layer (Fig. 2a), the electron density drops rapidly with decreasing temperature and, therefore, heat transfer is governed by heat conduction of the neutral species only. The wall is separated from the hotter part of the boundary layer by a relatively cold, non-ionized layer which acts more or less as a thermal insulator.

Substantially larger heat transfer rates are expected in the case of a "frozen" boundary layer (Fig. 2b). Heat transfer is enhanced by diffusion of charged particles to the wall and subsequent recombination. Since the wall is assumed to be

(a) Equilibrium boundary layer
(Catalytic or non-catalytic)

$$q = k_h \frac{dT}{dx}$$

(b) "Frozen" boundary layer
(catalytic wall)

$$q = k_h \frac{dT_h}{dx} + q_d$$

(c) "Frozen" boundary layer
(non-catalytic wall)

$$q = k_a \frac{dT_h}{dx} + q_d$$
$$+ k_e \frac{dT_e}{dx} + k_i \frac{dT_h}{dx}$$

q_d = diffusion of charged species with subsequent recombination at the wall

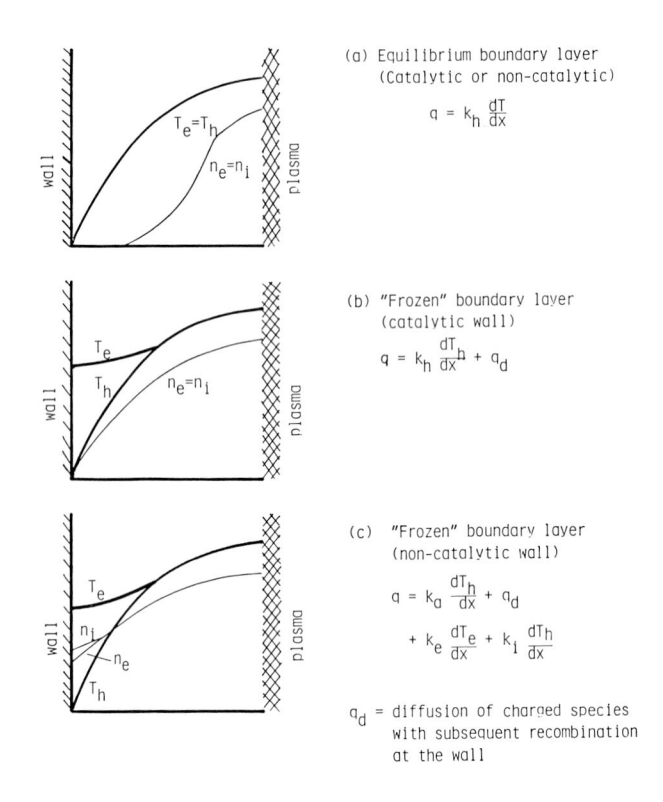

FIG. 2. Possible conditions in a continuum plasma-wall boundary layer.

catalytic, the charged particle densities drop to zero at the surface and the previously mentioned non-ionized layer in front of the wall shrinks substantially. Charged particle densities in the boundary layer are determined by diffusion in this situation. In the vicinity of the wall, electron temperature (T_e) and heavy particle temperature (T_h) tend to separate which, however, has no effect on heat transfer because the electron density drops to zero at the wall.

Finally, for the case of a "frozen" boundary layer in combination with a non-catalytic wall (Fig. 2c), the wall will assumed a finite negative charge. Close to the wall, electrons will be repelled and ions attracted. In a steady state situation electrons and ions will reach the wall at the same rate by ambipolar diffusion, i.e. the net surface charge will remain constant. Contrary to the previously discussed situation (Fig. 2b), there is now a finite charged particle density throughout the boundary layer, resulting in a substantially higher thermal conductivity due to the presence of free electrons. Heat transfer will be ennanced by two additional terms accounting for thermal conduction of electrons and ions.

Although the following considerations are related to the situation sketched in Fig. 2c, a significant modification is required. Since plasma sintering is primarily done at reduced pressures, the continuum approach for the boundary layer does no longer apply. In the following it will be replaced by an analysis based on free molecular flow. The mean free path lengths at reduced pressures (p ≈ 10 Torr) are of the same order of magnitude as the boundary layer thickness (≈ 1 mm).

The actual situation for pressures p ≥ 50 Torr may fall somewhere in between the two extreme cases of free molecular flow and continuum. Individual contributions by the neutral species, electrons, and ions have to be taken into account in order to calculate heat transfer to the wall in the case of molecular flow. For a catalytic wall, heat transfer will be dominated by electrons because of the high electron mobility. In the case of a non-catalytic wall, the surface assumes a net negative charge as previously explained. The corresponding electric field retards electrons and accelerates ions. For sufficiently strong electric fields ion bombardment will dominate the heat transfer process. In contrast to the previously discussed continuum situation, the ions falling freely through the collisionless boundary layer acquire high kinetic energies which they release at the wall. In a steady state situation, electrons and ions reach the surface at the same rate so the the negative charge on the surface remains constant. This implies that recombination of ions with electrons at the surface will also contribute to the heating of the wall.

The following analysis is based on a one-dimensional situation and the plasma is generated from a monatomic gas (for example argon) containing singly ionized species only, i.e. $n_e = n_i$. Heat fluxes due to atoms, ions and electrons may be expressed by [8-10]

Atoms:
$$Q_a = \frac{2n_a(k)^{3/2}(T_h)^{1/2}}{(2\pi m_a)^{1/2}} (T_h - T_w) \quad (3)$$

Ions:
$$Q_i = n_i \left(\frac{kT_h}{2\pi m_i}\right)^{1/2} (2kT_h)[1-\chi_p + 0.5\chi_p^2]$$
$$+ n_i \left(\frac{kT_h}{2\pi m_i}\right)^{1/2} (E_i) \exp(-\chi_p)$$
$$- \frac{2n_i(kT_w)^{3/2}}{(2\pi m_a)^{1/2}} \left(\frac{T_h}{T_w}\right)^{1/2} \quad (4)$$

Electrons:
$$Q_e = n_e \left(\frac{kT_e}{2\pi m_e}\right)^{1/2} (2kT_e) \exp\left(\chi_p \cdot \frac{T_h}{T_e}\right) \quad (5)$$

where the subscripts a, i, e, and h refer to atoms, ions, electrons, and heavy particles, respectively; T is the particle temperature, and T_w the wall temperature, k is the Boltzmann constant, m is the mass of the particles, and $\chi_p = e\phi_p/kT_h$ represents the dimensionless negative surface potential with ϕ_p as the surface potential itself.

4. RESULTS AND DISCUSSION

4.1 Estimated wall heat fluxes

Fig. 3 shows some calculated results for the case of an argon plasma at a pressure of p = 10 Torr and a wall tempera-

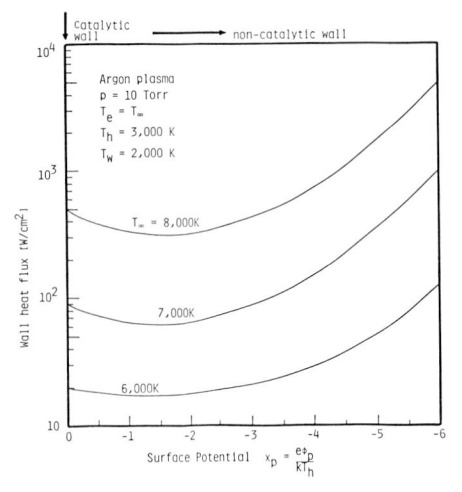

FIG. 3. Effect of surface potential on wall heat fluxes (molecular flow regime).

ture T_w = 2,000 K. The heavy particle temperature is assumed to be T_h = 3,000 K, and the electron temperature ($T_e = T_h$) may assume values from 6,000 K to 8,000 K.

As the surface potential starts to become negative, the wall heat flux first decreases, because the governing heat transfer by the electrons (Eq. 5) is reduced. For dimensionless surface potentials $\chi_p < -2$, the contribution of the positive ions becomes dominating, leading to a strong increase of the heat transfer. For the specified conditions the dimensionless potential reaches an equilibrium value of approximately $\chi_p = -5$. For this value of χ_p and a temperature of T_∞ = 7,000 K, the heat flux increases from 90 W/cm² to 360 W/cm² as shown in Fig. 3. This increase may even be larger if the large surface area of a green sample [5] is taken into account. As pointed out earlier such an increase of the heat flux (by a factor of 4 or more) for a non-catalytic wall can indeed explain the observed phenomena. It can be safely assumed that the surface of a green sample is non-catalytic, because the electrical conductivity is extremely small. As the sample is sintered, its electrical conductivity increases by approximately one order of magnitude [11] and, at the same time, the "active" surface area shrinks substantially.

4.2 Simulation of the plasma sintering process

Based on the results shown in Fig. 3, a numerical simulation has been undertaken for the plasma sintering process. Since the diameter of the sample is rather small, only axial variations will be considered. Because the heat fluxes will be decreasing during the sintering process, the heat transfer coefficient is, in a first approximation, assumed to be proportional to the porosity of the sample, i.e.

$$h = h_o\left[\left(\frac{h_g}{h_o} - 1\right)\frac{P}{P_g} + 1\right] \qquad (6)$$

where h_o and h_g are the heat transfer coefficients of the sintered and of the green sample, respectively. P_g is the initial and P the porosity of the sample during the sintering process.

For the simulation of the sintering process, a simple rate equation has been adopted [12] which reads as

$$\frac{dS_v}{dt} = \left(\frac{P_g - S_v}{1 - S_v}\right)(1.8 \cdot 10^7 - 2.37 \cdot 10^7 S_v^{0.514})^{0.53}$$

$$\times 5.37 \cdot 10^{10} \exp(-Q/81.5T) \qquad (7)$$

where S_v is the fractional volume shrinkage defined by the porosity

$$P = \frac{P_g - S_v}{1 - S_v} \qquad (8)$$

and Q is the thermal activation energy [J/kg].

The following calculation is based on the model sketched in Fig. 4. As the sample moves into the hot plasma region, it

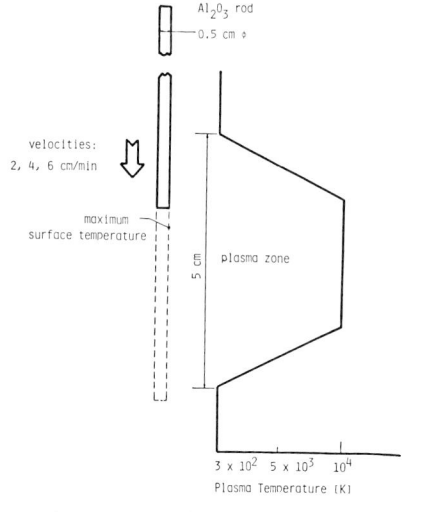

FIG. 4. One-dimensional model for the plasma sintering process.

receives heat from the plasma. Part of this heat is transferred by conduction to cooler regions and another part is lost by radiation.

The following situations have been simulated using the previously described model:
a.) Introduction of a green sample into the plasma with different velocities.
b.) A green sample introduced into the plasma is stopped after 1/3 of its length is completely sintered.
c.) An already sintered sample is moved through the plasma.
The data required for this simulation are listed in Table I.

It should be pointed out that the maximum plasma temper-

Material: Al_2O_3

Sample size: diameter 0.5 cm, length = 10 cm,
properties: powder size = 0.3 μm
density = 3,982 kg/m^3
specific heat = 1,500 J/kg
emissivity = 0.35
initial porosity = 52%
thermal conductivity $k/k_o = 0.92 - 1.34\ P$ for $P > 0.3$
(see ref. [14]) $k/k_o = \exp(-2.14\ P)$ for $P < 0.3$

Plasma: maximum temperature = 10^4 K
minimum gas temperature = 300 K
length = 5 cm

Sintering process:
activation energy = $6.2 \cdot 10^6$ J/kg
$= 4.5 \cdot 10^6$ J/kg (motion stopped)
heat transfer coefficient:

$$h = h_o\left[\left(\frac{h_g}{h_o}-1\right)\frac{P}{P_g}+1\right];\ h_o = 20\ W/m^2 K$$
$$h_g = 120\ W/m^2 K$$

propagation velocity of sample = 2, 4, and 6 cm/min.
P = porosity; subscript g = green sample
subscript o = sintered sample

Table I: Parameters for the simulation of plasma sintering.

ature, the length of the plasma zone, the heat transfer coefficient and its decrease during sintering are determined by the limiting temperatures observed in the experiments [4,5]. The heat transfer coefficient is assumed to be linearly related to the porosity (Eq. 6) and the activation energy in the sintering rate equation (Eq. 7) is assumed to be constant (6.2 x 10^6 J/kg) during sintering and it drops to a lower value (4.5 x 10^6 J/kg) whenever the motion is stopped. In reality, the activation energy is a function of temperature and porosity [13], i.e. it varies continuously during the sintering process.

Fig. 5 shows the temperature time histories of the sample temperatures. The temperatures of the green sample increases rapidly reaching a maximum temperature of approximately 2,200 K. As the motion of the sample is stopped, its temperature decreases from 2,200 K to approximately 1,600 K within 40 s. The fired sample temperature increases at a slower pace due to the reduced heat flux and it reaches a substantially lower maximum temperature close to 1,500 K.

Fig. 6 shows the temperature-time histories of the sintering process for different velocities of the sample. The rate of temperature rise as well as the maximum temperature increase with increasing propagation velocity of the sample. Also shown are results of the temperature-time histories for the same parameters, but with a constant, lower heat transfer coefficient. Because of the smaller heat fluxes, the rate of temperature increase is diminished and after the temperature reaches its maximum value it remains at this value.

5. CONCLUSIONS

All the puzzling observations described in the introduction can be explained by an analysis and simulation of the

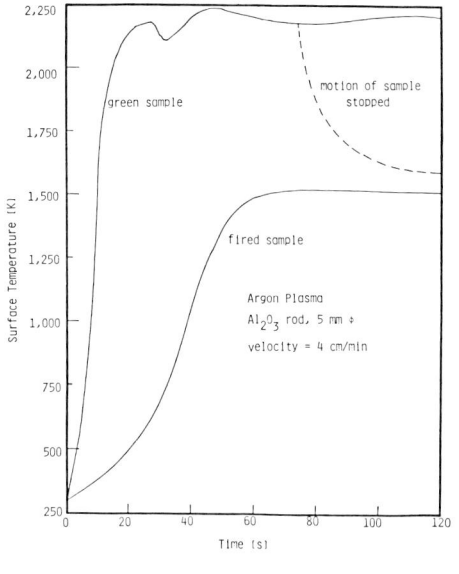

FIG. 5. Temperature histories of a green and of a sintered sample in a rf argon plasma.

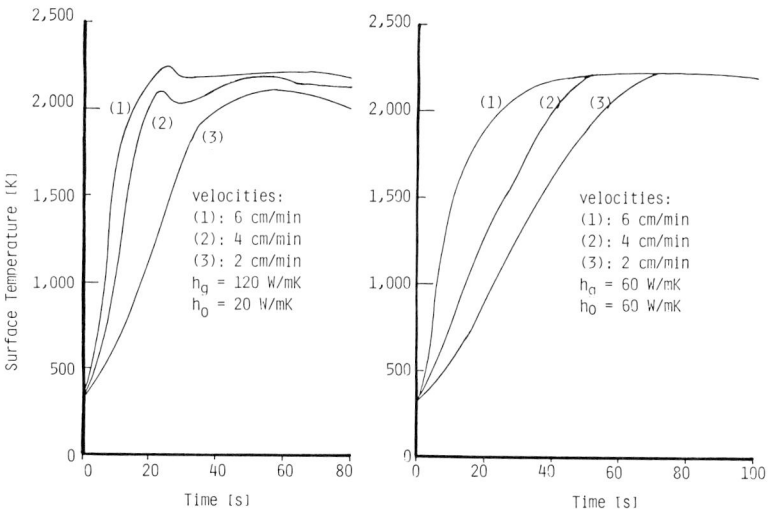

FIG. 6. Temperature histories of samples during the plasma sintering process with propagation velocities and heat transfer coefficients as parameters.

plasma heat transfer and sintering process at reduced pressures. The following conclusions may be drawn from the results of this study:

a.) Heat transfer from the plasma to the sample (Al_2O_3) depends strongly on the catalytic properties and on the area of the surface. In the case of a non-catalytic surface, the neg-

ative surface charge gives rise to an electric field which increases the rate and energy of the ions arriving at the surface, resulting in a substantial increase of the heat transfer compared to a situation with a catalytic surface.
b.) The high heating rate in combination with the low thermal conductivity of the green sample results in high surface temperatures ($\approx$ 2,200 K).
c.) The heating rate as well as the maximum temperature increase with increasing propagation velocity of the green sample in the plasma. This puzzling result is caused by the high rate of sintering which reduces the heat flux to the sample.
d.) The maximum temperature which a fired (sintered) sample assumes in the plasma is around 1,500 K, because the heat transfer rate to a fired sample is substantially lower than to a green sample.
e.) By stopping the propagation of the sample in the plasma, the surface temperature drops as sintering proceeds because of reduced heat transfer. The temperature drop from 2,200 K to the final temperature of approximately 1,600 K requires approximately 40 s.

6. REFERENCES

1. C.E.G. Bennett, N.A. McKinnon and L.S. Williams, Nature, 217, 1287-1288 (1968).
2. L.G. Cordone and W.E. Martinsen, J. Am. Ceram. Soc., 55, 380 (1972).
3. D.L. Johnson and R.A. Rizzo, Am. Ceram. Soc. Bull., 59, 467-472 (1980).
4. J.S. Kim and D.L. Johnson, Am. Ceram. Soc. Bull., 62, 620-622 (1983).
5. D.L. Johnson, Department of Material Science and Engineering, Northwestern University, private communication, 1983.
6. E.R.G. Eckert and E. Pfender, Advances in Heat Transfer, J.P. Hartnett and Irvine Jr., eds. (Academic Press, New York, 1967), 4, 270-272.
7. E. Pfender, Pure & Appl. Chem., 48, 199-213 (1976).
8. B.Y.H. Liu, K.T. Whitby and H.H.S. Yu, J. Colloid Interface Sci., 23, 367-378 (1967).
9. R. Godard and J.S. Chang, J. Phys. D, 13, 2005-2012 (1980).
10. N.N. Rykalin and A.A. Uglov, High Temp., 19, 404-411 (1981).
11. J.M. Brettell, Powder Metallurgy Internat., 13, 86-87 (1981).
12. H. Palmour III and D.L. Johnson, "Sintering and Related Phenomena", G.C. Kuczynski, N.A. Hooton and C.F. Gibbon, eds. (Gordon and Breach Sci. Pub., New York, 1967), 779-791.
13. H. Palmour III, M.L. Huckabee and T.M. Hare, 6th International Material Symposium (U.C. Berkeley, 1976), 308-319.
14. M.E. Cunningham and K.L. Peddicord, Int. Heat Mass Transfer, 24, 1081-1088 (1981).

IN-FLIGHT METAL EXTRACTION IN A NOVEL PLASMA REACTOR

K.J. REID,* J.J. MOORE,* AND J.K. TYLKO**
*Mineral Resources Research Center, University of Minnesota, Minneapolis, MN 55455; **Plasmatech Inc., P.O. Box 14055, Minneapolis, MN 55414

ABSTRACT

A new plasma reactor system based on a stable expanded conical plasma reaction zone has been used to examine the in-flight extraction of metals from mineral concentrates. The basic principles of the plasma system and details of the equipment are described. Results obtained from the treatment of taconite concentrate, low grade chromite concentrate and a copper-nickel concentrate are presented and discussed.

INTRODUCTION

1983 has been a vintage year for plasma with several major conferences and an increasing coverage in trade journals and the news media. A major plasma project has been funded by the state of Minnesota [1] and reviews of Plasma Processing [2], Plasma Melting/Remelting [3] and Plasma Extractive Metallurgy [4] are included in this conference. Also, during the year the National Materials Advisory Board initiated a study on plasma applications in the process industries and the Iron & Steel Society of AIME began work on a book covering the application of plasma technology in Process Metallurgy. This paper describes some of the research in plasma metallurgy carried out at the Mineral Resources Research Center (MRRC) at the University of Minnesota.

MRRC was established in 1911 by state legislation and has a primary responsibility for long-term mineral research related to the development and use of the mineral resources of the state. In 1979 a program in plasma metallurgy was initiated primarily to explore novel direct reduction routes for processing the abundant iron ore products mined from the Mesabi Range in the Northeastern part of the state. Reports on this work have been presented at the 5th ISPC Conference in Edinburgh and elsewhere [5,6,7]. A preliminary study of the in-flight plasma reduction of domestic chromite has also been reported [8] and an extension of this work is the subject of a current Ph.D study.

To the east of the Mesabi Range is the Duluth Gabbro formation which hosts several copper-nickel bearing mineral deposits. In recent years there has been considerable commercial activity to explore the potential for an operating mine. However, since the region lies close to the Boundary Waters Canoe Area, a unique geographical resource protected by Federal law, the state has made major investments in detailed environmental studies to determine the potential environmental impacts of mining and smelting activities. As part of an MRRC program to seek environmentally compatible technology, a three-pronged state-funded study was initiated to explore new smelting technology that would specifically minimize adverse environmental impacts. One of the routes explored was plasma flash smelting and preliminary results obtained from this work were reported at the ISPC-6 Conference in Montreal [9]. Further work on the plasma smelting of copper is currently being carried out as part of another Ph.D project. The present status of these three project areas is reviewed in this paper.

Although the literature on plasma chemistry is voluminous [10], there has been a disappointingly small number of commercial applications in process metallurgy other than melting or remelting where significant progress has been made in recent years [3]. One reason for this is the difficulty in establishing clear commercial benefits from the use of plasma and a recent detailed cost comparison of plasma smelting and coal based direct smelting [11] illustrates this point with the relative costs for pig iron delivered to Chicago shown in Table 1.

TABLE 1. Relative Costs for Pig Iron Delivered to Chicago

A. Technology Comparison	Minnesota	Canada	Brazil
Plasma Smelting			
Direct Smelting	1.10	0.96	1.07
B. Location Comparison	Minnesota	Canada	Brazil
Plasma Smelting			
Brazil Plasma Smelting	1.27	1.03	1.0
Direct Smelting			
Brazil Direct Smelting	1.24	1.14	1.0

From the technical viewpoint there are three basic approaches to the application of plasma technology in process metallurgy involving metallurgical reactions.

1) **Plasma Gas Heating**, in which a plasma torch is used to heat a process gas to temperatures above those attainable by conventional fossil fuel combustion. The high temperature gas is then used to convey the required reaction enthalpy to the reactants. Examples of this approach are the SKF PLASMARED [12] and PLASMASMELT [13] processes and the CRM PIROGAS [14] process.

2) **Molten Pool Transferred Arc Furnaces**, in which metal/slag reactions take place in the molten pool with the major energy input provided via the anode attachment zone on the molten surface. Examples of this approach include the molybdenite studies at Union Carbide [15] and McGill [16] and the chromite reduction work at MINTEK [17].

3) **In-Flight Plasma Reactors**, in which solids are introduced directly into the plasma medium and undergo metallurgical reactions while in flight. Examples of this approach are the TAFA zircon dissociation process [18] and the PLASMA HOLDINGS SSP technology [19].

The drive for exploring the application of these plasma systems in process metallurgy is influenced by several factors. One of the principal advantages of using plasmas in the reduction of mineral concentrates is the improved reaction kinetics available to effect the required reduction reactions at the elevated temperatures achieved by the plasma medium. This advantage is not merely due to the increased reaction rates obtained at elevated temperatures but, also, due to the more reactive species created within the plasma medium at these high temperatures.

Although plasma physics and plasma chemistry are well established scientific disciplines, there appears to have been relatively little activity until quite recently in the application of these disciplines in the less exotic areas of mineral processing and process metallurgy.

The reviews of Sayce [20], Hamblyn [21] and Maske and Moore [22] cover most of the practical applications and studies within extractive metallurgy up to 1982.

Another factor relates to the recent economic problems of conventional pyrometallurgical extraction operations which have accentuated the need for radically new approaches in extractive metallurgy. Any new technologies developed will be required to minimize capital costs, maximize energy efficiency, and have minimal adverse environmental impact. Although plasma processes have the potential to satisfy all of these objectives, the ultimate criterion will continue to be the commercial viability [11].

Of the three modes of plasma energy transfer available for process metallurgy applications, the gas phase and surface heat transfer modes rely principally on the thermal properties of the plasma. By contrast, the in-flight system has the potential to exploit the excited and ionic species that exist within the plasma volume itself. It is the in-flight mode that constitutes the main area of activity at MRRC. It should however be stressed that the optimal use of plasma for a required metallurgical reaction may involve the application of one or more of the energy transfer modes described. For this reason a 100-lb transferred arc melting/smelting furnace has recently been commissioned at MRRC and operated at power inputs up to 150 kW. Further details of the MRRC transferred arc furnace program will be reported at a later date.

SSP TECHNOLOGY

In order to achieve in-flight reactions there have been several approaches towards generating an expanded plasma volume [23,24,25] with the common general objectives of reducing the high velocity, viscosity and temperature gradients that exist in a conventional linear plasma jet and increasing the residence time of the feed materials within the plasma zone. One of these, the Expanded Precessive Plasma [25], introduced the concept of generating an expanded conical plasma by physically driving the tip of an inclined plasma torch along a circular path. The arc operated in the transferred mode to a stationary annular anode and thus generated a truncated conical volume of 'expanded' plasma, as illustrated in Figure 1. This device produced some interesting results in the 1970's but the mechanical driving mechanism limited the orbiting velocity to a maximum value of about 1500 rpm. In recent years there has been no work reported on in-flight reactions with the EPP system.

In 1978 the basic concepts for an improved expanded conical plasma were developed and led to the patents covering the Sustained Shockwave Plasma technology [19]. The basic conical plasma volume is created with a stationary cathode and anode and the arc is driven by external nonmechanical means [26]. By avoiding the mechanical constraints it has been possible to attain SSP orbiting velocities over the range 2,000 to 90,000 rpm.

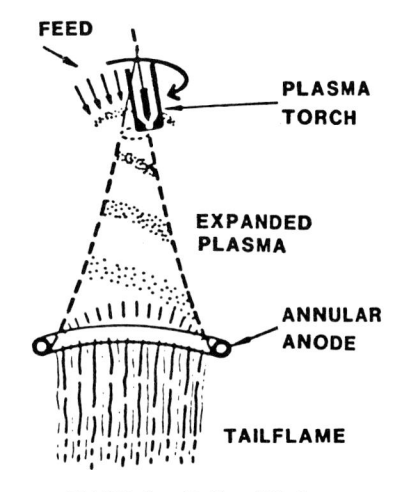

FIGURE 1. Tylko EPP System

The sustained shockwave plasma derives its name from the imposition of pulsations on the plasma which simultaneously generate pressure and temperature waves within the expanded plasma volume. These effects contribute to the rapid rate of reaction observed in the in-flight treatment of materials in the SSP reactor. A third factor of importance in relation to the effective utilization of the SSP conical volume is the ability to maintain a high particle loading within the plasma. These three basic principles, i.e., spinning, pulsation and loading, are exploited in the SSP system to facilitate plasma-particle interaction and in-flight metallurgical reactions and are interacting and to some extent synergistic.

By orbiting the plasma arc discharge two distinct fields are generated, an inner one, defined by the circumference of the plasma column orbit in which the charges are accumulating and an outer one in which they are dispersing. In slow orbiting speeds of only a few thousand rpm, the accumulation tends to be feeble and sporadic. With higher orbiting speeds, arc voltages in the range of 100-1000 volts and currents of 200-1000 amp., the accumulation effect becomes very distinct and well established. Substantially the whole truncated cone becomes filled with feebly ionized, highly turbulent plasma. The high angular orbiting speeds lower the normally very strong cohesion of the discharge column and thereby reduce the otherwise steep temperature and viscosity gradients. This results in a considerable improvement in the efficiency and uniformity of injection of the mineral particles into the plasma medium - an area of plasma technology which has produced difficulties in the past, causing poor utilization of the plasma effluents [25]. Formation of characteristic turbulence and microfields [26] results from the degree of variation of ionization within the plasma medium.

The second of the three mentioned characteristic features of SSP is the imposition of electrically produced pulsations. When an organized train of pulsations is introduced such as by periodic, short duration increase in power to the plasma column, a strong compression wave results, traveling radially outward through the ambient medium. Since the pulsation is periodic and takes place in axial confinement, some reflection at the walls takes place. With suitably chosen amplitude and mark space ratio, the sharply defined compression wave is followed by a continuous rarefaction. If the medium in which such phenomena take place contains particles, strong discontinuities in the form of shockwaves form around them. This also improves turbulence within the plasma and particle-plasma interaction. With high particle loading the cross-sectional area available for plasma is reduced and an improved cone stability is observed due to this constricting effect.

The advantages of applying the SSP principles in the processing of minerals may be summarized as:

1) effective utilization of whole plasma volume
2) no rotating parts
3) ease of scale-up
4) low refractory wear due to protection by the curtain of falling feedstocks
5) very short treatment time
6) ease of start-up and shut-down and
7) potential for utilizing a wide range of reductants and minerals.

DESIGN VARIABLES

The expanded SSP zone forms the heart of an SSP reactor system, as shown in Figure 2. There are several design variables which can influence the reaction sequence in an SSP reactor, the most important of which are as follows:

Feeder system,
Plasma zone geometry,
Secondary gas system,
Tailflame geometry,
Quenching, and
Gas/solids separation

Control of solids feed rate and uniform introduction into the reactor are important. Solids transport to the feed ports can be either by gravity from an overhead bin system or by pneumatic means depending on the physical characteristics of the material. In either case it is important to maintain a uniform flow rate. Solids can be introduced axially into the center of the plasma volume when the cathode configuration is annular or may be introduced at points on a circle surrounding the cathode for annular or point source cathodes. In the latter case injection can be discrete via a set of equally spaced feed ports or as a continuous annular curtain. For optimal loading it is also of importance to match the angle of solids injection with the geometry and operating conditions selected for the plasma zone.

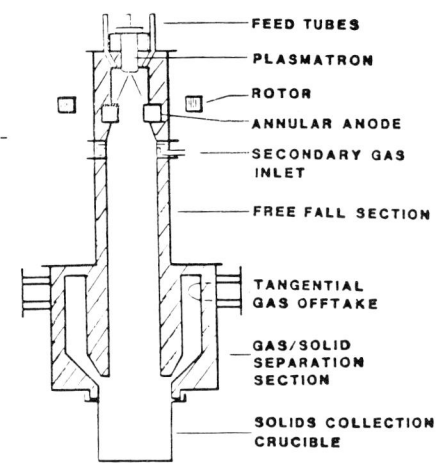

FIGURE 2. SSP Reactor Configuration

The geometry of the plasma zone can be varied by selecting the arc length, i.e., distance between cathode and anode, the diameter at the top of the cone by using either a point cathode or an annular cathode and the diameter of the base via the annular anode diameter. By these means geometries ranging from inverted cones through cylinders to long thin cones or short fat cones can be created.

In many cases the derived metallurgical reaction will require the supply of a secondary gas which is not involved in the establishment of the primary plasma volume. This gas can be introduced before the plasma zone, into the plasma zone, into the tailflame or below the tailflame.

Reactions continue within the tailflame volume below the anode and are influenced by the tailflame geometry. Residence time and hence extent of reaction can therefore be influenced by the length, diameter and shape selected for the tailflame section. In some cases a rapid quench is required to 'freeze' reactions or create metastable or amorphous product structures. This can be achieved by a variety of methods available to the design engineer such as gas or liquid injection, cold fingers and/or cooled walls.

The final operating zone in the reactor is the gas/solid separation section and three basic design variants are available. These are a simple radial port above the solids collecting crucible, a tangential offtake design, as shown in Figure 2, and an upward axial offtake similar in design to the vortex finder in a cyclone. The gas/solids separation design has a much lesser influence on the reaction sequence than the other variables discussed above and is more related to quench requirements and off gas dust loading criteria.

By careful selection of these basic design variables a wide range of rapid metallurgical reaction sequences can be accommodated.

LABORATORY FACILITIES

Work in the MRRC laboratories is based on four plasma power sources covering operating ranges of 1-15 kVA, 10-100 kVA, 50-250 kVA and 200-1400 kVA. The 50-250 kVA unit is a mobile diesel generator set and all others are supplied from mains power. DC output is provided by diode bridge rectifiers or SCR rectifiers and DC inductors. Experimental work is carried out in a range of SSP reactors with an open reactor and 100 kVA closed unit shown in Figure 3 and a second 100 kVA unit shown in Figure 4. The metal extraction work described in this paper used these two enclosed reactors. A much larger unit, approximately 25 feet tall and 4 1/2 feet diameter, is also available and was designed for in-flight cement reaction studies.

FIGURE 3. SSP Reactor Laboratory. a) Left, Enclosed 100 kVA reactor; b) Center, Open 40 kVA reactor; c) Right, Reactor control unit.

METAL EXTRACTION TESTWORK

Details of experimental results obtained with taconite concentrate [5,6,7], low grade domestic chromite [8] and copper-nickel concentrate [9] have been reported earlier. There are continuing programs in these three areas at MRRC and the main features of the results are presented and updated here. It should also be noted that a project to examine the reduction of ilmenite concentrates in the SSP reactor began recently and results will be published in due course.

For all testwork the feed materials were separately ground and then intimately mixed before being fed by gravity into the SSP reactor. Various reductants and stoichiometries were used. Solid products were collected and subjected to a range of analytical procedures including optical microscopy, scanning electron microscopy, electron probe microanalysis and chemical analysis. Chemical analysis was carried out on the bulk solid product and on selected components as appropriate. Results for the reduction of taconite

FIGURE 4. 100 kVA SSP Reactor

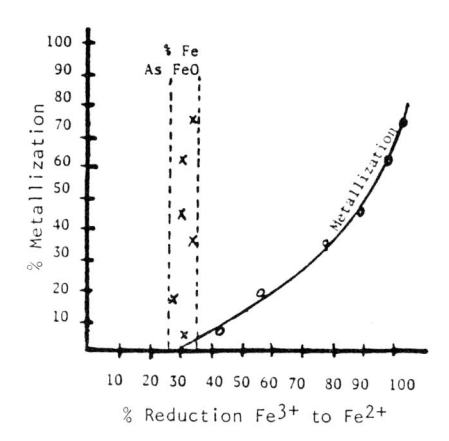

FIGURE 5. Degree of Metallization Versus Percent Reduction Fe^{3+} (Fe_2O_3) to Fe^{2+} (FeO)

TABLE II. Degree of Metallization Achieved with Taconite and Chromite Concentrates

Concentrate	Power	Feed Rate g/min	Degree of Metallization, % Fe	Cr
Taconite	60	76	74.7	-
Taconite	72	300	45.0	-
Chromite	30	100	46.0	7.3
Chromite	60	150	62.0	52.0

concentrate containing 6.23% SiO_2 and 66.81% total Fe are shown in Figure 5 and the degree of metallization achieved in the reduction of taconite and chromite concentrates is shown in Table II. Point analyses of metallic particles obtained in these runs are shown in Table III. The chromite feed material had a chromite to iron ratio of 1.53 and assayed 43.8% Cr_2O_3, 22.8% FeO, 18.3% Al_2O_3 and 11.4% MgO.

TABLE III. SAM Point Analysis of Metallic Particles

Feed	Wt % Fe	C	Cr
Taconite	98.2	1.8	-
Chromite	55.2	12.8	32.0

For the copper-nickel concentrate runs the feed material contained 20.7% Cu, 0.44% Ni, 34.7% Fe and 31.6% S and initial tests were done under an oxidizing atmosphere aimed at producing a high strength SO_2 off-gas. For this work the highest degree of metallization achieved was 95.7% and some individual copper particles were assayed at 99.7% Cu. Subsequent tests were carried out under reducing conditions with a carbon reductant and using CaO or Na$_2$CO$_3$ as sulfur scavengers, thereby avoiding sulfur bearing off-gas. To date the highest degree of metallization achieved under oxidizing conditions has not been matched, the corresponding figure under reducing conditions being 75.9%. Some data from the copper test runs are given in Tables IV and V. Examples of photomicrographs obtained from the three programs are shown in Figure 6.

TABLE IV. Copper Smelting. EPMA Analyses of Various Products

Product	Analysis, %			
	Cu	Fe	Ni	S
Metallic Veins	25	74	6	35
Iron Rich Metallics	1.6	76	0	0
Copper Rich Metallics	95	0.3	3.4	1.4
High Purity Copper	99.7	0.3	0	0

Maximum observed copper metallization = 95.7%

TABLE V. Copper Reduction Results, -325# Feed

		Flow Rate g/min	Analysis of +50μm Metallic Flakes		
			Cu Total	Cu Metal	% Metallization
A.	Using CaO + C	72	13.6	1.2	8.8
		163	14.7	5.2	35.4
B.	Using Na$_2$CO$_3$ + C	72	14.3	2.8	19.6

Maximum observed copper metallization = 75.9%

Results from the testwork to date can be summarized as follows:

  i. Plasma volume stability is improved by addition of mineral particles.
  ii. Reduction was improved with higher volatile reductants.
  iii. With low volatile reductants reduction was improved by the introduction of air or oxygen.
  iv. In all studies reactions appear to be initialized along intercrystallite boundaries within the mineral structure.
  v. Fine metallic globules form initially within the mineral particles and subsequently coalesce and grow to form larger metal globules.
  vi. Metal products exhibited a wide variation in structure and composition.
  vii. In the oxide reduction studies no sulfur was detected in any metal products.
  viii. In the taconite work the presence of silica resulted in the formation of fayalite (Fe$_2$SiO$_4$).
  ix. In some treated chromite particles topochemical reaction zones were observed with low-melting magnesium-rich silicates on the outside and high-melting natural chromite spinel structures in the center.
  x. There is a limiting particle size which can be completely reacted under given SSP reactor conditions.

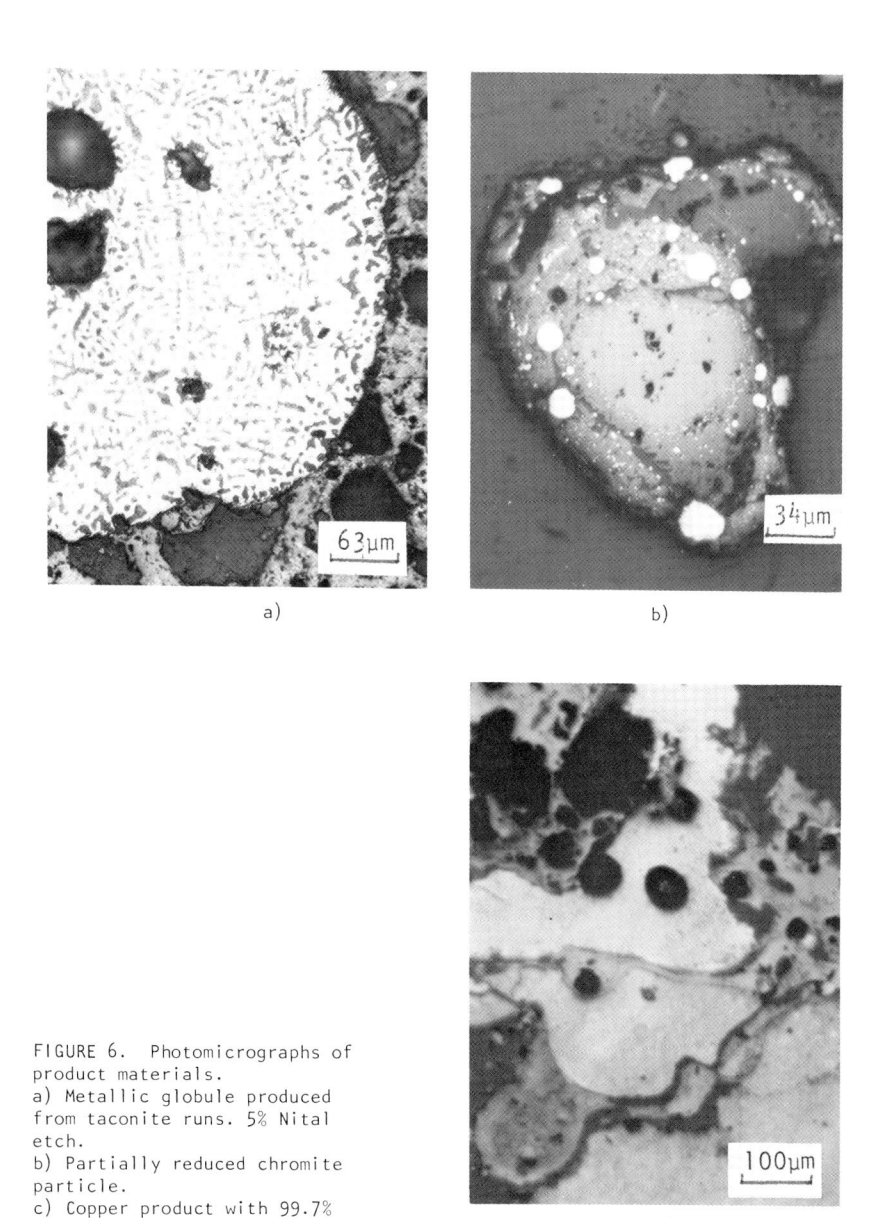

FIGURE 6. Photomicrographs of product materials.
a) Metallic globule produced from taconite runs. 5% Nital etch.
b) Partially reduced chromite particle.
c) Copper product with 99.7% copper particle.

  xi. In the copper studies multiphase products were obtained including separate copper-rich and iron-rich metallics.
 xii. In all cases medium to high levels of metallization have been achieved.
xiii. Separation of metallics from residual slag can be achieved using simple mineral processing separation methods.

xiv. In the oxidizing sulfide work SO_2 off-gas concentrations over 45% were attainable.

DISCUSSION

Reaction mechanisms based on both conventional reaction kinetics and those conditions thought to exist within the SSP reaction zone have been postulated in previous papers [6,8,9] for the in-flight plasma treatment of taconite, chromite and copper concentrates. In essence, these are concerned with

i) the selective reduction of oxides based on their thermodynamic stability, e.g., for chromite, in the order: $FeO \cdot Cr_2O_3 \cdot MgO \cdot Al_2O_3$

ii) the kinetics associated with a two-stage process for taconite reduction, i.e., $Fe_2O_3 \cdot FeO + C = 3FeO + CO$ and $FeO + C = Fe + CO$, but unlike conventional reduction of Fe_2O_3 the second reduction stage ($FeO \rightarrow Fe$) was found to start before complete consumption of the Fe_2O_3.

iii) the overall reaction for chalcopyrite

$$8CuFeS_2 + 21\ O_2 \rightarrow 8CU + 2FeO + 2Fe_3O_4 + 16SO_2$$

iv) the initial formation of low melting point silicates as soon as the minerals interact within the plasma medium

v) the simultaneous outgassing and formation of microplasma channels along intercrystallite boundaries in the mineral thereby producing reactive channels along which metallics are nucleated and subsequently coalesce.

vi) the fact that the reaction sequences take place within 100 ms was qualitatively explained by the presence of the highly activated molecular, gaseous and ionic species present and the high temperatures achieved within the SSP medium.

At such elevated temperatures contributions to the reaction sequence can be made by

a) dissociation of oxides which occurs in the range 2,600 to 5,100 K for Fe_3O_4, MgO, Al_2O_3, Cr_2O_3, SiO_2 and FeO in order of increasing temperature

b) molecular forms of carbon, C_1 through C_5, which have been spectroscopically observed at temperatures above 2000°C [27]

c) dissociation and reassociation of diatomic and complex gases [28], and d) ionization of gaseous species of carbon and carbon monoxide [28].

In addition to these considerations the SSP is in a nonequilibrium condition in which electron temperatures can be an order of magnitude higher than ion temperatures. These conditions make any study of reaction mechanisms and kinetics extremely speculative but, nevertheless, offer a qualitative explanation of why rapid in-flight reduction is possible within the SSP reactor.

These nonequilibrium and varying conditions also provide an explanation for the different microstructures and carbon diffusion levels found in the metallics produced from the same plasma experiment. These metallics range from hypereutectoid steel to high chromium iron compositions [29] which, on subsequent separation by conventional mineral processing techniques, could be briquetted and used in the production of alloy steels and irons. In the copper extraction work the presence of different metallic species also indicates the range of conditions and varying physical forces brought to bear on the feed material.

The continuing programs in these areas are focused on obtaining more precise data over a wider range of primary variables including feed grade, feed rate, feed size, reductant type, power level, etc., in an attempt to identify optimal conditions for complete metallization and to further elucidate the prevailing reaction mechanisms.

CONCLUSIONS

The in-flight SSP treatment of taconite chromite and copper concentrates has been demonstrated with the achievement of metallization levels up to 74.7% Fe from the taconite concentrate, 62.0% Cr from the chromite concentrate and 95.7% Cu from the copper concentrate. Since particle resident time in the plasma zone is of the order of 100 ms the substantial extent of reaction indicates the high rates of reaction and favorable kinetics achieved within the sustained shockwave plasma. The versatility of the plasma reactor for metallurgical processing has been demonstrated by testwork involving both oxide and sulfide systems.

ACKNOWLEDGMENTS

The support for this work provided by several different organizations is gratefully acknowledged. Supporting organizations are the Legislative Commission on Minnesota Resources, the federal Title III Mineral Institutes program, Plasmatech Inc., and Metallurgy Inc. The authors and MRRC are also grateful to Plasma Holdings N.V. for making technical support and details of the SSP technology available for these studies.

REFERENCES

1. G.J. McManus, Plasma Could Hold the Key to One-Step Steel Processing, Iron Age, October 3, 1983.
2. J. Szekely, Overview of Plasma Processing, Proceedings MRS Symposium on Plasma Processing & Synthesis of Materials, Boston, November 15-17, 1983, Elsevier.
3. W.C. Roman, Thermal Plasma Melting/Remelting Technology, Proceedings MRS Symposium on Plasma Processing & Synthesis of Materials, Boston, November 15-17, 1983, Elsevier.
4. W.H. Gauvin, Plasmas in Extractive Metallurgy. Proceedings MRS Symposium on Plasma Processing & Synthesis of Materials, Boston, November 15-17, 1983, Elsevier.
5. K.J. Reid, J.J. Moore and J.K. Tylko, Reduction of Taconite in the Sustained Shockwave Plasma (SSP). Proceedings 5th ISPC, IUPAC, Edinburgh, August 1981.
6. K.J. Reid, J.J. Moore, Plasma Reduction of Minnesota Taconite Concentrates. Proc. 54th Annual Minnesota Sec. AIME Meeting, Duluth, January 1981, University of Minnesota.
7. K.J. Reid, J.K. Tylko and J.J. Moore, The Application of the Sustained Shockwave Plasma (SSP) in Process Metallurgy. Second World Congress of Chemical Engineering, Montreal, Canada, October 1981.
8. J.J. Moore, K.J. Reid and J.K. Tylko, In-Flight Plasma Reduction of Domestic Chromite. Journal of Metals, p 43, August 1981.
9. K.J. Reid, M. Murawa and N.M. Girgis, Plasma Smelting of Copper-Nickel Concentrate for Minimal Environmental Impact. Proceedings 6th International Symposium on Plasma Chemistry ISPC-6, Montreal, 1983.
10. Bibliography of Plasma Chemistry. International Union of Pure and Applied Chemistry.
11. K.J. Reid and P. Kakela, Direct Iron Smelting: The Influence of International Location on the Competitive Posture of Two New Technologies. Journal of Metals. December 1983.
12. S. Santen, Plasma Technology Gives New Lease of Life to Swedish DR Plant. Iron & Steel International, 53, No. 1, 1980.

13. S. Santen, Plasma Smelting. Proceedings 6th International Symposium on Plasma Chemistry ISPC-6, Montreal, 1983.
14. N. Ponghis, R. Vidal and A. Poos, PIROGAS - A New Process Allowing Diversification of Energy for Blast Furnaces. High Temperature Technology, Vol 1, No. 5, August 1983, p 275-281.
15. J.H. Harrington, Reduction and Dissociation of Molybdenum Compounds in a Transferred Plasma Arc. 4th Int. Symp. on Plasma Chemistry, Zurich, August 1979.
16. W.H. Gauvin, G.R. Kubanek and G.A. Irons, The Plasma Production of Ferromolybdenum - Process Development and Economics. J. Metals, January 1981, pp 42-46.
17. T.R. Carr, N.A. Barcza, K.U. Maske and J.F. Mooney, The Design and Operation of Transferred-Arc Plasma Systems for Pyrometallurgical Applications. Proc. 6th International Symposium on Plasma Chemistry ISPC-6, Montreal, July 1983.
18. M.L. Thorpe and P.H. Wilks, Electric-Arc Furnace Turns Zircon Sand to Zirconia. Chemical Engineering, Nov 15, 1971, pp 117-119.
19. J.K. Tylko to Plasma Holdings N.V., U.S. Patent 4,361,441.
20. I.G. Sayce, Plasma Processes in Extractive Metallurgy, Advances in Extractive Metallurgy, Inst. Min. Met., p 241, 1972.
21. S.M.L. Hamblyn, Plasma Technology and Its Application to Extractive Metallurgy, Minerals Sci Engng, Vol 9, No. 3, July 1977, p 151.
22. K.J. Maske and J.J. Moore, The Application of Plasmas to High Temperature Reduction Metallurgy. High Temperature Technology, August 1981, p 51.
23. D.R. McRae, R.G. Gold, C.D. Thompson, W.R. Sandall, Ferrovanadium Production by Plasma Carbothermic Reduction of Vanadium Oxide, 34th Electric Furnace Conference, Iron and Steel Society of AIME, Dec 1976, St. Louis.
24. D.K.A. Donyina, J.D. Lavers, A.McLean, R.S. Segsworth, Plasma Processing of Ferromanganese Slags. 37th Electric Furnace Conference, Iron and Steel Society of AIME, Dec 1979, Chicago.
25. J.K. Tylko, Expanded Precessive Plasmas, IUPAC Plasma Chem. Symposium, Kiel, West Germany, 1973.
26. Frank-Kamenetskii, Lektsii po Fizike Plazmy Atomizdat, Moscow, 1964.
27. B.R. Brontin, The Application of Plasmas to Chemical Processing, R.F. Baddour, R.S. Timmis, ed., Oxford: Pergamon Press, 1967.
28. T.B. Reed, ibid., pp 26-28.
29. J.J. Moore, K.J. Reid, J.M. Sivertsen, Production of Metastable, Metallic Particles Directly from the Mineral Concentrate in a Plasma Reactor. Proceedings MRS Symposium on Plasma Processing and Synthesis of Materials, Boston, Nov 15-17, 1983, Elsevier.

PLASMA-MELTED AND RAPIDLY-SOLIDIFIED POWDERS

RICHARD F. CHENEY
GTE Products Corporation, Chemical and Metallurgical Division, Hawes Street, Towanda, PA 18848

ABSTRACT

A novel process is described for making rapidly solidified powders. In its simplest form, powders are melted in a high-temperature plasma and the resulting high-velocity spray is quenched into a fluid medium. Rapid cooling rates are achieved by free fall in inert gases, by quenching into liquid argon or nitrogen, or by impacting against a chilled, moving substrate. Any composition can be processed as long as it or its constituents are available in powder form. When necessary, alloy or composite components are agglomerated to 15-45µm diameter prior to plasma synthesis. The agglomeration is typically achieved by spray drying. Powders with melting points up to and including tungsten have been melted.

INTRODUCTION

A highly specialized plasma synthesis technology has been developed which is applicable to the production of a wide range of powders [1-5]. The original motivation was to develop processes for the manufacture of hardsurfacing powders having high melting points. It was necessary to form spherical powders (for good flow) with controlled particle size and narrow distribution (for homogeneous melting). These novel processes have very adequately met those original objectives and have the added bonus of rapid solidification of the melted powder product.

The Plasma Melting and Rapid Solidification (PMRS) process can be applied to any composition as long as its constituents are available in powder form. The desired starting powders are agglomerated by spray drying to achieve flow, size uniformity, and chemical homogeneity. An illustration of how this technology would be applied is shown in Figure 1. The resulting agglomerates would be passed through a plasma flame, melted, cooled rapidly, and collected, all under an inert atmosphere. Details of these operations are given in the following paragraphs.

AGGLOMERATION

Agglomeration is accomplished by spray drying a slurry consisting of the desired powders and a binder in either an aqueous or organic liquid [6-12]. Spray drying is an economical, commercial process often used for such items as dried milk, instant tea and coffee, dyes, detergents and pharmaceuticals. The agglomerates are formed by atomizing a slurry of powder and binder into a chamber where the spherical drops are contacted by hot gas, causing the carrier liquid to quickly evaporate. The dried powder falls to the bottom of the chamber where it is collected. The moist gas is exhausted from a side port and any entrained particles are removed by scrubbing or filtration.

Examples of spray-dried, sintered agglomerates are shown in Figures 2 and 3. Figure 2 illustrates 15-60 µm molybdenum agglomerates made from 1-5 µm molybdenum powder. Figure 3 shows two types of agglomerates of tungsten carbide/tungsten with 12% cobalt. In one case, a 1 µm tungsten carbide was used and in the other, a 6 µm tungsten carbide. The cobalt size was about 0.8 μm and it is not distinguishable in the figure.

Commonly used binders are organic materials such as paraffin, polyvinyl alcohol, or methyl cellulose [13-15]. However, inorganic binders are sometimes attractive, e.g. ammonium molybdate is used with molybdenum (16). Since the binders are soluble in the suspending liquid, it is easy to classify the product of spray drying to the desired range and to return the out-of-size material to the slurry tank for respraying.

Agglomerate-size control is further achieved by adjusting the spray drying conditions and the composition of the slurry. Average agglomerate sizes from 20 μm to 300 μm are possible. The particle size range may be as narrow as 50 μm for the finer sizes.

While aqueous liquids are suitable for many compositions, others, for example aluminum, require the use of organic solvents such as acetone, ethanol, hexane, and methanol. Because these liquids are flammable, an inert spray-drying gas must be used, usually nitrogen. The drying systems are closed circuits so that the organic liquid can be condensed and recycled. The drying gas is scrubbed and recycled also [17-21].

One of the advantages of spray drying is the ability to limit the temperature to which the solids are exposed. For example, the maximum temperature which an agglomerate would typically experience is about 10 °C less than the outlet temperature of the dryer. This results from the evaporative cooling effect of liquid leaving the agglomerate surface during drying. If water were being evaporated, the agglomerate would probably experience a temperature of about 110 °C. With organic solvents, temperatures are generally lower.

Another major advantage of spray drying is the homogeneity of the product. Elements of widely varying sizes and densities can be uniformly dispersed in a liquid slurry. The atomization and rapid drying ensures that this homogeneity is retained in the agglomerate. For example, extremely fine graphite (<0.5 μm) and cobalt (<1 μm) can be intimately mixed with tungsten carbide (1 μm to 20 μm). Carbon adjustments as small as 0.02% are possible at an average carbon level of 6.10%.

Particle statistics also play a role in insuring homogeneity. For example, a 30 μm spherical agglomerate can be shown to contain more than 10,000 particles of 1 μm diameter. However, the same agglomerate would contain only about 62 particles of 6 μm diameter. Our experience indicates the agglomerate/particulate diameter ratio should always be 5/1 or larger. Figure 3 illustrates this ratio very well. The Tungsten carbide/tungsten/cobalt agglomerates shown were made from different tungsten carbide particle sizes.

Slurry formulation plays an important role in achieving particle homogeneity. In addition to the carrier liquid and binder, specific additives may be useful. Surfactants, wetting, or suspending agents can enhance particle suspension and help prevent solids from settling in the slurry. Deflocculents can prevent the clustering of fine particles, and antifoaming agents can minimize foaming during mixing [22].

The final step in agglomerate formation is to remove the binder (dewax) and strengthen the dewaxed agglomerates (sinter). Most binders are removed by heating at relatively low temperatures, e.g. below 550 °C in furnace atmospheres appropriate to the compositions being processed, e.g. hydrogen, inert gas, or vacuum. After removing the binder, agglomerates are further strengthened by heating to a temperature high enough to allow particle-to-particle bonding by surface diffusion.

For example, the molybdenum and the tungsten carbide/cobalt agglomerates were heated at 1050 °C in hydrogen. After heating, the agglomerates can be blended, vibrated, and otherwise handled without breakup.

PLASMA MELTING

This step is performed by feeding the agglomerates through a plasma flame. The agglomerates are entrained in a carrier gas and fed into the plasma such that melting and solidification occur without interparticle contact. The plasma-melted particles are spherical and dense with excellent flow and packing properties. An example is the molybdenum powder shown in Figure 4.

If the agglomerates consist of more than one type of particle, of more than one metal or ceramic component, or any combination of these, these materials will react or alloy during melting to produce prealloyed or composite powder. An example is the tungsten carbide-cobalt system shown in Figure 5. The agglomerate consisted of tungsten carbide, tungsten and cobalt as shown in Figure 3b. After melting, the microstructure consists of tungsten carbide particles (dark gray), a matrix of Co-W-C (light gray) and a few patches of residual cobalt (white).

The plasma conditions can be varied to control the powder product. Plasma temperature is determined by the type of gas and by the power input. The plasma velocity is controlled by the gas flow rate, composition, and nozzle diameter. Speeds from subsonic (800 ft/sec) to Mach 2 can be obtained [23-25]. The point at which the powder is injected is also important. The temperature attained by the powder is affected by the agglomerate diameter, the residence time in the plasma and the flame temperatures which the particles experience.

In the process of melting, the particles are accelerated to nearly the velocity of the plasma gas. The solidified particles are collected in a cooled chamber under inert atmosphere or partial vacuum. The atmosphere is generated by the plasma itself although supplemental gas may also be used.

Cooling rates are very high after plasma melting because of the high droplet velocity and small droplet size. Cooling can be further enhanced by gas jets designed to impinge upon the molten drops as they exit the plasma flame. Data are unavailable yet for the cooling rates attained in our equipment, however the particle diameters are smaller and their velocities are higher than for gas-atomization processes which claim cooling rates of 10^4-10^5 °C/sec. [26].

There are important advantages to the PMRS process in addition to the ability to create unique compositions and microstructures. The melting process does not require a container, thereby eliminating a potential source of refractory contamination. Energy savings result from heating only the product in the absence of the usual, large, furnace structures. Added energy savings, economics and higher purity result from the fact that only the desired size fractions need to be melted. Classification, if needed, is done after spray drying.

The Plasma-Melting and Rapid-Solidification (PMRS) technique has the capability for processing a wide variety of materials. The feed for PMRS need not be agglomerates but can be any desired particulate, e.g. prealloyed gas-atomized powders or naturally occurring minerals.

APPLICATIONS

Most of the past production applications have been powders for hardsurfacing, including tungsten, tungsten carbide-cobalt, and molybdenum, as discussed earlier. However, many custom powders have been made. For example, solid-solution alloys have been produced of tungsten-molybdenum and of tungsten-rhenium. Based on our experience, only the PMRS process has the capability for producing powders of such high-melting alloys.

Very recently, pilot quantities of molybdenum-5% cobalt were produced by the PMRS process with the objective of creating a supersaturated solid solution. Cobalt has almost no solubility in molybdenum at room temperature. Figure 6 shows the Mo-Co PMRS powder and microstructure. It also shows the microstructure of pure PMRS molybdenum for comparison. X-ray diffraction studies showed no trace of either cobalt or cobalt-molybdenum compounds in the PMRS powder and the hardness, 330 DPH, was much higher than that of the pure PMRS molybdenum, 175 DPH. The mean powder diameter was 23 μm.

Another recent PMRS powder was Mo-Ti-Zr-C. This is a precipitation-strengthened alloy which was PMRS processed to determine the effect on the powder properties and articles made from the powder, which has a mean size of 23μm and a hardness of 320 DPH. Evaluations of the Mo-Co and Mo-Ti-Zi-C alloys are presently underway at GTE and at AMMRC, the U.S. Army Materials and Mechanics Research Center.

SUMMARY

A process for the plasma synthesis of powders has been described in which rapid solidification is an intrinsic feature. The PMRS process is broadly applicable and for many materials it is economically attractive. The powders produced can be finer in size (100% <44μm) than those made by other methods, e.g., atomization or rotating electrode processing. Since a crucible is not necessary, there is no contamination from refractory compounds.

LIST OF REFERENCES

1. Houck, D. L., "Powders for Plasma Spray Coatings", Progress in Powder Metallurgy, Vols. 34 and 35. 1978-1979. p 207-216.
2. Houck, D. L., R. F. Cheney, (GTE, Chemical and Metallurgical Division) and R. F. Cribbs, (Sealed Power Corporation), "Properties of Oxygen-Bearing Molybdenum Coatings", ITSC 1980 Proceedings of the 9th International Thermal Spraying Conference, p 365-371.
3. Houck, D. L., "Techniques for the Production of Flame and Plasma Spray Powders", Modern Developments in Powder Metallurgy, Vols. 12 to 14.
4. Houck, D. L., "New Powder Developments for Non-Transferred Plasma Arc Spraying", The International Journal of Powder Metallurgy and Powder Technology, Vol. 18, No. 1, p 69-79.
5. Houck, D. L. and R. F. Cheney, "Method for Producing Prealloyed Chromium Carbide/Nickel/Chromium Powders and Properties of Coatings Created Therefrom", Presented at 10th International Thermal Spray Conference Poster Session, Essen, West Germany, May 2-6, 1983.
6. Masters, K. Spray Drying Handbook, 3rd Edition, George Godwin Limited, London, John Wiley and Sons, New York, 1979.
7. Sherrington, P. J. and R. Oliver, Spray Drying, from Granulation, a Monograph in Powder Science and Technology, Heyden, New York, 1981, p 122-139.

8. Perry, R. H., C. H. Chilton, Spray Dryers, Chemical Engineers Handbook, 5th Edition, McGraw Hill, 1973, Section 20, p 58-63.
9. Katta, S. and W. H. Gauvin, Some Fundamental Aspects of Spray Drying, American Institute of Chemical Engineers Journal, Vol. 21, No. 1, 1975, p 143-152.
10. Marshall, Jr., W. R. and E. Seltzer, Principals of Spray Drying, Chemical Engineering Progress, Vol. 46, No. 10, 1950, p 501-508; Vol. 46, No. 11, 1950, p 575-584.
11. Duffie, J. A. and W. R. Marshall, Jr., "Factors Influencing the Properties of Spray Dried Materials", Chemical Engineering Progress, Vol. 49, No. 8, 1953, p 417-413; 480-486.
12. Houck, D., Spray Drying, Chapter to be published in American Society for Metals Handbook, 9th Edition, Vol. 7, Powder Metallurgy.
13. Houck, D., "Spray Drying", Chapter to be published in American Society for Metalsd Handbook, 9th Edition, Vol. 7, Powder Metallurgy.
14. Masters, K., Spray Drying Handbook, 3rd Edition, George Godwin Limited, London, John Wiley and Sons, New York, 1979.
15. Dittrich, F. J., "Flame Spray Powders and Process", U. S. Patent 3,617,358.
16. Powers, J. A., M. B. MacInnis, and D. J. Port, "Flame-Spray Powder of Cobalt-Molybdenum Mixed Metal Agglomerates Using Molybdenum Salt Binder and Process for Producing Same", U. S. Patent 4,011,073, Assignee - GTE Sylvania Incorporated.
17. Masters, K., Spray Drying Handbook, 3rd Edition, George Godwin Limited, London, John Wiley and Sons, New York, 1979.
18. Jensen, A. O. and K. Masters, "Spray Dryer for Producing Tungsten Carbide Products", Bulletin F-125, Niro Atomizer.
19. "Closed Circuit Spray Drying Systems", Anhydro Bulletin 1342.
20. "Closed Cycle Systems", Bowen Engineering Bulletin.
21. "Spray Drying Processing Plants", Anhydro Bulletin A-2-ENG and personal communication Jullian Tunno (Anhydro), March 28, 1983.
22. F. J. Dettrich, U. S. Patent 3,617,358, "Flame Spray Powder and Process".
23. Ingham, H. S. and A. J. Fabel, "Comparison of Plasma Flame Spray Gases", Welding Journal, February 1975.
24. Gal-Or, B., "Plasma Spray Coating Processes", Physico-Mathematical Characterization, Journal of Engineering for Power, Vol. 102, July 1980, p 589-593.
25. Chang, C. W. and J. Szekely, "Plasma Applications in Metal Processing", Journal of Metals, February 1982, p 57-64.
26. "DARPA Expands Research into Rapid Solidification Technology", Metal Powder Report, September 1982.

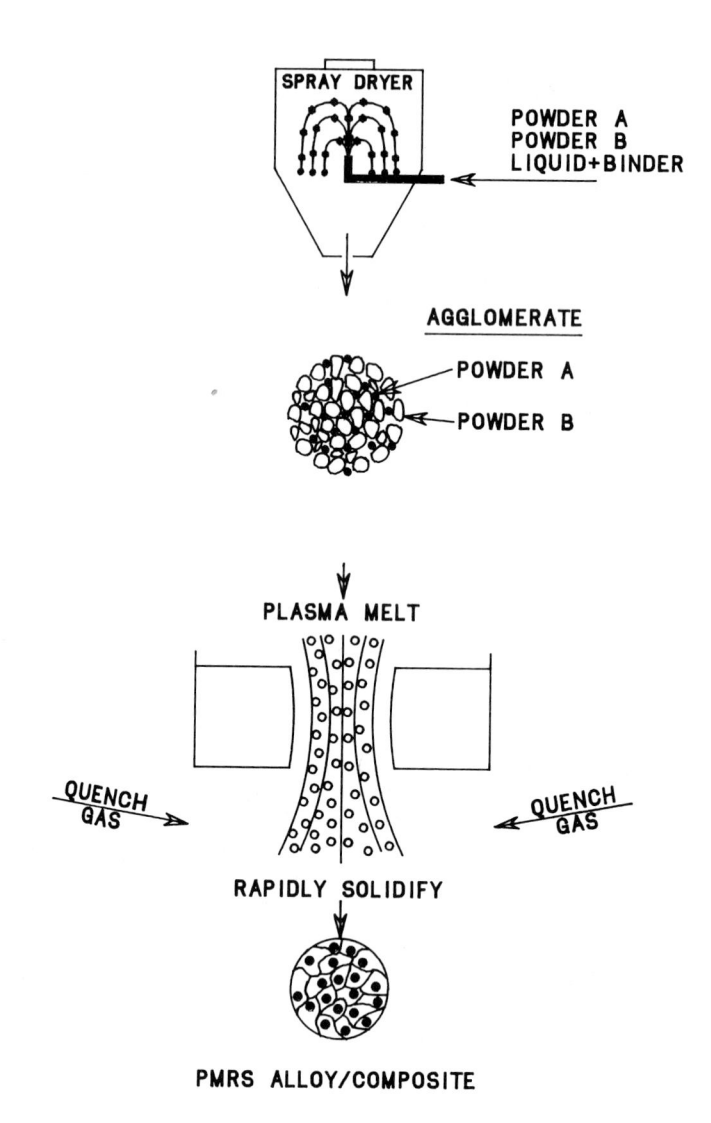

Figure 1 - Plasma Melted and Rapidly Solidified (PMRS) Powder Process

Figure 2 - Spray-Dried And Sintered Agglomerates
(15-60μm) made from 1-5 μm Molybdenum
Powder

Using 1 μm WC
(a)

Using 6 μm WC
(b)

Figure 3 - Spray-Dried and Sintered Tungsten Carbide/
Cobalt/Tungsten Agglomerates

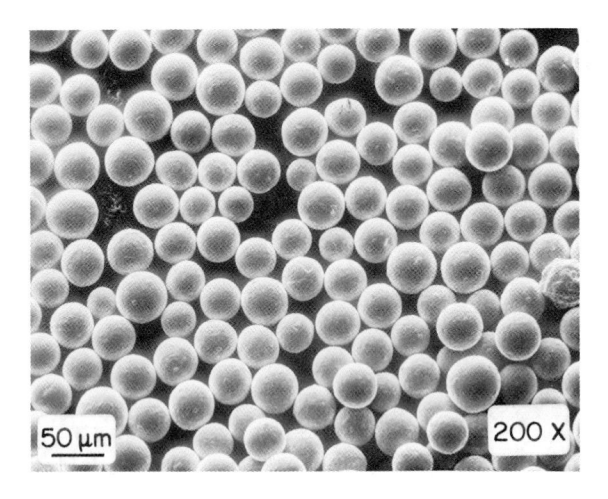

Figure 4 - Plasma-Melted and Rapidly-Solidified Molybdenum Powder

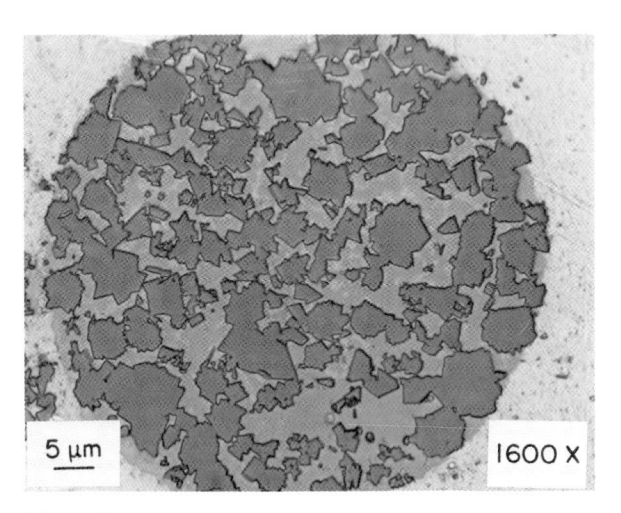

Figure 5 - Light Micrograph of Plasma-Melted and Rapidly-Solidified WC-12Co. Sample was unetched showing Cobalt-rich matrix surrounding Tungsten Carbide particles

Molybdenum-5 Cobalt

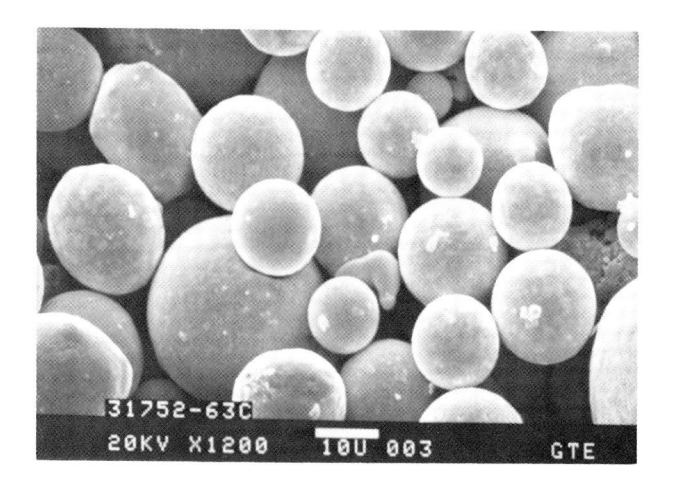

Molybdenum

Molybdenum-5 Cobalt

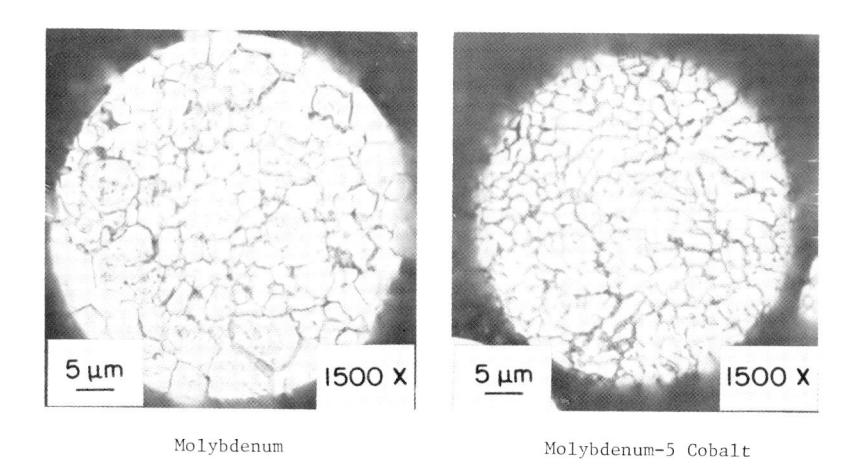

Figure 6 - Plasma melted and rapidly solidified (PMRS) Mo-5Co powder and light microstructures of PMRS Mo and PMRS Mo-5Co

THE EFFECT OF STRUCTURE ON THE THERMAL CONDUCTIVITY OF PLASMA SPRAYED ALUMINA

H.C. Fiedler
Metallurgy Laboratory, General Electric Company, Corporate Research and Development, Schenectady, New York 12301

ABSTRACT

Plasma sprayed deposits of alumina are normally of the metastable gamma phase, which transforms to alpha upon heating to an elevated temperature. The poor thermal conductivity of gamma phase is improved by transforming to alpha, but the improvement is restricted by the fragmented nature of the structure, a consequence of alpha being a denser phase than gamma.

Alpha phase has been reported to form when deposits are made on preheated substrates. It was found that reducing the distance between gun and deposition surface, thereby raising the temperature of the latter, resulted in transforming gamma to alpha, and as the temperature continued to rise, alpha formed directly from the liquid. The thermal conductivity of alpha formed directly from the liquid is 0.27 Watt/cm K at 323 K, which approaches the thermal conductivity given for 99.5% pure, 98% dense, polycrystalline alpha alumina.

INTRODUCTION

It has been known for some time that flame sprayed deposits of alumina are of a metastable phase [1], and that the metastable phase, gamma, transforms to alpha upon subsequent heating to a high temperature [1,2]. Ault [1] determined the density, the coefficient of thermal expansion and the thermal conductivity of the gamma phase. The latter ranged from 0.050 Watt/cm K at 473 K to 0.027 at 1373 K, which is substantially less than for alpha alumina. He observed that the gamma phase converted to alpha within minutes at 1733 K, but was unchanged after three days at 1273 K.

Hurley and Gac [3] determined the crystal structure of plasma sprayed alumina deposits, their thermal diffusivity and the heat treatments required to transform the structure to alpha phase. From X-ray diffraction patterns they concluded that the as-deposited structure was primarily eta phase, which is metastable, and unmelted alpha powder particles. The differences in the diffraction patterns of gamma and eta are subtle, and most researchers identify this metastable phase as gamma.

The thermal diffusivity of their deposit was 0.007 cm^2/sec. at 310 K, which is a factor of ten smaller than the value for alpha alumina [4]. Heating for one hour at temperatures of 1473 K and above resulted in complete transformation to alpha. Since alpha is denser than the metastable phase [5], the transformation results in a volume shrinkage which fragments the structure. Because of the fragmentation, the thermal diffusivity increased by only a factor of four after annealing at 1200°C but continued to increase after annealing at higher temperatures, which they attributed to the closing of cracks.

Hurley and Gac [3] associated the low thermal diffusivity of the eta (or gamma) phase to the relative disorder of the metastable phase. The reduction they calculated in the mean free path for the scattering of phonons was consistent with the observed large decrease in diffusivity. Cracks and pores in the plasma sprayed deposit would also adversely affect diffusivity, but the effect is minor as compared to crystal structure.

McPherson [6] considered the question of why the metastable rather than the equilibrium phase is formed in plasma sprayed deposits of alumina. He concluded that homogenous nucleation at considerable undercooling results in the formation of gamma rather than alpha because of its lower liquid-solid interfacial energy. The interfacial energy reflects the difference in structure between the crystalline phase and the liquid, and the nucleus which forms would be expected to be the phase with structure most nearly approaching that of the liquid.

The work to be reported demonstrates that dense deposits of alpha alumina with high thermal conductivity can be made by sufficiently reducing the distance between gun and deposition surface, thereby increasing the temperature of the latter. The microstructures suggest that the alpha phase is then formed directly from the liquid rather than by the transformation of gamma.

PROCEDURE

The powder sprayed was METCO 105SFP aluminum oxide. This powder typically is -20+5 microns in size and 99.5% pure.

Two plasma spray processes were used to make deposits. One was in air and the other was a low pressure process done in a reduced pressure (60 torr) inert gas atmosphere. The operating conditions for the processes are given in Table I.

The air plasma gun was manually held and the distance between gun and deposition surface varied between 5 and 15 cm. Since it was recognized that spray distance might have an effect on properties, the samples were divided before thermal conductivity measurements into three groups: those sprayed at more than 8 cm, those sprayed at about 8 cm and those sprayed at less than 8 cm. Although the division was not rigorous, it did assist in understanding the results.

TABLE I. Plasma Spray Conditions

	Air	Low Pressure
Plasma Gun	Metco 3MB-GH	EPI 03CA-80
Power Settings	52V, 550A	53V, 1300A
Primary Gas	Argon, 460 slm	Argon, 179 slm
Secondary Gas	Hydrogen, 26 slm	Helium, 48 slm
Powder Feed Gas	Argon, 12 slm	Argon, 15 slm
Powder Feed Rate	840 g/h	---
Spray Distance	5-15 cm	30 cm

The deposits were made by spraying the end of a 1.27 cm diameter copper bar. The disks as-deposited were 1.0 to 1.3 mm thick and were subsequently ground to a thickness of 0.63 mm, with material being removed from both sides.

The thermal diffusivity measurements were made by the Thermal Physics and Chemistry Laboratory, Re-Entry Systems Operation, General Electric Co. The procedure involved supplying a thermal burst to one side of the disk with the Nd/glass laser and detecting the transient thermal response on the back side with an InSb infrared detector.

RESULTS

Table II gives the density, thermal diffusivity and thermal conductivity of 14 different disks. The density values of the disks were determined from the dimensions and weight. The thermal conductivity is the product of density, diffusivity and specific heat, with the value for the latter taken as 0.197 Cal/gm K [7]. Also listed is the phase identification as determined by the X-ray diffraction spectra from both sides of six disks.

The sintered 96% alumina disks have the highest thermal conductivity of the samples measured, about 0.16 Watt/cm K. These disks consist of alpha grains about 20 microns in diameter with particles of a glass phase dispersed throughout the structure.

Disks 4 through 8 were air plasma sprayed. The disk made with a distance of 8 cm between gun and deposition surface and the two made with a distance greater than 8 cm all have low thermal conductivities, and are primarily gamma phase. Disk 7 (Fig. 1) has a microstructure that is typical of air sprayed deposits with low thermal conductivity: lamellar grains of gamma alumina with unmelted alpha alumina particles. The X-ray diffraction spectrum from a face of this disk is shown in Fig. 2.

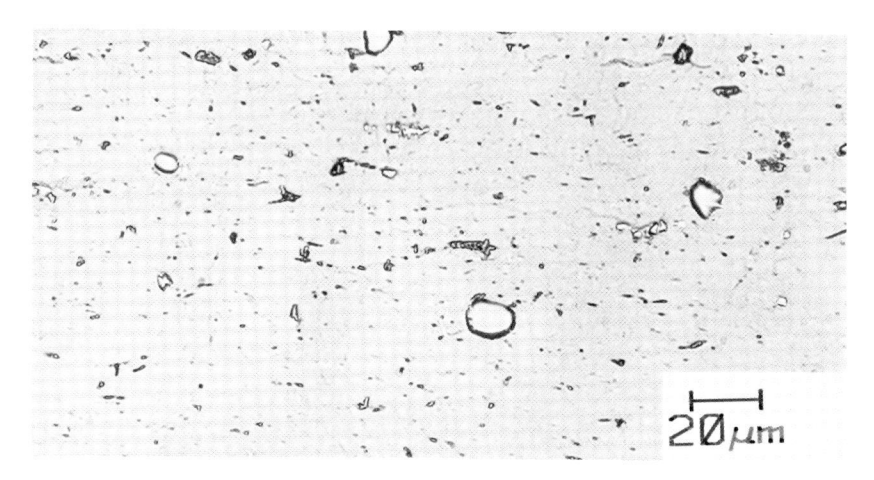

FIG. 1. Cross-section of Disk 7, made by air plasma spraying. Metastable gamma and theta are the major phases. The white particles are unmelted alpha powder particles.

TABLE II. Thermal Conductivity of Disk Samples at 323 K

Disk	Material	Spray System	Spray Distance, cm	Density	Thermal Diffusivity cm^2/sec	Thermal Conductivity Watt/cm K	Phases
1.	Sintered 96% Al_2O_3	---	---	3.88	0.0486	0.155	α
2.	Sintered 96% Al_2O_3	---	---	3.85	0.0510	0.161	
3.	Sintered 96% Al_2O_3	---	---	3.84	0.0528	0.167	
4.	105SFP Al_2O_3	Air	>8	3.23	0.0109	0.029	
5.	105SFP Al_2O_3	Air	>8	3.23	0.0092	0.0245	
6.	105SFP Al_2O_3	Air	<8	3.44	0.0348	0.099	major α, traces δ, Θ
7.	105SFP Al_2O_3	Air	~8	3.36	0.0120	0.033	major γ and Θ, trace α
8.	105SFP Al_2O_3	Air	~8	3.54	0.0218	0.0635	major α, trace γ
9.	105SFP Al_2O_3	LP	30	3.29	0.0120	0.032	
10.	105SFP Al_2O_3	LP	30	3.32	0.0113	0.031	
11.	105SFP Al_2O_3	LP	30	3.29	0.0119	0.032	
12.	105SFP Al_2O_3	LP	30	3.33	0.0131	0.036	major σ and γ
13.	105SFP Al_2O_3	LP	30	3.35	0.0134	0.037	
14.	105SFP Al_2O_3	LP	30	3.39	0.0143	0.040	

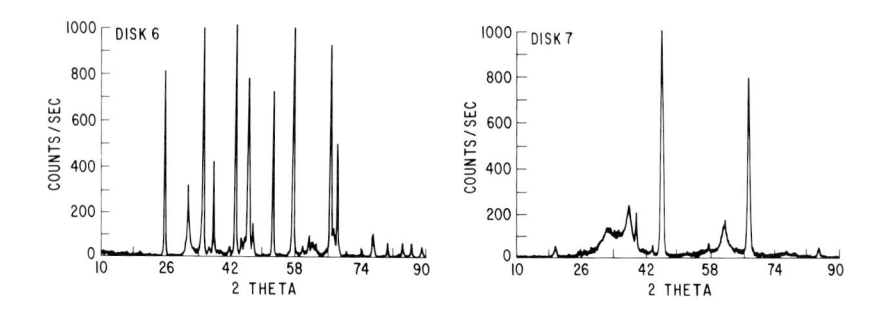

FIG. 2. X-ray diffraction spectra of two of the disks. Alpha alumina is the predominant phase in the spectrum at the left; gamma alumina is the predominant phase in the spectrum at the right.

Disk 8, sprayed at a distance of about 8 cm, has double the thermal conductivity of Disks 4, 5 and 7, and the major phase is alpha rather than gamma. The photomicrograph in Fig. 3 shows a highly fragmented structure as compared to Fig. 1.

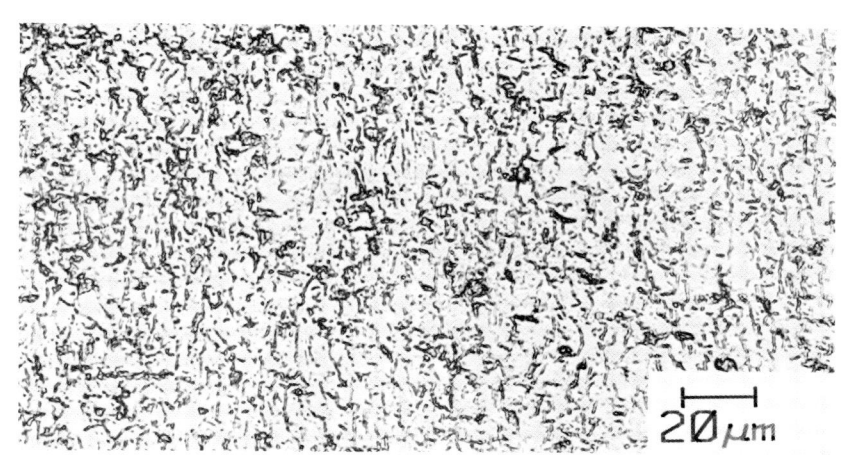

FIG. 3. Cross-section of Disk 8, made by air plasma spraying. The as-deposited structure was gamma, which from being heated in the plasma transformed to the fragmented alpha phase seen in the photomicrograph.

Disk 6, sprayed at a distance of less than 8 cm, has the highest thermal conductivity of the plasma sprayed samples, and alpha alumina is the only major phase. The X-ray diffraction spectra from the two sides of this disk are essentially identical; the spectrum from one side is shown in Fig. 2. The microstructure (Fig. 4) shows an abrupt change in morphology about midway through the thickness of the sample. Half of the disk consists of a fragmented alpha phase, as comprised the entire cross-section of Disk 8 (Fig. 3), but the other half consists of a columnar alpha structure.

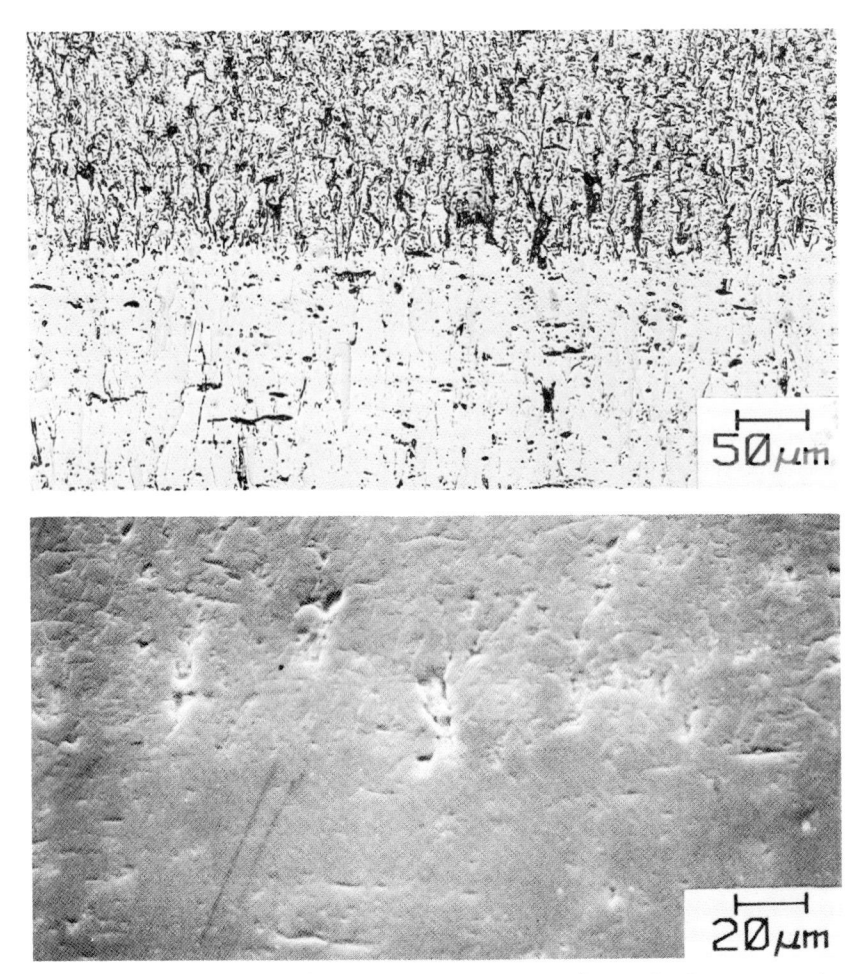

FIG. 4. Photomicrograph (top) and SEM micrograph (bottom) of cross-section of Disk 6, made by air plasma spraying. The fragmented structure in the top half of each micrograph is alpha resulting from the transformation of gamma. The columnar structure in the bottom half is also alpha, but it formed directly from the melt.

Two sets of disks were low pressure plasma sprayed and examples are given in Disks 9 to 11 and 12 to 14. The thermal conductivities are low, consistent with the structure which has major amounts of both alpha and gamma. The relatively large fraction of alpha is due to the many unmelted alpha powder particles.

DISCUSSION

At least two investigations have reported forming alpha alumina by spraying on to a preheated substrate. Eichhorn, Metzler and Eysel [8] sprayed on to a substrate preheated to 1373 K and found only alpha lines in the X-ray diffraction spectrum of the deposit. However, it is not possible to tell from their paper whether alpha formed directly or by the transformation of gamma.

Huffadine and Thomas [9] flame sprayed alumina onto mandrels heated to a series of temperatures and found that heating to 1673-172 K was required for making dense alpha alumina deposits. The grain structure of their deposit viewed perpendicular to the deposition surface consisted of columnar shaped grains similar to those seen in the bottom half of the photomicrograph in Fig. 4.

In these experiments, the temperature of the deposition surface was affected by the distance from the air plasma spray gun. Disks formed by spraying at more than 8 cm were primarily gamma, whereas Disk 8, sprayed at 8 cm was primarily alpha. However, it is apparent from the fragmented microstructure in Fig. 3 that the alpha resulted from the transformation of gamma.

Disk 6 was deposited with a shorter distance from gun-to-deposition surface than was Disk 8, and this surface can be assumed to have reached a higher temperature. A reasonable interpretation of the structure of Disk 6 is that the deposit not only became hot enough to transform the gamma to alpha, but eventually became hot enough to form alpha directly. Based on the Huffadine and Thomas [9] experience, that temperature was probably in excess of 1673 K. The particles of molten alumina form lamellar shaped deposits upon impinging on the surface, and each deposit is comprised of columnar-shaped grains with their long axis perpendicular to the deposit surface [10,11]. It is likely that the large columnar grains seen in Fig. 4 resulted from the growth of these grains in the temperature gradient established by the plasma.

Knowing the thermal conductivity of Disk 6, the thermal conductivity of the fragmented alpha alumina and the volume fraction of the two forms of alpha alumina, the thermal conductivity of the columnar alpha alumina can be calculated. The thermal conductivity of a material with two components in series is related to the thermal conductivities of the two components by the relation [12]:

$$k_m = \frac{k_1 k_2}{k_1 V_2 + k_2 V_1}$$

where k is thermal conductivity and V is volume fraction. In the layered structure of Disk 6 (Fig. 4), the volume fraction of fragmented alpha, V_1, is 0.53 and the volume fraction of columnar alpha, V_2, is 0.47. The thermal conductivity of the fragmented alpha, k, is 0.063 as measured on Disk 8. Since k_m of Disk 6 is 0.099, the equation gives k_2 = 0.27 Watt/cm K. This conductivity is 80% greater than that of the sintered 96% alumina, and approaches the literature value of 0.33 Watt/cm K given for 99.5% pure, 98% dense, polycrystalline alumina at 50°C [13].

Mackay and Muller [14] have reported the thermal conductivity of plasma sprayed alumina to range from 0.27 Watt/cm K at 323 K to 0.18 Watt/cm K at 473 K. They gave no details of their spray procedure other than that they used METCO 105 alumina powder, which is 96% pure. No X-ray diffraction identification was made of the deposit, which was 0.86 mm thick. However, their reported pycnometer density of 3.61 g/cm^3 is more compatible with that of gamma alumina (3.65 g/cm^3) than with that of alpha alumina (3.98 g/cm^3). (Densities for these phases from Ref. 5.) Ault [1] reported a true density for his gamma phase deposit of 3.60 g/cm^3.

CONCLUSIONS

1. Plasma sprayed deposits of alumina normally consist primarily of a metastable phase or phases.

2. Gamma alumina, the most prevalent of these metastable phases, has a thermal conductivity lower than that of sintered 96% alpha alumina by about a factor of five.

3. Alpha alumina was formed directly from the liquid by raising the temperature of the surface, a consequence of decreasing the distance from gun to deposition surface.

4. The thermal conductivity of alpha formed directly from the liquid, 0.27 Watt/cm K, is:

    a. about a factor of ten greater than that of gamma phase,
    b. about a factor of four greater than that of alpha made from transforming gamma,
    c. about 80% greater than that of sintered 96% alpha alumina, and
    d. approaches that of 99.5% pure, 98% dense, polycrystalline alumina.

ACKNOWLEDGEMENTS

We are indebted to J.D. Hanson and J. Bueche of the Re-Entry Systems Operation for the thermal conductivity measurements.

C.R. Rodd made the photomicrographs; J.R. Ross made the SEM micrograph. The X-ray diffractograms were provided by D.W. Marsh. Valuable contributions to this study were made by P.A. Siemers, D.M. Gray, S.F. Rutkowski and J.R. Rairden.

REFERENCES

[1] N.N. Ault, J. Am. Ceram. Soc., 40 69-74 (1957).
[2] V.S. Thompson and O.J. Whittemore, Jr., Ceram. Bull., 47, 637-41, (1968).
[3] G.F. Hurley and F.D. Gac, Am. Ceram. Soc. Bull., 58, 509-11 (1979).
[4] Y.S. Touloukian, R.W. Powell, C.Y. Ho and M.C. Nicolaou, Thermophysical Properties of Matter, Thermal Diffusivity, 10, (IFI/Plenum, 1973) p. 378.
[5] E. Ryshkewitch, Oxide Ceramics: Physical Chemistry and Technology, (Academic Press, New York, 1960).
[6] R. McPherson, J. Mater. Sci., 1973, 8, 851-858 (1973).
[7] Y.S. Touloukian, Thermophysical Properties of Matter, Specific Heat of Nonmetallic Solids, 5, (IFI/Plenum, 1970) p. 25.
[8] F. Eichhorn, J. Metzler and W. Eysel, Metalloberflaeche, 26, 212-13, (1972).
[9] J.B. Huffadine and A.G. Thomas, Powder Met., 7, 290-299 (1964).
[10] Volker Wilms and Herbert Herman, Thin Solid Films, 36 251-62 (1976).
[11] S. Safai and H. Herman, Advances in Surface Coating Technology, London, 1978, Paper 5.
[12] W.D Kingery, Introduction to Ceramics, (John Wiley and Sons, New York, 1960) p. 501.
[13] Y.S. Touloukian, R.W. Powell, C.Y. Ho, P.G. Klemens, Thermophysical Properties of Matter, Thermal Conductivity, 2, (IFI/Plenum, 1970), p. 119.
[14] T.L. Mackay and M.L. Muller, Ceram. Bull., 46, 833-35 (1967).

PLASMA SPRAY PROCESSING OF CERAMIC OXIDES

N. RAVI SHANKAR, H. HERMAN,* AND R. K. MAC CRONE**
*Dept. of Materials Science and Engineering, State University of New York, Stony Brook, NY 11794; **Materials Engineering Department and Department of Physics, Rensselaer Polytechnic Institute, Troy, NY 12181

ABSTRACT

We report some measurements of unique physical properties that arise in plasma-sprayed alumina. It is suggested their origin is the O_- species formed on the surface of the particles during the fabrication process.

INTRODUCTION

Plasma spraying involves the melting and rapid deposition and solidification of a wide range of materials.[1] The technique, in its various forms, is a well-established technological process used to form protective coatings of ceramics and metallic alloys. Research has been directed at the highly metastable states and the imperfections resulting from the unique mode of material build-up.[1] In particular, the microstructures and morphologies of the deposit constituents has been related to plasma spray process parameters.

On an atomic level there exists a host of questions and effects which are of importance, and which have largely not been addressed; such as the quenching-in of excess point defects and the formation of other atomic scale configurations of interest. These entities are expected, in the case of ceramics, to give rise to interesting electrical, magnetic, and optical properties. For some time we have been addressing these effects on an ad hoc basis.

In this paper we describe some results of dielectric measurements on plasma-sprayed alumina; that is, the dielectric behavior of alumina coatings sprayed onto steel. The results show dielectric behavior which is not typical of this normal bulk alumina.

Also measured are the EPR spectra of the same batch of alumina powder, but plasma-sprayed into water rather than onto a metal substrate. These measurements reveal the presence of the oxygen species O^-, presumably absorbed (adsorbed?) on the surface of the spray-processed powder particles.

EXPERIMENTAL PROCEDURE

Plasma Spraying

Plasma-sprayed coatings were formed from pre-fused commercial alumina of a particle size -325 mesh, obtained from Muscle Shoal Mineral. The spray process was carried out in accord with accepted industrial practice. For the coated specimen, the steel substrate was grit blasted with alumina grit prior to coating. The substrate was pre-heated to approximately 150°C prior to application of the coating. For plasma spraying directly into water, the torch was maintained at a distance of 25 cm over a container of distilled water. The sprayed particles were collected and oven dried.

Dielectric Measurements

The dielectric measurements were made using a G.R. 1615 Capacitance Bridge operating at 1 kHz. The electrodes were asymetrical, one consisting of the steel substrate, the other consisting of silver paste. A guard ring was used. The values of the capacitance were found to be independent of the ac applied voltage, as well as the dc bias up to 1000 v/cms: this is taken to indicate that no surface barrier effects are present.

For dielectric measurements, the specimens were contained in an airtight shielded metallic container with a dynamic over-pressure of dry nitrogen.

EPR Measurements

The EPR measurements were made in a home-made honodyne spectrometer operating at 9.27 gHz at a power level of -20 dbm. The modulation was at 4.0 kHz with a peak-to-peak amplitude ∿1.0 Oe.

RESULTS

Microstructure

Figure 1 shows a typical transmission micrograph of the plasma-sprayed coating. Of interest are the presence of a two-phase structure. X-ray diffraction patterns of this material indicate that the major phase present is β-alumina. The two phases are metastable to temperatures up to 1000 K., a temperature in excess of any used in this work.

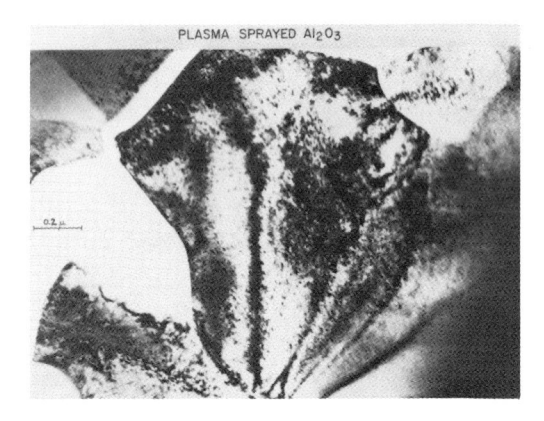

FIG. 1: Micrograph of the plasma-sprayed alumina.

Dielectric Measurements

The results of the dielectric measurements are shown in Figures 2a and 2b. The imaginary part of the dielectric constant was converted into conductivity in this case here, where there is no evidence for dispersion and is thus a physically appropriate reduction of the data.

The value of the dielectric constant measured here is between 7.0 and 12.0, which compares very closely with that of a-alumina, with literature values between 6.0 and 12.0[2] and 8.0 for sapphire. Thus we are measuring a very reasonable value, the dielectric constant being related in the first instance to the type of atoms present, which in this case must correspond closely to a-alumina.

However, the dielectric constant is seen to include a smaller component which increases with temperature. This cannot be due to dipolar reorientation, which has a 1/T temperature dependence. Likewise, thermal expansion will result in an apparent decrease in dielectric constant. This

FIG. 2a

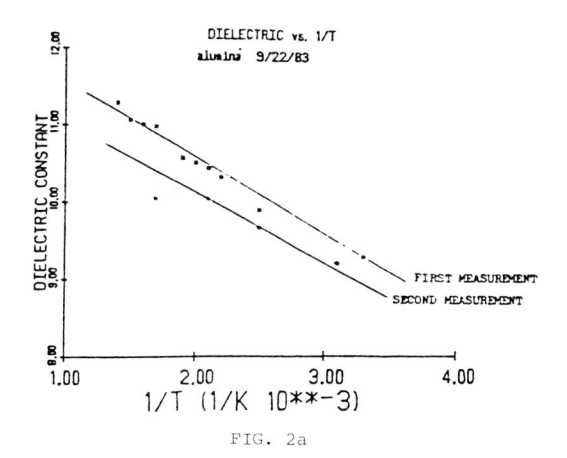

FIG. 2b

FIG. 2: Dielectric constant and resistivity of plasma-sprayed alumina as a function of temperature.

rather unexpected behaviour must result from a state induced by the plasma spraying itself. We will speculate on a possible physical origin subsequently in the discussion.

The conductivity of the plasma-sprayed coating shows a thermally activated process with an activation energy of 0.49 eV up to 360°C, and another with an activation energy of 1.2 eV at higher temperatures. We interpret these as conduction from ionic motion. The fact that there is no detectable polarisation at the temperature of the knee, that is, no corresponding dielectric knee is seen in Figure 2a, is consistent with the assertion.

EPR Measurements

The EPR measurements of the plasma-sprayed powder collected into water are shown in Figure 3. This spectrum is very intense, and very stable, hardly showing any variation over many months. The spectrum at 20 K is characterized by three g-values: $g_1 = 2.00$, $g_2 = 2.11$, and $g^3 = 2.17$.

FIG. 3a

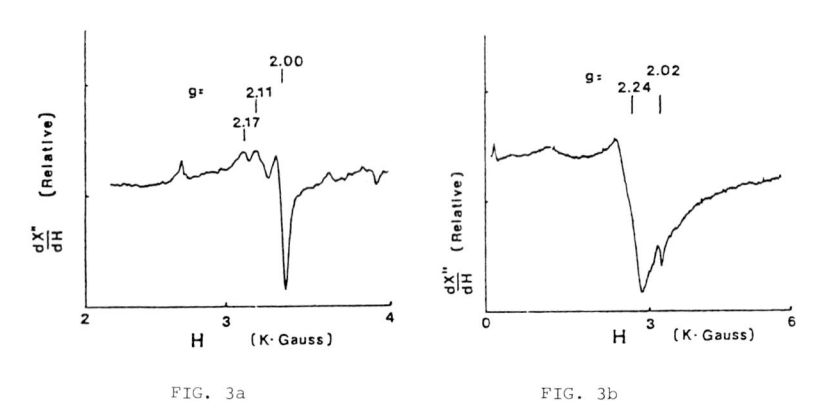

FIG. 3b

FIG. 3: EPR spectrum of plasma-sprayed alumina at 20 K and at 300 K.

Since the Al^{+3} ion is non-magnetic, the spectrum must arise from some other species. The most likely candidate is an oxygen-charged species incorporated onto the surface during the flight of the powder particle to the substrate, and when quenched into water.

DISCUSSION

The EPR spectra of the $\overline{O_2}$ and O^- surface species are well known, and are reviewed by Lunsford.[3]

Essentially, the $\overline{O_2}$ is characterized by three g values, one close to 2.00, one with a g value less than 2.0 and one with a g value greater than 2.0. On the other hand, the O^- is characterized also by three g values, one close to 2.0, but the other two g values are greater than two. The spectrum in Figure 3 are consistent with the latter specie in our powder, namely O^-.

Explicitly, the expressions for the g values are[3]

$g_{zz} \approx g_e$

$g_{xx} = g_e + 2\lambda/\Delta E_1$

$g_{yy} = g_e + 2\lambda/\Delta E_2$

with the quantities E_1 and E_2 as defined above. From these expressions and the observed g values we find $E_1 = .165$ eV and $E_2 = .255$ eV, assuming a spin orbit coupling coefficient of $L = .014$ eV.[4] With FTIR instrumentation, these energy transitions should be easily detectable, an experiment we plan in the future work.

At very low temperatures, we observe an additional weaker line with a g value of 1.95. We interpret this as the third g value of the O_2^- species, present in our powder in much lower concentrations.

As mentioned above, a linear increase in the dielectric constant with temperature is not characteristic of a dipolar process in thermal equilibrium. Although other mechanisms with an increasing polarisability with temperature are known, it seems reasonable to associate our observation with the clearly one unique characteristic of these plasma-sprayed materials, namely the presence of large amounts of O^- incorporated onto (or possibly, into) the alumina particles.

It can be shown, using free energy considerations, that the polarisability of an electron in such a system is given by

$$\Delta X = Np^2/E(T)$$

where p is the first order dipole matrix element and $E(T) > kT$, (in the case, $p = \langle p_z/ex/p_x \rangle$, see insert). We explain the temperature dependence of ΔX through the observed decrease in the value of E with increasing temperature, deduced from the increase in the g values with increasing temperature, namely

$T = 10\,K:\quad g_1 = 2.00 \quad g_2 = 2.11 \quad g_3 = 2.17 \quad E_1 = .165\,eV \quad E_2 = .255\,eV$

$T = 300\,K:\quad g_1 = 2.02 \quad g_2 = 2.24 \quad \sim g_3 \quad E_1 = .117\,eV$

We would like to emphasize that the explanation here is tentative, and that further work need be done to establish a quantitative relation between the concentrations of O_- determined from dielectric and EPR data respectively, and correspondingly the inferred dependence of E(T) on temperature.

The thermally activated loss shown in Figure 2a appears not to be associated with the dipolar process just discussed. We interpret this as simply the thermally activated conductivity from an ionic impurity in the alumina. This conductivity is not due to the migration of intrinsic lattice ions since their activation energies reported between 4.3 and 6.0 eV.[5-8] We plan to perform x-ray fluorescence analysis to determine the species responsible.

CONCLUSION

It has been shown that the extreme temperatures encountered in the plasma-spraying process can produce metastable phases, in this case β-alumina, with reasonable dielectric properties, the loss being primarily due to the thermally activated motion of ionic impurities.

Apparently, the process also incorporates the O^- species, probably on the surface.

ACKNOWLEDGEMENTS

We would like to acknowledge helpful discussions with G. Korenowski, and B. Homan and K. Sheng for experimental assistance. The financial support of this project is received, with gratitude, from Teledyne-Stillman under their Trap. program.

REFERENCES

1. S. Safai and H. Herman, "Plasma Sprayed Material" in Vol. 20, Tr. on Mat. Sc. and Tech., Academic Press, NY, 1981, pp. 183-214.

2. W. H. Gitzen, "Alumina as a Ceramic Material", American Ceramic Society, Columbus, OH, 1970.

3. J. H. Lunsford, Catalysis Reviews, 8(1), pp. 135-157, 1973.

4. P. H. Kasai, J. Chem. Phys., 43, pp. 3322, 1965.

5. A. E. Paladino and W. D. Kingery, J. Chem. Phys., 37(5), pp. 957, 1962.

6. R. M. Cannon and R. L. Coble in "Deformation of Ceramic Materials", edited by R. C. Bradt and R. E. Tressler, Plenum Press, NY, p. 61, 1975.

7. P. A. Lessing and R. S. Gordon, J. Mater. Sci., 12, p. 2291, 1977.

8. R. C. Rossi, J. D. Buch and R. M. Fulrath, J. Amer. Ceram. Soc., 53, p. 629, 1970.

AN EXPERIMENTAL STUDY OF POWDER MELTING DURING LOW PRESSURE PLASMA DEPOSITION

M. PALIWAL AND D. APELIAN
Department of Materials Engineering, Drexel University,
Philadelphia, PA 19104

ABSTRACT

Low pressure plasma deposition (LPPD), a recent advancement in plasma spray metallizing, is currently being developed for high performance materials applications. An experimental study of particle melting within the plasma jet was pursued to identify the effect of the process variables and the material properties on the resultant deposit. In tandem the experimental results have been utilized in the development and verification of a mathematical model for the melting of powder particles during the process.

Two binary iron based model alloys - Fe-20 wt% Mn and Fe-20 wt% Cu - were plasma sprayed using Ma 2.4 and Ma 3 guns. Two different methods were used to evaluate the degree of particle melting within the plasma jet. The first method intercepts the particle path in the plasma jet with a glass slide, whereas in the second method the powder particles are collected in free flight using a powder collector (resolidification of powder particles occurs before they impact the collector walls). The droplets which impacted on glass slides and the collected (using the powder collector) powder particles were studied for mode and degree of powder particle melting using scanning electron microscopy. "Sweet spot" deposits (with no relative motion between the plasma gun and the substrate) were also made for the two model alloys in different size ranges using the Ma 2.4 and Ma 3 guns. The resulting deposits were metallographically evaluated. Mode and degree of particle melting injected under different process variables as well as the resultant deposit structures are presented.

INTRODUCTION

Ever since Duwez's [1] pioneering work in 1960 it has been known that unique structure/properties can be achieved by rapid solidification. Though cooling rates of 10^4-10^6 K/s can now typically be achieved in powder atomization and melt spinning processes (or variations thereof) [2-6], it has not been possible to impart these unique properties to near net shape structures. The traditional powder metallurgy route, i.e. consolidation and sintering alters the desirable as-quenched structure. In contrast, plasma arc metallizing, a process which has been used for the last 25 years has been known to be capable of producing near net shape structures [7]. The powder particles injected into the high temperature plasma gas are melted and subsequently upon impacting the substrate resolidify at high cooling rates [7-12]. The resulting structures may exhibit amorphous phases, metastable phases and extended solid solubility [10-14].

Plasma spraying has not been thought of as a viable technology for producing rapidly solidified near net shape structures since the process has traditionally been carried out under ambient conditions resulting in deposits containing inherent defects - oxide inclusions and low densities.

Recent developments in the field of plasma arc spraying, namely the introduction of the low pressure plasma deposition (LPPD) process [15] has resulted in a renewed interest in plasma processing. The LPPD process has been used to deposit superalloys, in general the properties are superior to those achieved via the P/M or casting routes [16,17]. The LPPD process is currently being investigated as a process for forming near net rapidly solidified metallurgical structures and has been named RSPD-Rapid Solidification by Plasma Deposition [18].

The addition of a low pressure environment to the plasma spray process has introduced several advantages, making possible deposits of highly alloyed materials which can be used for high performance applications. Some of these advantages are: i) inert environment; ii) pre-heat/cleaning ability of plasma transferred arc; iii) process flexibility for spraying alloy particles having different melting ranges, and iv) deposition size/uniformity. Work is now underway to evaluate and optimize the plasma deposition process for producing high quality superalloy deposits which can be used as preformed RSR materials or near net shape structures. This is particularly beneficial in alloy systems which are difficult to cast and forge.

In order to successfully and repetitively produce deposits of good quality, the basic controlling mechanisms of the RSPD process must be understood. Variables which affect the deposit structure include: i) plasma jet enthalpy and velocity (determined by gun design, power level, and arc gas); ii) powder characteristics; iii) powder injection and iv) substrate variables (distance, temperature, thermal properties and geometry) [19-22]. The objective of our study has been to isolate the process of melting and solidification through experimental techniques and to study each process separately. The experimental results are used to identify the mechanisms of melting and solidification during RSPD. These results have also been utilized for the purposes of development and verification of our activities in process modeling the melting phenomena [23].

Prior to impacting the substrate, a completely melted particle is termed a droplet. However, the degree of particle melting depends on the heat gain during the particle's journey through the plasma plume. The degree of particle melting can be broadly categorized and is shown in Fig. 1 as a) unmelted particles which may fracture on impact, b) partially melted particles which fracture on impact, c) partially solid particles (droplets with small amount of solid left) which splat at impact and d) fully molten particles (droplets) which splat on impact.

The predominant requirement for the production of dense high-strength deposits via RSPD is that a large fraction of the particles be in a molten state when they impinge on and impact the substrate (or the previously deposited particles...). Isolating the melting and solidification phenomena in the plasma spray process is a difficult task. The degree of particle melting, the mode of powder particle melting, the final splat thickness, the particle cooling rate, and the quenched microstructures can be masked by the recrystallization caused by the extensive heat flux from the plasma jet. Recrystallization of the deposits, especially in simple alloys, can be quite extensive due to the low contamination in the low pressure plasma deposits. In addition to deposit recrystallization, the other confounding variables in the process are the non-uniform powder sizes, non-uniform heat flux to the particles, and non-uniform powder trajectories.

EXPERIMENTAL PROCEDURE

Two model alloy systems were used in this study: Fe 20wt% Mn (melting range = 1703 to 1723 K) and Fe-20wt% Cu (melting range = 1369 to 1733 K). The argon atomized powders were sieved and the -400 mesh was air classified. Batches of different particle sizes were used for the experiments. These will subsequently be designated by their average particle size, i.e., FeMn 29 refers to Fe-20wt% Mn powder with an average particle size of 29 μm. Powders used in the study were FeMn 41, FeMn 29, FeMn 16, FeCu 31 and FeCu 14.

The powders were plasma sprayed using two commercially available plasma guns; both employing internal powder injection. Characteristic parameters for these two guns (referred to as Ma 3 and Ma 2.4 guns) are listed in Table I.

TABLE I: Characteristic Parameters of the Two Plasma Guns Used for the Experiments

Plasma Gun	Operating Conditions				Calculated Conditions at the Nozzle Exit		
	Gas	Gas Flow Rate, lpm	Chamber Pressure, Torr	Gun Power, kW	Gas Temperature, K	Gas Velocity, m/s	Stagnation Temperature, K
Ma 3	Ar-He	Ar:142 He: 38	40	80	4139	3437	10,000
Ma 2.4	Ar-He	Ar:123 He: 38	40	72	5718	2978	10,576

Three different experimental approaches were pursued simultaneously to yield and verify different aspects of the phenomenon that occur during the particles flight and impact on the substrate. These experiments which are referred to as i) glass slide experiments, ii) particle collector experiments, and iii) sweet deposit experiments, respectively, are described subsequently.

i) <u>Glass Slide Experiments</u>: Glass microscope slides were positioned at various distances from the plasma gun exit, low powder feed rates (<50 gram/min) were set, and the plasma gun was rapidly (>30 cm/sec) traversed over the slides. The gun to substrate distance was varied from 20 to 45 cm at 5 cm intervals. The slides intercepted varying numbers of powder particles, but all the deposits were less than a single particle thick without major overlaying of particles. The glass slides were then viewed via both optical and scanning electron microscopy (SEM).

ii) <u>Powder Collector Experiments</u>: Another technique to sample the powders from the plasma jets was a powder collector whose aperture could be positioned in the plasma jet. The collector was a 45 mm ID by 95 cm long stainless steel tube topped with a 90° stainless steel cone with a 3 mm aperture port drilled in the center, and a stainless steel collector cup at the bottom. This collector was inserted into the jet at 40 cm axial distance to collect powders over a 5 min deposition time. During this time powders deposited onto the cone surface and a small fraction entered the collector aperture and cooled during flight before impacting the collector bottom 95 cm downstream.

iii) <u>Sweet Deposit Experiments</u>: "Sweet-spot" (where the plasma gun is not traversed) deposits onto substrates were also made to study the microstructural features (unmelted particles, splat thickness, porosity, etc.) of actual built-up deposits. These deposits were made onto 15 cm x 15 cm plates positioned at an axial distance of 41 cm from the plasma gun. Using spraying times of 1-2 minutes, 2-4 mm thick deposits were made on preheated substrates (800-900°C).

RESULTS

a. <u>Glass Slide Experiments</u>

The results of the glass slide experiments are used to study the morphological changes in individual particles upon impacting a substrate. The individual particles shown in Fig. 1 indicating various levels of melting and splatting resulted from the glass slide experiments. On a more macroscopic level glass slides from plasma deposition run will exhibit these four kinds of particles/droplets in varying amounts, since particles from a large cross-section of the plasma jet are deposited onto the glass slide. These experiments thus give an indication of overall melting of particles across the plasma jet cross-section. The glass slide experiments were used to compare powder melting as a function of distance along the plasma jet axis for combinations of powder alloy, powder size and the plasma gun. In all these experiments, an improvement in the degree of particle melting was found as the gun to substrate distance was increased from 20 to 40 cm. The particles are heated up rapidly within a distance of 30 cm and less, from the nozzle exit. Increasing the distance from 30 cm to 40 cm causes an increase in the degree of particle melting; the effect however is only marginal.

The glass slide experiments can also be used to qualitatively compare the effect of different plasma guns (or operating conditions) for similar size powders or the effect of powder size for a specific gun. These experiments have shown that the lower power Ma 2.4 gun results in more complete melting of powder particles. This is better seen in the results of the sweet deposit experiments.

b. <u>Powder Collector Experiments</u>

SEM micrographs of plasma sprayed and collected, different size powders of Fe-20wt% Mn, and Fe-20wt% Cu alloys are shown in Fig. 3. The starting material, i.e., the argon atomized powders in all cases had a spherical morphology. Compared to the starting material (not shown), the plasma sprayed and collected FeMn 29 (Fig. 3a and 3b), FeCu 31 (Fig. 3c) show excessive amount of fragmented particles. The FeMn 29 powder plasma sprayed using the higher enthalpy Ma 2.4 gun shows a larger fraction of fragmented particles than the same powder sprayed using the Ma 3 gun. A comparison of the powders sprayed using the Ma 2.4 gun shows a higher amount of fragmentation and smaller spherical powders for the wider freezing range alloy Fe-20wt% Cu (Fig. 3c) as compared to the Fe-20wt% Mn powders (Fig. 3b).

The observed powder fragmentation can occur during the heating stage: when the particles are being heated by the plasma jet or during the resolidification stage which occurs during the powder collection. However, the smaller size, FeMn 16 (Fig. 3d) and FeCu 14 (Fig. 3e and 3f) plasma sprayed and collected powders have a spherical shape. Since the finer particles would be more completely melted in the plasma jet, the fragmentation cannot be due to resolidification. It can then be concluded that the observed fragmentation occurred during the heating stage of the powders. Fragments from these fractured particles can subsequently melt within the plasma jet. These fragmented and melted particles can be seen in the collected powders as spherical particles which are smaller in size than the original powders (Fig. 3c).

Some fluffy type very fine (2µm and less) particles can also be seen in the smaller powders (FeMn 16 and FeCu 14) plasma sprayed using the higher enthalpy Ma 2.4 gun. The finer particle - higher enthalpy gun combination causes surface evaporation of the particles. The finer fluffier particles seen in these micrographs (Fig. 3e and 3f) are believed to have formed from this vapor phase. The vaporized alloys would be highly reactive and can

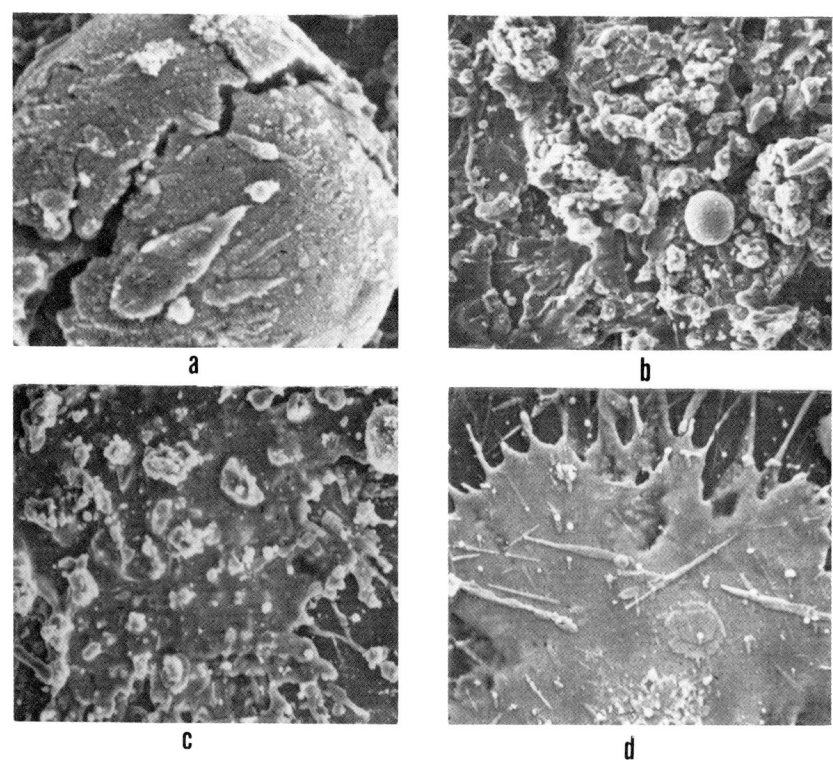

Figure 1: SEM Photographs of Fe 20 Wt% Mn Powders showing varying amount of melting and splatting upon impacting a substrate.

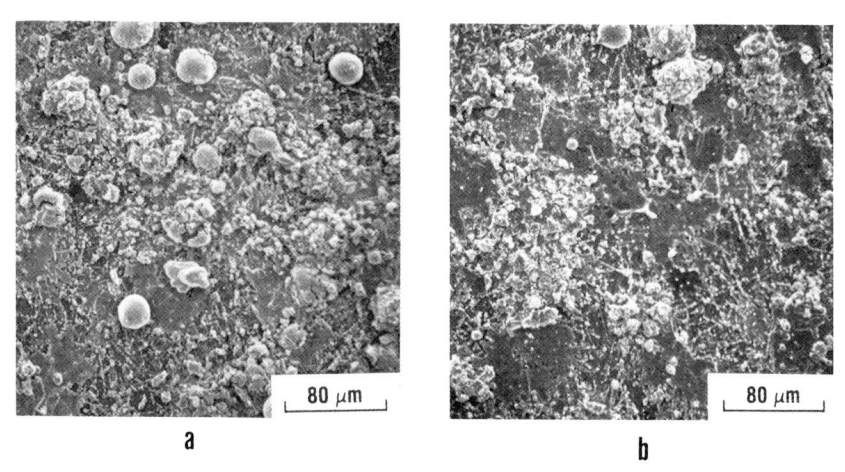

Figure 2: SEM micrographs of Fe Mn 41 powders plasma sprayed onto glass slide substrates using the Ma 2.4 gun. Gun to substrate distance - a) 20 cm, b) 40 cm.

a) Fe Mn 29, Ma 3 gun

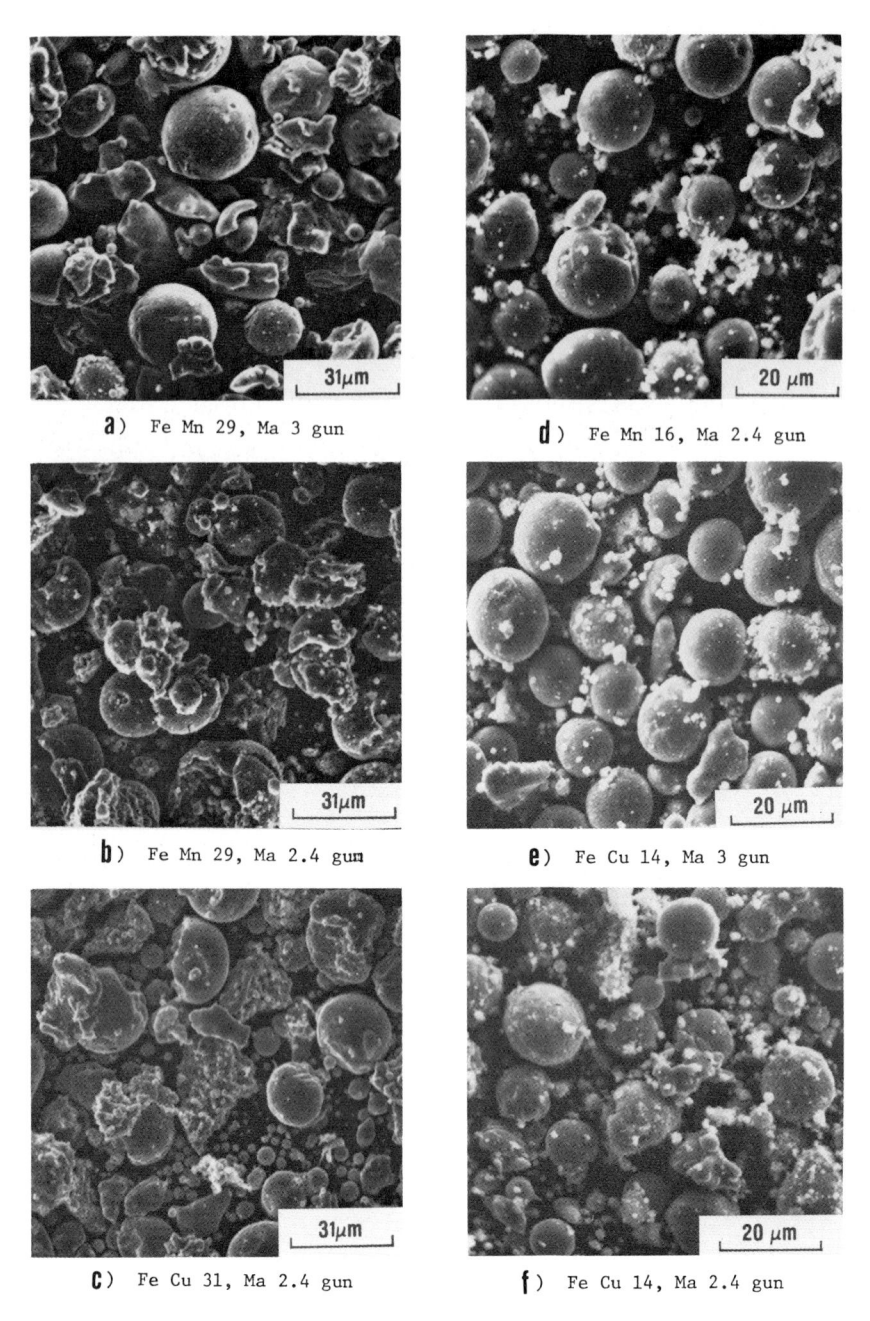

b) Fe Mn 29, Ma 2.4 gun

c) Fe Cu 31, Ma 2.4 gun

d) Fe Mn 16, Ma 2.4 gun

e) Fe Cu 14, Ma 3 gun

f) Fe Cu 14, Ma 2.4 gun

Figure 3: SEM photographs of plasma sprayed and collected powders.

combine with the trace amounts of reactive gases in the Ar-He mixture. That this is indeed the case cannot be categorically stated since the starting powders itself is a source of oxygen (surface absorbed oxygen) because it was not kept under inert gas environment between the times of the atomization and plasma spraying processes.

A higher magnification view of a fracturing particle can be seen in Fig. 4. This fragmentation can also be seen in the deposit of the larger size FeMn 41 powder plasma sprayed using the Ma 2.4 gun; Fig. 5.

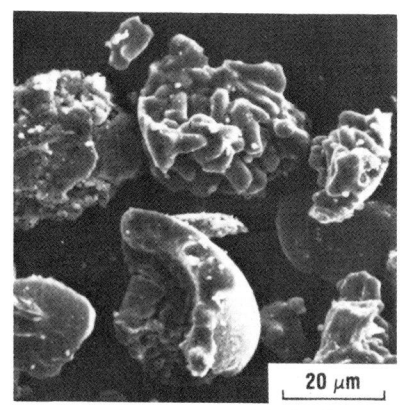

Figure 4: Interdendritic fracture of Fe Mn 41 particles plasma sprayed using the Ma 2.4 gun.

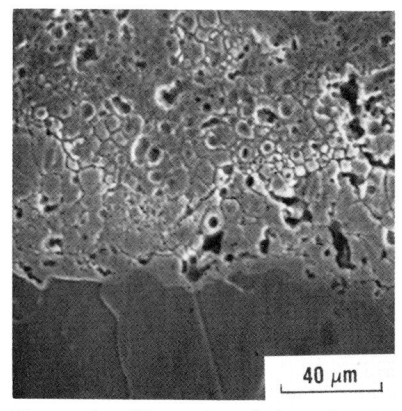

Figure 5: "Sweet Spot" deposit of Fe Mn 41 powders plasma sprayed using the Ma 2.4 gun.

c. Sweet Spot Deposits

Microstructures of the sweet deposits made using different powder batches and gun combinations (gun-to-substrate distance = 41 cm) are shown in Figs. 6a to 6e. The FeMn 29 powders with the Mach 3 gun are essentially unmelted but are heated to a temperature sufficient to cause deformation of the particles on impact (Fig. 6a). The deposit made using the Mach 3 gun and the FeMn 16 powder (Fig. 6b) has a structure containing few particles that have retained their original shape and consists of particles that were partially but sufficiently melted so as to splat upon impact. Fragments of the solid remains of the partially melted particles can be seen spread across the splats throughout the structure. Figure 6c shows the microstructure of the deposit for FeMn 29 powder plasma sprayed using Mach 2.4 gun. The particles are well melted and splatted.*

The deposits of the FeCu 31 and FeCu 14 powders, plasma sprayed with the Mach 3 gun, are shown in Figs. 6d and 6e respectively. The Fe-20wt% Cu alloy has a melting range of 1369 to 1733 K. Thus, for approximately the same average particle sizes as the Fe-20wt% Mn and the same plasma gun, the Fe-20wt% Cu alloy shows a higher fraction of melted particles. A number of unmelted particles are, however, clearly seen in the deposit of FeCu 31 made using the Mach 3 gun (see Fig. 6d). FeCu 14 powder plasma sprayed with the

---

* The powder collector experiments had shown FeMn 29 powder plasma sprayed using the Ma 2.4 gun to be fragmented but not melted. This is not a discrepancy in results. The gas flow conditions near the substrate/powder collector for the two experiments are different, which in turn effect the upstream conditions. Lack of one to one correspondence between the results of the two experiments is not surprising.

194

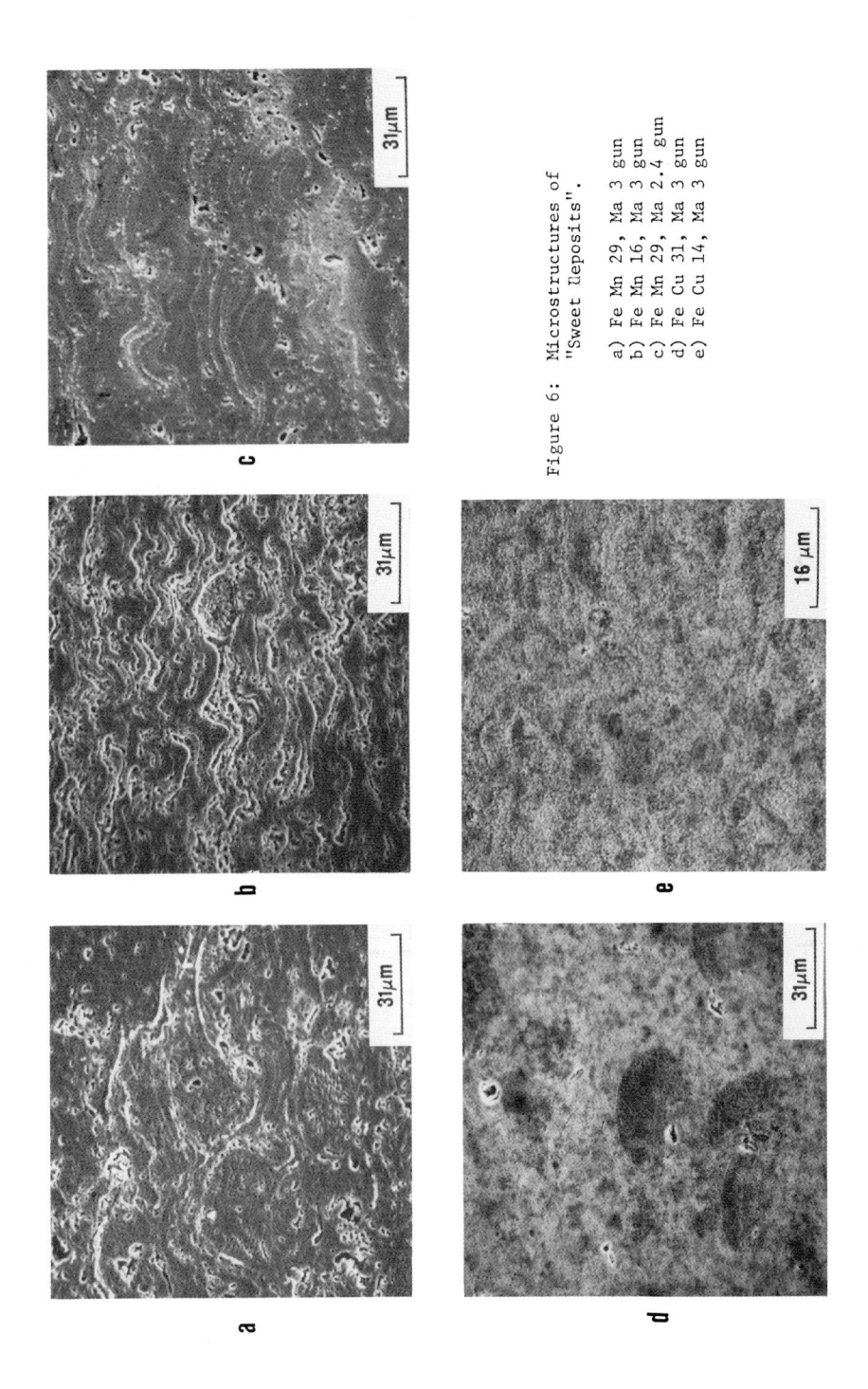

Figure 6: Microstructures of "Sweet Deposits".

a) Fe Mn 29, Ma 3 gun
b) Fe Mn 16, Ma 3 gun
c) Fe Mn 29, Ma 2.4 gun
d) Fe Cu 31, Ma 3 gun
e) Fe Cu 14, Ma 3 gun

Mach 3 gun also depicts a structure with traces of unmelted particles; however, these are more difficult to identify - Fig. 6e.

CONCLUSIONS

Based upon the results of these experiments, it can be concluded that:

1. The major part of the heating of powder particles occurs within a distance of 31 cm (12 in.). Increasing the gun distance from 31 cm (12 in.) to 41 cm (16 in.) causes a marginal increase in the degree of heating. Gun-to-substrate distances should thus be chosen based on consideration of heat flux from the plasma gas to the substrate rather than that of particle melting.

2. Powder melting using the EPI Mach 3 gun is rather inefficient. Powders with an average diameter of 29 µm of the model alloy Fe-20 wt% Mn could not be melted using the Mach 3 gun. Injected Fe-20 wt% Mn powders with an average diameter of 16 µm could be heated up sufficiently so as to splat on impact but were not completely melted.

3. Fe-20 wt% Mn particles in the size range 325-400 mesh (average size = 41 µm) could not be melted using the EPI Ma 2.4 gun. In contrast to the Ma 3 gun, Fe-Mn 29 and Fe-Mn 16 both could be melted using the Ma 2.4 gun. It can therefore be concluded that guns having lower velocity and higher enthalpy characteristics should be used to optimize particle melting.

4. Alloyed powders with dendritic structure, when plasma sprayed, are fragmented into smaller irregular particles. This fragmentation is caused by the interdendritic melting of the segregated structures during heating of the powder particles. These fragmented particles can then melt completely (seen in the collected powders as spherical powders which are smaller in size than the as-received powders).

ACKNOWLEDGMENTS

The research work reported herein was sponsored by DARPA Contract No. F33615-81-C-5156 and the authors would like to thank Dr. Steve Fishman of DARPA; Air Force Project Monitor, Dr. H.C. Gegel; and Dr. Paul Siemers of General Electric Research and Development for their continued support.

REFERENCES

1. W. Klement, R. Willens and P. Duwez: Nature, 187, 869 (1960).

2. M. Lebo and N.J. Grant: Met. Trans., 5, 1547 (1974).

3. N.J. Grant in: Rapidly Quenched Metals III, Vol. 2, B. Cantor, ed. (Metals Society, London 1976) p. 172.

4. P.R. Holiday, A.R. Cox and R.J. Patterson in: Rapid Solidification and Processing -- Principles and Technologies, R. Mehrabian, B.H. Kear and M. Cohen, eds. (Claitors Publishing Division, Baton Rouge, 1978) p. 246.

5. H.H. Liebermann: Coaxial Jet Melt-Spinning of Glassy Alloy Ribbons, Technical Information Series, Report No. 80 CRD 117 (G.E. Corporation, Schenectady, New York 1980).

6. E.M. Breinan and B.H. Kear in: Rapid Solidification Processing Principles and Technologies, R. Mehrabian, B.H. Kear and M. Cohen, eds., (Claitors Publishing Division, Baton Rouge, 1978) p. 87.

7. D.R. Mash and I.M. Brown: Matls. Engg. Quarterly, Feb. 1964, p. 18.

8. S. Safai and H. Herman in: Ultra Rapid Quenching of Liquid Alloys, Vol. 20 of Treatise on Materials Science and Technology, H. Herman, ed. (Academic Press, New York, 1981) pp. 183-214.

9. V. Wilms and H. Herman: Proc. Eighth International Thermal Spraying Conference (American Welding Society, Florida 1976) p. 236.

10. R. McPherson: J. of Materials Science, 15, 3141 (1980).

11. M. Moss: Acta Metallurgica 16, 321 (1968).

12. K.D. Krishnanand and R.W. Cahn in: Rapidly Quenched Metals, N.J. Grant and B.C. Giessen, eds., (MIT, Cambridge, Mass., 1976) p. 67.

13. M. Moss, D.L. Smith and R.A. Lefever: Appl. Phys. Lett., 5, 120 (1964).

14. B.C. Giessen, et al.: Metall. Trans. 8A, 364 (1977).

15. E. Muehlberger: Proc. Seventh International Thermal Spraying Conference, (The Welding Institute, London 1973) p. 245.

16. M.R. Jackson, J.R. Rairden, J.S. Smith and R.W. Smith: J. of Metals, 33 (11), 23 (1981).

17. M.R. Jackson, R.W. Smashey and L.G. Peterson in: Proc. Rapid Solidification Processing, Principles and Technologies III, R. Mehrabian ed. (National Bureau of Standards, Gaithersburg, Dec. 6-8, 1982) p. 198.

18. P.A. Siemers: Rapid Solidification Plasma Deposition (RSPD) for Fabrication of Advanced Aircraft Gas Turbine Components, Contract F33615-81-C5156, Interim Reports, Jan. 15, 1982, Apr. 15, 1982, July 15, 1982, Final Report Dec. 31, 1982. Air Force Systems Command, AFWAL, Wright Patterson Air Force Base, OH.

19. R.W. Smith, M. Paliwal and D. Apelian, in: Proc. Rapid Solidification Processing and Technologies III, R. Mehrabian, ed., (National Bureau of Standards, Gaithersburg, Dec. 6-8, 1982) p. 105.

20. T.L. Cheeks, M.E. Glicksman, M.R. Jackson and E.L. Hall, in: Proc. Rapid Solidification Processing Principles and Technologies III, R. Mehrabian, ed., (National Bureau of Standards, Gaithersburg, Dec. 6-8, 1982) p. 118.

21. R.W. Smith, D.V. Rigney and J.R. Rairden in: Proc. Rapid Solidification Processing Principles and Technologies III, R. Mehrabian, ed., (National Bureau of Standards, Gaithersburg, Dec. 6-8, 1982) p. 468.

22. A.M. Ritter and M.R. Jackson in: Proc. Rapid Solidification Processing Principles and Technologies III, R. Mehrabian, ed., (National Bureau of Standards, Gathersburg, Dec. 6-8, 1982) p. 270.

23. D. Wei, D. Apelian, M. Paliwal and S.M. Correa: "Melting of Powder Particles in a Low Pressure Plasma Jet", Annual Meeting of Materials Research Society. Paper No. L4.6, Boston, MA, Nov. 14-17, 1983.

MELTING OF POWDER PARTICLES IN A LOW PRESSURE PLASMA JET

D. WEI*, D. APELIAN*, S. M. CORREA** AND M. PALIWAL*
*Department of Materials Engineering, Drexel University,
Philadelphia, PA 19104; **General Electric Company,
Corporate Research & Development, Schenectady, NY 12345.

ABSTRACT

A numerical model has been developed to predict the temperature profile of particles injected in a D.C. plasma jet. The equations governing particle melting were applied to spherical powders of binary model alloys. Thermophysical properties of the gas and the powder material have been taken to be temperature dependent. In the proposed model, the latent heat of melting was taken into account by introducing apparent enthalpy as a function of the fraction of liquid formed which can be derived from equations describing non-equilibrium melting. The temperature and velocity profiles of the plasma jet used in this analysis are for a free jet (without target interference) and were calculated using the parabolic Navier-Stokes equation with a K-ε turbulence model. Correction factors have been introduced to take into account non-continuum effects encountered in the low pressure environment and the results show that both heat and momentum transfer between the plasma gas and the injected particles are reduced.

INTRODUCTION

In the plasma spraying process a D.C. plasma jet is used as a heat source to melt and accelerate the injected powder particles which subsequently impinge and solidify on a given substrate. Conventional Atmospheric Plasma Sprayed (APS) deposits which have been used for coating applications have inherent defects such as oxide inclusions and low deposit densities which preclude their use as high performance structural components. With the introduction of the Low Pressure Plasma Deposition (LPPD) process or the Vacuum Plasma Spraying (VPS) process, fine grained, homogeneous and nearly fully dense deposits are produced. Here the plasma jet is expanded into a low pressure environment (∿ 40-80 torr) which results in high plasma jet velocities - Ma 2 to Ma 3. Compared to APS, the spray distances are longer and the plasma jet and the spray pattern are broader. In addition, the LPPD process allows the use of a transferred arc between the gun and the substrate thus enabling one to heat and clean the substrate prior to deposition (1-3).

In this paper, an integrated process model has been developed for the LPPD process, capable of predicting the plasma gas temperature and velocity, as well as particle velocity and thermal history. The melting of injected particles in a plasma jet has been previously addressed for the LPPD process, however the model considered a continuum mechanics approach of heat and momentum transfer between the particles and the plasma (4). In this paper, the non-continuum effects which occur during LPPD are incorporated in the analysis.

PROCESS MODEL

The integrated process model which predicts the temperature of the powder particles during their flight through the plasma jet is based on models describing the plasma gas temperature and velocity, injected particle trajectory and the heat transfer between the particles and the plasma gas. These models are briefly described below.

Plasma Gas Temperature and Velocity

A mathematical model for the free jet flow from a D.C. plasma gun has been developed which predicts the plasma gas temperature and velocity profiles. The model is fundamentally based on the Favre averaged continuity parabolized Navier-Stokes equations with a transitional K-ε turbulence model and is described in detail by Correa (5).

Momentum and Heat Transfer Between Plasma Gas and Particles

Particle velocity can be calculated using the one dimensional equation for particle acceleration.

$$\frac{dV_p}{dt} = \frac{3}{4} \frac{C_D}{D} \frac{\rho_g}{\rho_p} (V_g - V_p) |(V_g - V_p)| \qquad (1)$$

In the LPPD process, due to the low pressure environment and supersonic plasma velocities, the non-continuum effects on the heat and momentum transfer to the particle become significant and should be taken into account. In addition, at high plasma jet velocities the gas compressibility should also be taken into consideration when calculating the plasma jet temperature and velocity profiles.

The relative importance of the non-continuum and compressibility effects on the particle-plasma gas interactions can be characterized by the aid of a dimensionless parameter, the Knudsen number. For an ideal gas, the Knudsen number as a function of the Mach number, Ma, Reynolds number, Re and the ratio of specific heats, γ is given by:

$$Kn = \sqrt{\frac{\pi \gamma}{2}} \cdot \frac{Ma}{Re}$$

Based upon the Knudsen number, the flow regime can be classified as continuum or non-continuum flow; the non-continuum region is further subdivided into three regimes - slip flow, transitional flow and free molecular flow. The different flow regimes, according to two different classifications are shown in Table I (the two classifications are not entirely consistent with each other since the choice of the boundary between any two regions is somewhat arbitrary).

Table I: Classification of the different flow regimes.

Flow Regime	Classification #1 (After Schaaf and Chambre, Ref. 6)	Classification #2 (After Devienne, Ref. 7)
Continuum Flow	Re < 1: Ma/Re < 0.01 Re > 1: Ma/√Re < 0.01	$Kn < 10^{-3}$
Non-continuum Flow		
• Slip Flow	Re < 1: 0.01 < Ma/Re < 0.1 Re > 1: 0.01 < Ma/√Re < 0.1	$10^{-3} < Kn < 0.25$
• Transitional Flow	Re < 1: 0.1 < Ma/Re < 3 Re > 1: 0.1 < Ma/√Re < 3√Re	$0.25 < Kn < 10$
• Free Molecular Flow	Ma > 3Re	$Kn > 10$

In this analysis the drag coefficient in supersonic flow regime was calculated using the formulations suggested by Vallerani (8):

- For $3.3 \times 10^{-1} > Kn_\infty > 3.3 \times 10^{-4}$

$$C_{D,NC} = C_{D,C} + (0.98 + 0.55 f) \sqrt{Kn_\infty} - (0.152 + 0.3 f) Kn_\infty \qquad (2)$$

where $C_{D,C} = 0.9 + ((0.24/Ma) - 0.04)$

$$f = 3.2 (6.7 \phi)^{0.15/(0.45 + \phi)}$$

$$\phi = x^2/Ma^3$$

and $x = \sqrt{T_p/T_{g_\infty}}$

- For $3.3 > Kn_\infty > 3.3 \times 10^{-1}$

$$C_{D,NC} = C_{D,fm} = 0.93 \, g/Kn_\infty + 0.06 \, g^2/Kn_\infty^2$$

where $C_{D,fm} = [(2-a_n) + a_t + \frac{2}{3}\sqrt{\frac{2\pi}{\gamma}} \cdot \frac{x}{Ma}][\frac{1}{3} \cdot (2+\sqrt{1+\frac{1}{Ma^2}})\sqrt{1+\frac{1}{Ma^2}}]$

and $g = 0.96 + 0.0165 \, Ma/x$

The drag coefficient for particles moving at low Reynolds and Mach numbers was calculated using an approximate solution suggested by Phillips (9):

$$C_{D,NC} = \frac{15 - 3C_1 Kn + C_2(8 + \pi a)(C_1^2 + 2)K_n^2}{15 + 12C_1 Kn + 9(C_1^2 + 1) + 18C_2(C_1^2 + 2)Kn^3} C_{D,C}$$

where $C_1 = (2-a)/a$ and $C_2 = 1/(2-a)$; 'a' being the accommodation coefficient which typically has a value in the range 0.8 to 1.0.

The heat flow from the plasma gas to the injected particles is given by: $q = h(T_g - T_s)$. However, for a gas moving at high velocities, an additional factor becomes important. One must take into account the resultant friction and stagnation of the plasma gas near the injected particulate body. This is called the aerodynamic heating effect and is incorporated into the heat flux balance by accommodating the temperature of the gas interacting with the moving particulate, i.e., $q = h(T_r - T_p)$ where T_r is now defined as the recovery temperature, and is expressed by a dimensionless parameter called the recovery factor r $[r = T_r - T_p)/(T_s - T_p)]$ (10). In this work the recovery factor used was obtained from experimental results reported by Drake and Backer (11).

Heat transfer to spheres and bodies in subsonic flow has been investigated experimentally over a wide range of Mach and Reynolds numbers (12,13); however, little experimental data for heat transfer to bodies in supersonic flow exists. In this study, an experimental correction factor (14) for heat transfer to bodies in subsonic flows in the transitional flow regime has been extended to the supersonic flow regime:

$$Nu_{NC} = \left(1 + \frac{Nu_{fm}}{Nu_c}\right)^{-1} \cdot Nu_{fm} \qquad (4)$$

where Nu_{fm} is the Nusselt number in free molecular flow (15) and Nu_c is the Nusselt number in continuum flow (16). In addition one must correct the heat transfer coefficient, h, by taking into account the steep temperature gradients around the moving particulate bodies. This was done by correcting h by the factor $[\rho_\infty \mu_\infty / \rho_s \mu_s]^{0.6}$.

Heat Conduction Within the Particle with Phase Change

In evaluating the heat conduction, temperature gradient within the particle and the temperature dependent properties of both the gas and the powder particle are considered. The governing differential equation and boundary conditions for a pure material during the various stages of melting, i.e. (i) all solid; (ii) liquid + solid; (iii) all liquid or (iv) evaporating surface, have been utilized (17,18). In addition, the melting model takes into account the fact that during melting of alloyed particles a mushy region exists where both solid and liquid phases are present. An enthalpy method developed by Voller (19) has been used to take into account the mushy region during melting; the temperature is related to the enthalpy by

For a Pure Material

$$T = H/C_p \quad \text{when} \quad H \leq \int_0^{T_{mp}} C_p \, dT$$

$$T = T_{mp} \quad \text{when} \quad \int_0^{T_{mp}} C_p \, dT < H \leq \int_0^{T_{mp}} C_p \, dT + L$$

$$T = (H-L)/C_p \quad \text{when} \quad H > \int_0^{T_{mp}} C_p \, dT + L$$

and

For an Alloy

$$T = H/C_p \quad \text{when} \quad H \leq \int_0^{T_{sol}} C_p \, dT$$

$$T = (H - f_L L)/C_p \quad \text{when} \quad \int_0^{T_{sol}} C_p \, dT < H \leq \int_0^{T_{liq}} C_p \, dT + L$$

$$T = (H-L)/C_p \quad \text{when} \quad H > \int_0^{T_{liq}} C_p \, dT + L$$

It has been assumed that the enthalpy changes linearly in any time interval and thus the melting front has been estimated by linear interpolation.

COMPUTED RESULTS

Plasma Gas Temperature and Velocity

The model was subsequently used to calculate the plasma gas temperature and velocity profiles for two supersonic conditions at the nozzle exit -

Ma 2.4 and Ma 3; the results have been presented previously (4).

Plasma/Particle Interaction - Non-Continuum Effects

In low pressure plasma spraying, it is found that the Knudsen number ranges from 0.5 to 5 when injecting particulates in the size range of 10-50 μm. The non-continuum effects on heat and momentum transfer are substantial. The ratio of heat flux and drag coefficient between continuum flow regime and transitional flow regime are calculated and are shown in Figures 1 and 2, where $Q_{O,C} = h(T_r - T_p)$ is calculated assuming continuum flow regime and $Q_{O,NC}$ is the heat flux for Knudsen numbers when non-continuum effects are taken into account. Similarly, $C_{D,C}$ is the drag coefficient for continuum flow and $C_{D,NC}$ is the drag coefficient calculated using formulations suggested by Vallerani (8) when non-continuum effects are taken into consideration. The results indicate that both momentum and heat transfer can be reduced by as much as a factor of 0.05 to 0.8.

Particle Velocity

The particle velocity as a function of axial distance was calculated for 10 μm, 30 μm, and 50 μm tungsten particles (Figure 3) for the plasma condition which has been used to experimentally evaluate the particle velocity and reported by Frind et al. (20). The calculated velocity of 50 μm particle is lower, the error being 21% at the first reported data point and 10% and lower for distances from 12 to 30 cm, than the reported experimental value. For 10 μm particles, the experimental value for the particle velocity is reported only at a distance of 5 cm from the nozzle exit; the calculated value agrees with the single experimental data point available.

The computed results for particle velocities of 18 μm and 30 μm Fe-20 wt% Mn particles as a function of plasma jet axis using non-continuum approach are shown in Figure 4. The calculations indicate that the momentum transfer occurs during the initial stages of particle flight (up to 15 cm) and the introduction of the non-continuum effects leads to particle velocities which are significantly lower than those predicted by continuum mechanics.

The dwell time corresponding to the distance beyond which plasma gas temperature is lower than 1708°K (the solidus temperature of Fe-20 wt% Mn alloy) of the two different size particles for both of the continuum and non-continuum mechanics approach are listed in Table II. As can be seen, the dwell time calculated for the LPPD conditions by including non-continuum effect are much higher than those calculated using continuum mechanics only.

Table II. Comparison of the Dwell Times Calculated Using the Continuum and Non-Continuum Approaches.

Plasma Gun	Spray Distance, cm ($T_g = T_p$ solidus)	Dwell Times, ms			
		Dia. = 18 μm		Dia. = 30 μm	
		C*	NC[†]	C*	NC[†]
LPPD, Ma 3.0	15	0.2	0.53	0.29	0.72
LPPD, Ma 2.4	18	0.25	0.79	0.36	1.08

* Continuum mechanics approach.   [†] Non-continuum mechanics approach.

Particle Temperature Profile

The melting model gives the temperature distribution within the particle as a function of distance along the plasma jet axis. These computations were carried out for particles of the model alloy, Fe-20 wt% Mn (melting range = 1708-1733°K), for two supersonic guns - Ma 2.4 and Ma 3.

Results of these computations (taking into account the non-continuum effects) are shown in Figure 5 as the mean particle temperature vs. distance along the plasma jet axis for two different size particles.

According to the theoretical results the heating of the powder particles is more efficient with the Ma 2.4 gun than with the Ma 3 gun. The Ma 2.4 gun has a lower velocity, a higher enthalpy as well as a lower operating power level than the Ma 3 gun. Powders of 30 μm size can be melted using the Ma 2.4 gun but these are not melted using the Ma 3 gun. Furthermore, 18 μm particles are superheated when injected with the Ma 2.4 gun but are only partially melted using the Ma 3 gun as shown in Figure 5. These results are in agreement with the experimental data reported for particle melting in the LPPD process (21). It should be noted that the reason the temperature of the 18 μm particles (using the Ma 3 gun) does not increase significantly after a distance of 6 cm is that these particles are in the mushy state and thus the heat input is being absorbed as latent heat of fusion.

DISCUSSION

The developed model has been used to theoretically evaluate (i) plasma jet temperature and velocity; introducing non-continuum effects to evaluate (ii) particle velocity and (iii) particle temperature for two different operating conditions. The calculated particle velocity calculations (Figure 3) shows good agreement with the experimental data reported by Frind et al (20). Because comprehensive experimental data is lacking it is difficult to assess the accuracy of the computations. The integrated model as yet does not take into account all the complexities of the process. The plasma jet model which is based on conditions that are relevant for atmospheric spraying (subsonic flow) conditions has been extended to plasma jets which are initially supersonic. Specifically, the model neglects compressibility effects which should be taken into account for jets and flow which are supersonic or approaching sonic velocity; similarly the model neglects abrupt variations in plasma jet properties (velocity, pressure, density, temperature, etc.) across the shockwave as the flow changes from supersonic to subsonic.

Likewise, due to lack of experimental data on heat transfer under supersonic conditions, in the presented model, the empirical formulations for heat transfer in subsonic flow have been extended to supersonic flow. In addition the contribution of ion-electron recombination on the total heat transfer has been neglected. Furthermore, since there is no reliable data in the literature on the thermal and momentum accommodation coefficients of plasma heat transfer - an accommodation coefficient of 0.8 was assumed in this analysis. We are presently refining the model and addressing the deficiencies.

The results of the presented analysis, however are in much better agreement with the reported experimental particle velocities (20) and temperatures (21) than those based on a model using a continuum approach (4). It is evident therefore that for the process conditions of the LPPD process non-continuum effects play an important part and should be taken into account.

The theoretical analysis also indicates that the advantages of the LPPD process (i.e. higher plasma and particle velocities, broader spray patterns, cleaner deposits and transferred arc capabilities, etc.) have been achieved at the cost of the higher particle heating efficiency of the conventional (atmospheric) plasma spraying process. Clearly, an optimum exists between the two extreme conditions - atmospheric spraying (subsonic jets) and the LPPD process operating at supersonic velocities - Ma 2 to Ma 3. Work is presently ongoing to further refine the model such that one can analytically optimize the plasma jet operating conditions.

LIST OF SYMBOLS

a = Accommodation coefficient
C_D = Drag coefficient
h = Heat transfer coefficient
H = Enthalpy
T = Temperature
t = Time
D = Diameter of particle
ρ = Density
μ = Viscosity
L = Latent heat of fusion
C_p = Specific heat
f_L = Fraction of liquid

Subscripts

g = Plasma gas
p = Particle surface
S = Stagnation state
C = evaluated by continuum mechanics
fm = evaluated by free molecular flow
NC = evaluated by non-continuum approach
∞ = Free stream condition
sol = Solidus
liq = Liquidus
mp = Melting point
n = Normal direction
t = Tangential direction

Dimensionless Numbers

Re = Reynolds Number
Ma = Mach Number
Kn = Knudsen Number
Nu = Nusselts Number
r = Recovery factor

ACKNOWLEDGEMENTS

The research work reported herein was sponsored by DARPA Contract No. F33615-81-C-5156 and the authors would like to thank Dr. Steve Fishman of DARPA; Air Force Project Monitor, Dr. H.C. Gegel; and Dr. Paul Siemers of General Electric Research and Development for their continued support.

REFERENCES

1. E. Muehlberger: in "Proceedings of the 7th International Thermal Spray Conference", (Welding Institute, Abington, Cambridge, U.K., 1974), p. 245.
2. M.R. Jackson, J.R. Rairden, J.S. Smith and R.W. Smith: J. of Metals, 33(11), 23 (1981).
3. S. Shanker, D.E. Koenig and L.E. Dardi: J. of Metals, 33(10), 13 (1981).
4. D. Wei, S.M. Correa, D. Apelian and M. Paliwal: "6th International Symposium on Plasma Chemistry", Vol. 1, Montreal, Canada (1983), p. 83.
5. S.M. Correa: "6th International Symposium on Plasma Chemistry", Vol. 1, Montreal, Canada (1983), p. 77.
6. S.A. Schaaf and P.L. Chambre: in "Fundamentals of Gas Dynamics", (H.W. Emmons, ed.), High Speed Aerodynamics and Jet Propulsion, Vol. 3, (Princeton University Press, Princeton, New Jersey, 1958), p. 587.

7. M. Devienne: "Frottement et échanges thermiques dans las gaz raréfiés", Gauthier-Villars, Paris, 1958.
8. E. Vallerani: "A Review of Supersonic Sphere Drag from the Continuum to the Free Molecular Flow Regime", AGARD Conference Proceedings No. 124, on Aerodynamic Drag.
9. W.F. Phillips: Physics Fluids 18, 1089 (1975).
10. H.A. Johnson and M.W. Rubesin: Trans. A.S.M.E., 71(5), 447 (July 1949).
11. R.M. Drake, Jr. and G.H. Backer: Trans. A.S.M.E. 74(7), 1241(Oct. 1952).
12. G.S. Springer: Heat Transfer in Rarefied Gas, in "Advances in Heat Transfer", Vol. 7, J.P. Hartnett and T. Irvine, eds. (Academic Press, New York 1971), pp. 163-218.
13. L.L. Kavanau: Trans. A.S.M.E., 77(5), 617 (July 1955).
14. F.S. Sherman: in "Rarefied Gas Dynamics", J.A. Laurmann, ed., Vol. 2, (Academic Press, Inc., New York, 1983), p. 228.
15. A.K. Oppenheim: J. Aeronaut Sci. 20(1), 49 (1953).
16. W.E. Ranz and W.R. Marshall, Jr.: Chem. Eng. Progress, 48, 173 (1952).
17. T. Yoshida and K. Akashi: J. Appl. Phys., 48, 2252 (1977).
18. J.K. Fiszdon: Int. J. Heat and Mass Transfer, 22, 749 (1979).
19. V. Voller and M. Cross: Int. J. Heat and Mass Transfer, 24, 545 (1981).
20. G. Frind, C.P. Goody and L.E. Prescott: "6th International Symposium on Plasma Chemistry", Vol. 1, Montreal, Canada (1983), p. 120.
21. D. Apelian and M. Paliwal: "Material Research Society 1983 Annual Meeting (November 14-17, 1983).

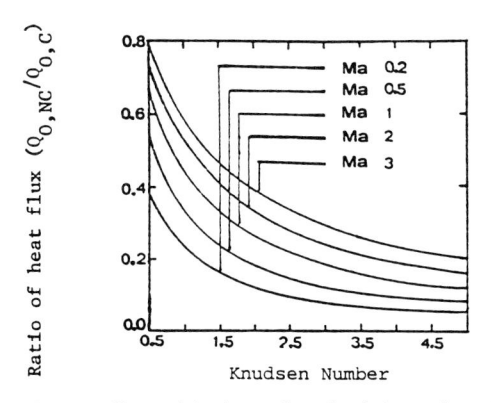

Fig. 1: Ratio of heat flux with ($Q_{O,NC}$) and without ($Q_{O,C}$) non-continuum effect for different Kundsen numbers. (T_g=5000K, T_r/T_g=0.1, a=0.8).

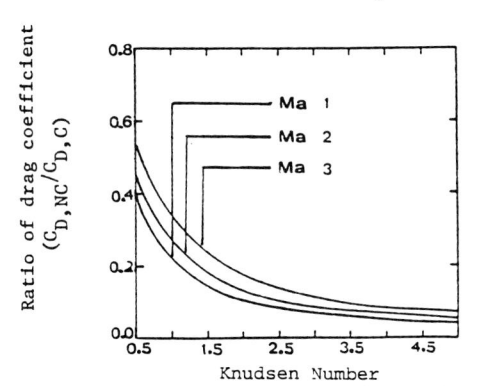

Fig. 2: Ratio of drag coefficient with ($C_{D,NC}$) and without ($C_{D,C}$) non-continuum effect for different Knudsen numbers. (T_g=5000K, T_p/T_g=0.1, a=0.8)

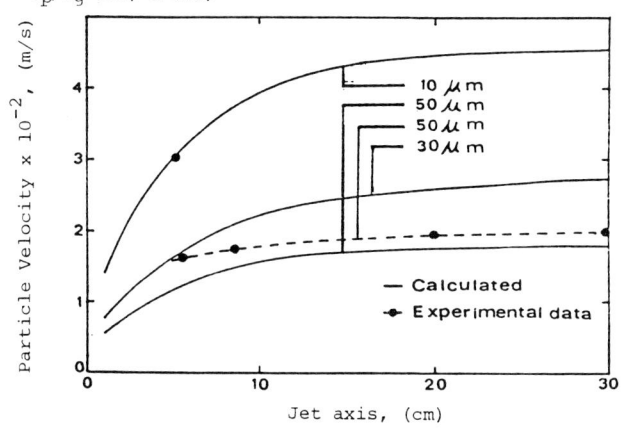

Fig. 3: Calculated and experimental (Ref. 20) tungsten particle velocities along the plasma jet axis.

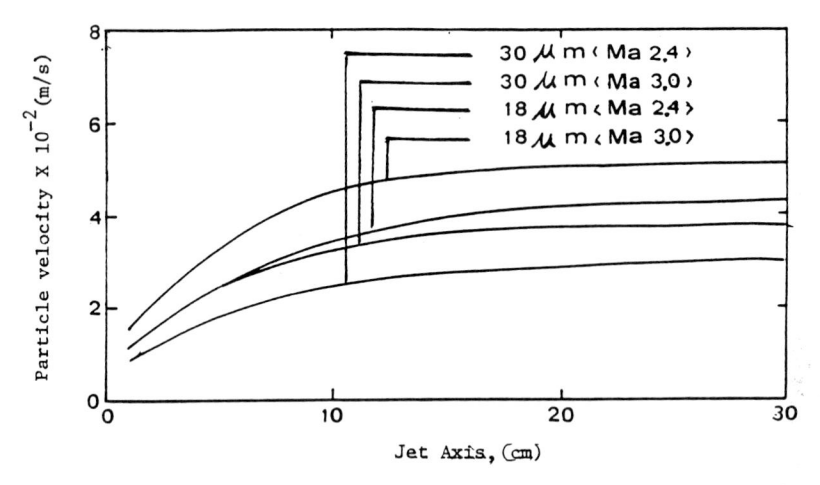

Fig. 4: Calculated velocity of Fe-20 wt.% Mn particles along the plasma jet axis.

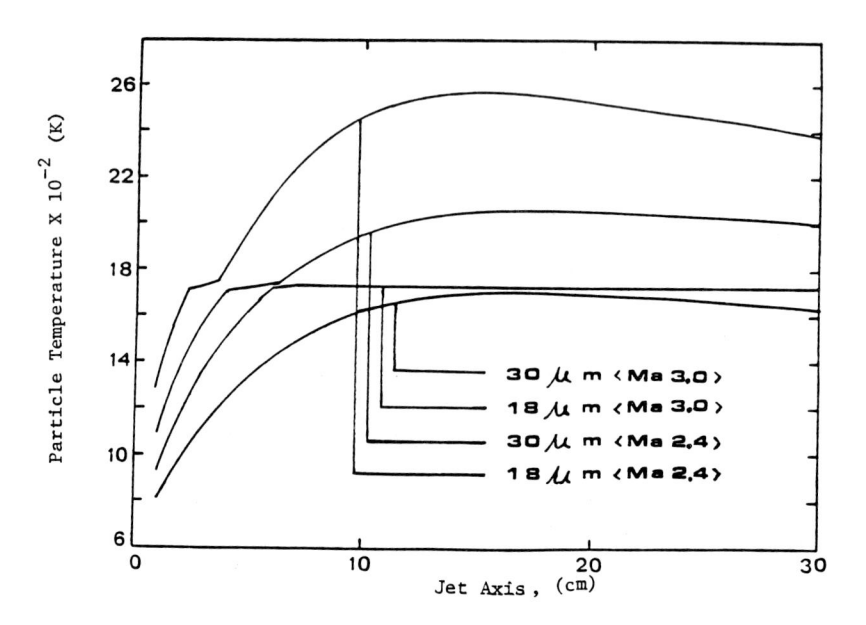

Fig. 5: Calculated mean temperature of Fe-20 wt.% Mn particles along the plasma jet axis.

MICROPROCESSOR CONTROL OF THE SPRAYING OF GRADED COATINGS

R. KACZMAREK*, W. ROBERT*, J. JUREWICZ*, M.I. BOULOS* AND S. DALLAIRE**
*Department of Chemical Engineering, University of Sherbrooke
Sherbrooke (Québec), Canada, J1K 2R1; **National Research Council Canada
Industrial Materials Research Institute, 75 de Mortagne, Boucherville
(Québec), Canada, J4B 6Y4

ABSTRACT

A systematic study was made of the characteristics of a microprocessor controlled powder feeder in view of its utilization for the deposition of graded coating. The unit tested had the capability of feeding three different powders at rates which could be continuously varied up to about 40 g/min. $MgOZrO_2$ was deposited on a stainless steel substrate using a series of thin (Ni/Cr)-Zirconia coats as intermediate layers. The results showed that such a technique could provide a valuable means of achieving a precise and effective control on the composition of the intermediate layers which can have important effects on the final quality of the deposit.

1. INTRODUCTION

Plasma spray-coating has immerged over the last few years as an excellent technique which can be used for the deposition of a wide range of protective coatings on an industrial scale. These could be applied to a substrate either to restore and attain a desired dimension, or to improve its erosion or corrosion resistance, or to create a thermal barrier for high temperature operation.

To a large extent, the success of a thermal spraying operation depends not only on the appropriate selection of the coating material to achieve the desired surface properties, but also on the compatibility between the properties of the material deposited and that of the substrate. Specifically, large differences between the coefficients of thermal expansion of the coating and that of the substrate can cause high internal stresses which are often at the origin of the premature failure of the coating.

One way of overcoming such a difficulty is to build up the coating by the successive deposition of thin layers of gradually varying composition [1]. Such coating are known as "graded coating". In a metal-ceramic system, it is not unusual for a 800 μm coating to be built up of a number layer with a composition varying from 100% metal applied directly to the substrate to 100% ceramic for the top coat.

Unfortunately due to technological reasons, most of the graded coatings used so far [2-5] have been limited to the deposition of two or three layers of a pre-mixed powder which results in a rather poor graduation of the composition of the coating as shown schematically in Figure (1-a). An obvious desirable alternative is to increase the number of layer deposited as much as possible (6 or even 8) and to gradually change the composition of each of these layers as shown in Figure (1-b). Such a task, however, necessitates a tight control on the thickness of the layers deposited each time, and, for practical reasons, finding an alternative to the use of pre-mixed powders.

The present investigation addresses itself specifically to the question of the control of the composition and the powder feed rate in the plasma spray-coating operation. Incorporating microprocessor technology, a powder feeder was designed and developed to achieve a programmable, fine-grain, variations of the composition of the coating. The system was tested using a NiCrAl alloy - $MgOZrO_2$ ceramic system deposited on a stainless steel substrate.

2. EXPERIMENTAL TECHNIQUES

2.1 Plasma Torch

The d.c. torch used was of standard design using a water-cooled thoriated tungsten cathode and a 6 mm i.d. water-cooled copper anode. Details of the plasma torch and spraying set-up are given in Figure 2.

2.2 Powder-Feeding System

The principal design feature of the powder feeding system used in the present investigation is essentially that, rather than using pre-mixed powders, we opted to individually feed each powder at a closely controlled rate in such a way as to create, in situ, the desired composition of the deposited layer.

A schematic of the powder-feeding system used is given in Figure 3. This consisted essentially of two parallel screw feeders driven by individually microprocessor-controlled stepper motors. Repeated calibration of each of these powder feeders showed a close and reproducible relationship between the motor pulse rate (or r.p.m.) and the powder feed rate. The latter, however, was also found to depend on such parameters as the amplitude of the vibration of the powder feeder, the carrier gas flow rate as well as the density and the particle size distribution of the powder. Throughout our experiments, powder-feeder vibration level was kept constant. Individual calibration curves were made for each of the powders used with different carrier gas flow rates.

Specific attention was given to the way in which the powders were injected into the plasma. Four distinct alternatives were identified and individually tested. These are schematically illustrated in Figure 4. In the arrangement given in Figure 4-a, the two powders to be simultaneously deposited are mixed at the exit of the powder feeder upstream of the powder transport line. In the arrangement shown in Figure 4-b on the other hand, the two powders are mixed only prior to their injection into the plasma. Unfortunately, in either case, it is not possible to achieve a close control on the injection velocity of each of the powders to be deposited. With the slightest difference between the densities, or the particle size distribution of the two powders used, substantial segregation can take place resulting in the deposition of the powders on two different locations on the substrate.

The arrangement shown in Figure 4-c seemed at first as an appropriate solution allowing an individual control of the injection velocity of each powder. It was, however, particularly difficult to adjust the powder injection conditions in such a way to insure that the powder introduced at the different points around the anode are deposited simultaneously on the same spot on the substrate. Considerable success was achieved in this regard using the powder injection arrangement shown in Figure 4-d which was used in most of our subsequent experiments.

2.3 Torch-Substrate Movement

The coatings were applied to 70 x 70 mm square stainless steel substrates. In order to achieve a uniform thickness of the deposit, it was necessary to move both the torch and the substrate as shown schematically in Figure 5. The torch movement was controlled using a cam system which maintained for most of the torch sweep a linear velocity on the substrate of about 300 mm/s. The substrate, on the other hand, was moved in the plane perpendicular to that of the torch using a screw feeding system at velocity close to 30 mm/s. Throughout the spraying operation the substrate was cooled using an air jet directed towards its back.

2.4 Materials

Commercial NiCrAl and MgOZrO$_2$ powders were used for spraying. The NiCrAl powder, composed essentially of a nickel-chromium alloy with an aluminium constituent, produced high bond strength, readily machinable coatings which are resistant to oxidation at high temperatures. The MgOZrO$_2$ powder was, on the other hand, a simple blend of 76% ZrO$_2$ and 24% MgO. More information about the powders used is given in Figure 6. The substrate was 304 stainless steel plate, 3 mm thick which was sandblasted prior to spraying.

3. RESULTS AND DISCUSSION

A photomicrograph of the graded coating deposited using premixed powders is shown in Figure 7. In this case, the first layer constituted of a NiCrAl bond coat. The intermediate coat was a commercially available blend of NiCrAl/MgOZrO$_2$. The final coat was composed of MgOZrO$_2$. It is to be noted that in the intermediate layer, the NiCrAl particles are more or less uniformly spread in the ceramic matrix. There is also an important apparent porosity in the deposit which are the result of the particular spraying conditions used.

Figure 8 presents a cross section of a graded coating obtained using two-point powder injection, one for the NiCrAl alloy while the other for the MgOZrO$_2$ ceramic. The mixing between the two powders was carried out in this case in situ. The results indicate that the alloy and the ceramic powders were not deposited simultaneously on the same spot. As shown in Figure 8, such a situation can lead to the formation of a statified rather than a uniformly graded deposit.

It is interesting to note that a slight adjustment of the injection conditions can completely change this situation as shown in Figure 9 where the two powders were deposited simultaneously on the same spot on the substrate with the subsequent formation of a reasonably well graded coating.

The gradual variation of the composition of the coating obtained in this case was further confirmed using electron microscopic analysis. Due to the discreate and relatively large size of the particles in the deposit, the analysis had to be carried out on relatively large surface areas of the cross section. These were defined by a rectangle of dimensions 50 x 150 μm. Based on the semi-quantitative composition profiles obtained for Ni, Cr, Zr and Mg which are shown in Figure 10, it can be concluded that the present powder feeding technique offers a very powerful tool for achieving a good control on the composition profile in a graded coating. Whether such coatings will indeed have substantially improved the performance properties of the coating remains to be seen.

4. CONCLUSIONS

A microprocessor-controlled powder feeder has been developed and tested for the deposition of graded coatings. The results indicate that while in situ mixing of two powder constituants during the spray-coating operation offers the added flexibility of achieving any desired composition, special care has to be taken to insure that both powder are injected in the plasma jet in such a way as to be simultaneously deposited on exactly the same spot on the substrate. Otherwise important stratification effects can be observed in the deposit.

ACKNOWLEDGMENT

This work was partially supported by the National Research Council of Canada, Industrial Materials Research Institute, Industrial Materials Research Institute, through contract agreement 09SD 31255-1-5015. The support by the Ministry of Education of the Province of Quebec and the Natural Sciences and Engineering Research Council of Canada is greatfully acknowledged.

REFERENCES

[1]. A Plasma Flame Spray Handbook, Final Report, Report No. MT-093 March (1977), Naval Ordnance Station, Louisville, Kentucky, 40214.

[2]. S. Stecura, Two-Layer Thermal Barrier Coating for High Temperature Components, Ceramic Bulletin, Vol. 56, No. 12 (1977).

[3]. C.M. Taylor, R.C. Bill, Temperature Distributions and Thermal Stresses in a Graded Zirconia/Metal Gas Path Seal System for Aircraft Gas Turbine Engines, AIAA 16th Aerospace Science Meeting, Huntsville, Alabama, January 16-18, (1978).

[4]. A.F. Erickson, J.C. Nablo, C. Panzera, Bonding Ceramic Materials to Metallic Substrate for High-Temperature Low-Weight Applications, ASME Publication, 78-WAIGT-16 (1978).

[5]. R.L. Newman, W.G. Spicer, Thermal Conductivity of Mixed-Composition Plasma-Sprayed Coatings, AIAA Journal, Technical Notes, Vol. 11, No. 3, March (1973).

(a)

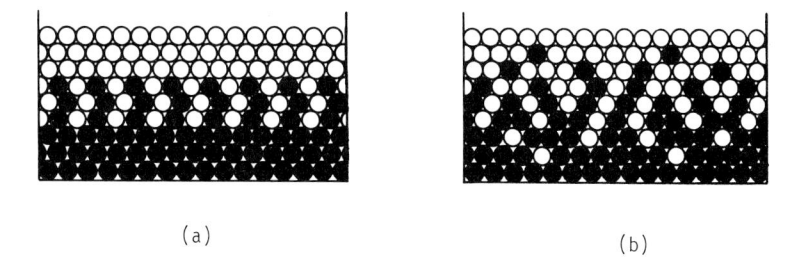

(b)

Figure 1. Schematic of the possible compositions of graded coatings

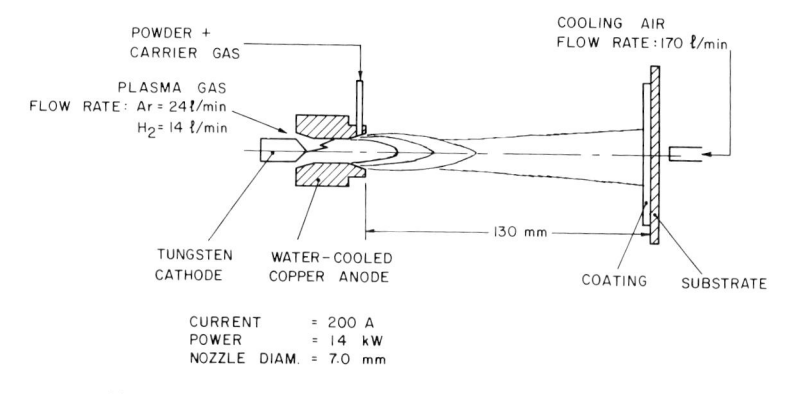

Figure 2. Experimental set-up showing the torch and substrate arrangement

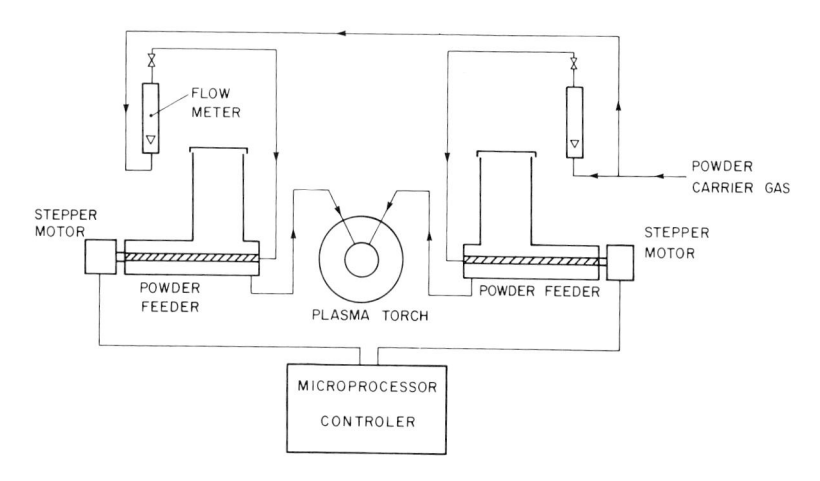

Figure 3. Schematic of the microprocessor controlled powder-feeding system

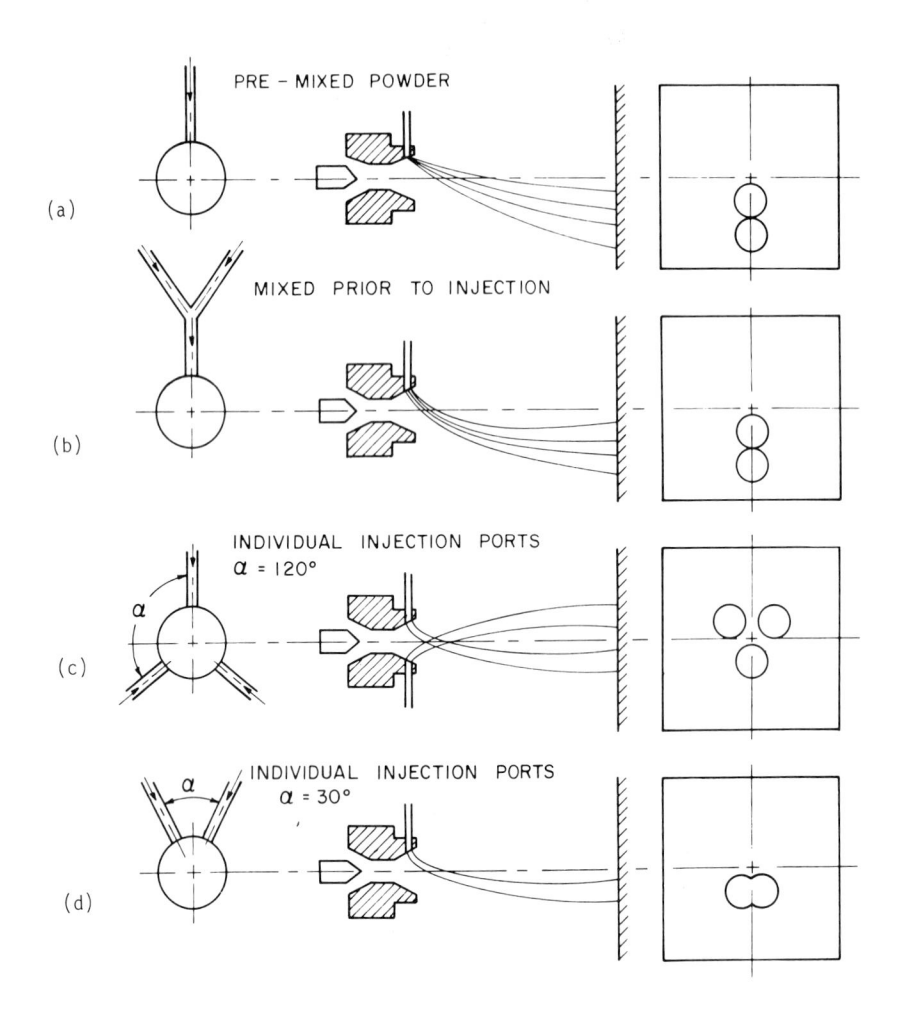

Figure 4. Different powder injection techniques

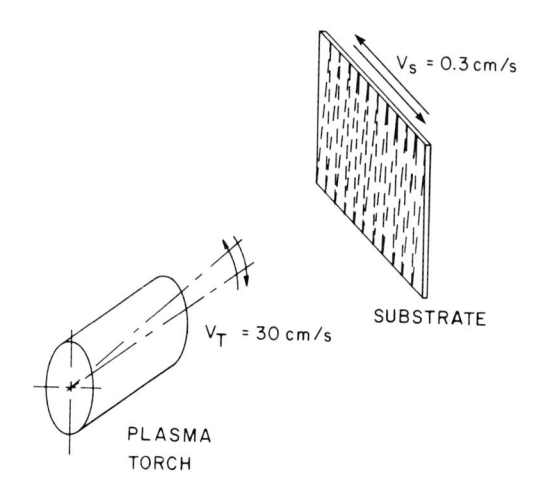

Figure 5. Schematic of the torch substrate movement

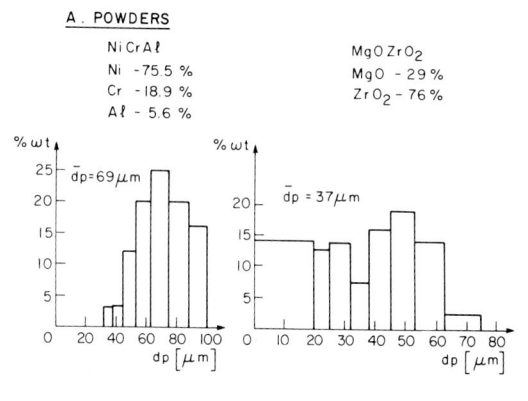

A. POWDERS

NiCrAl
Ni -75.5 %
Cr -18.9 %
Al - 5.6 %

MgO ZrO$_2$
MgO - 29 %
ZrO$_2$ - 76 %

$\bar{d}p = 69 \mu m$

$\bar{d}p = 37 \mu m$

B. SUBSTRATE

MATERIAL - STAINLESS STEEL 304
DIMENSION - 70mm x 70mm
THICKNESS - 3.1 mm (1/8")

Figure 6. Materials used for spraying

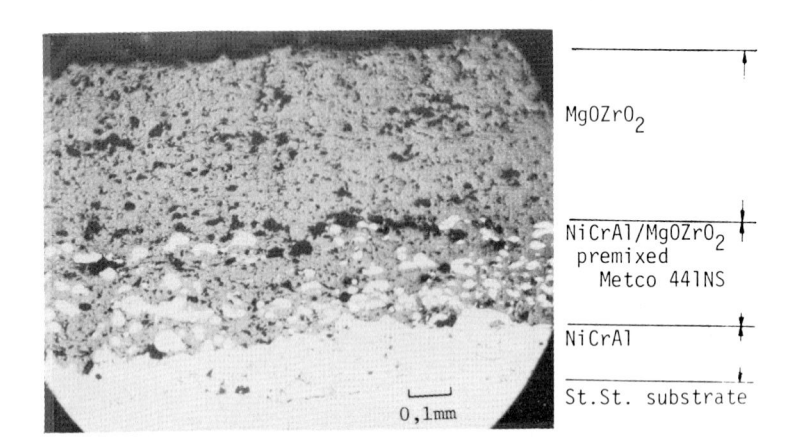

Figure 7. Micrograph of a graded coating obtained using a premixed powder as an intermediate layer

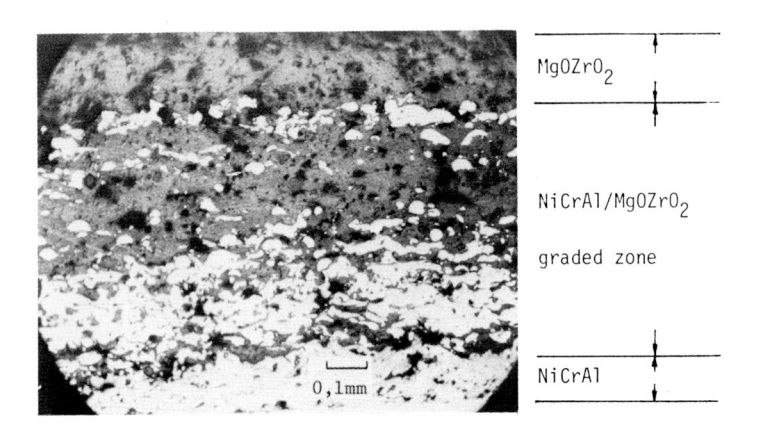

Figure 8. Micrograph of a graded coating obtained with the powder constituents simultaneously deposited at different spots on the substrate ($\alpha=60°$)

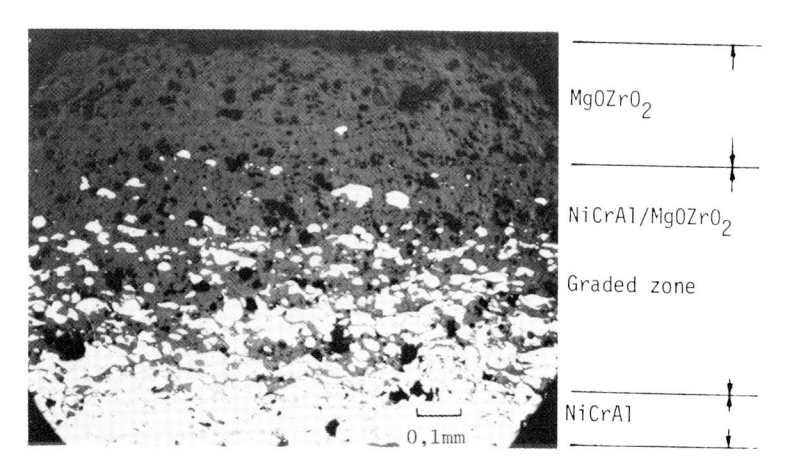

Figure 9. Micrograph of a graded coating obtained with the powder constituants simultaneously deposited on the same spot on the substrate (α=30°)

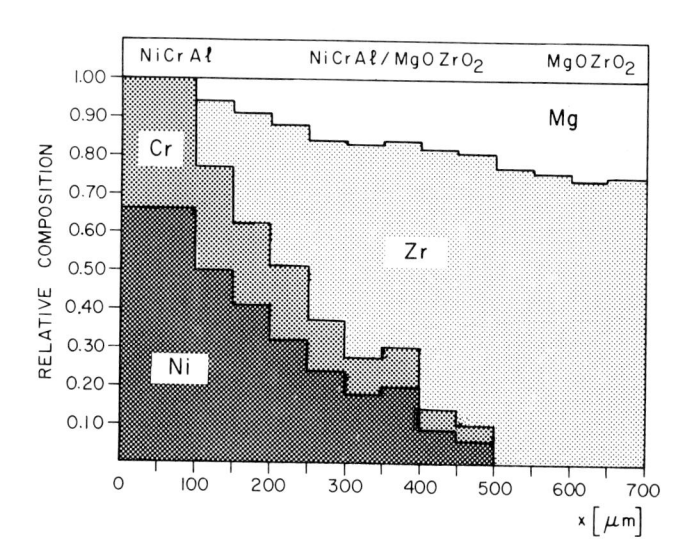

Figure 10. Composition profile across the graded coating

CONSOLIDATION OF NICKEL BASE SUPERALLOYS POWDER
BY LOW PRESSURE PLASMA DEPOSITION

R.W. Smith
L.G. Peterson
W.F. Schilling
General Electric Co. Gas Turbine Division Schenectady, New York 12345

INTRODUCTION

Powder Metallurgy

Powder Metallurgy (P/M) technology has seen an important series of advances in the past twenty-five or so years. It has progressed from a press + sinter methodology (which still has important uses) to one of being a complete processing technology where, ultimately, the structure and properties of totally new materials can be synthesized. A key step in this evolution was the application of P/M techniques to high performance materials such as nickel-based superalloys. This effort began in the early 1960's and has resulted in several important commercial applications such as P/M gas turbine disks and shafts. Attention has now turned to the potential use of Rapid Solidification Technology (RST) as a means of furthering the span of materials which can be made available for a wide variety of engineering applications. Rapidly solidified nickel, cobalt, titanium, aluminum and copper based materials are all the subject of extensive research and development at the present time (1).

Consolidation of rapidly solidified materials also remains an area of significant interest. To date, more or less conventional process techniques have been utilized to consolidate rapidly solidified materials which are in particulate, filament, or sheet form. While a large amount of variation exists, the consolidation methods are generally comprised of the following steps:

o   Canning or containerization
o   Degassing
o   Densification

Achievement of final shape generally requires some post-densification machining.

To-day's P/M processing methods for consolidation of high performance materials are not without problems. For example, improper selection of canning materials can result in unwanted metallurgical reactions between the can material and the alloy which is to be densified. These include modification of the surface composition due to interdiffusion or the formation of unwanted phases such as carbides or intermetallic compounds. Improper powder handling procedures can result in the incorporation of metallic or non-metallic inclusions in the densified material which can affect mechanical properties. Incomplete degassing can produce thermally induced porosity on subsequent high temperature exposure which can also affect metallurgical stability and mechanical properties.

While it is likely that conventional consolidation techniques will be in use for quite a time to come, the search for improvements over these methods has already begun. The ideal solution to many of the issues already outlined may be techniques which employ spray forming methods to achieve both rapidly solidified structures and full densification in a one-step manner.

Plasma Deposition

Plasma Deposition, a spray forming technique, has been used to consolidate metal powders. The powders are injected into a plasma jet, typically formed in a D.C. plasma arc gun, where they are heated to a range of temperatures and accelerated to high velocities. When these heated powder particles impact the substrate, some are molten and others are in various degrees of melting depending on the alloy. The energy of impact flattens out the particle through liquid or liquid/solid shear deformation. Millions of particles impact the substrate one on top of another consolidating the original powder form. The rapid solidification that occurs after impact or inclusion of the original solid particles not fully melted maintains the very fine microstructure and good compositional uniformity typical of other types of P/M consolidation.

One difficulty of previous plasma deposition techniques has been the deposits, high oxide content when spraying reactive alloys such as superalloys. These type of depositions were done in air where large amounts of oxygen was entrained into the process. Recent developments in plasma deposition processing include introducing plasma spraying into a low pressure, inert gas environment which significantly improves the properties of nickel-base superalloys. The properties have been shown to have equivalent strengths of superalloys powders consolidated by other techniques [2]. The addition of a low pressure environment to the plasma spray process has introduced several advantages, making possible deposits of highly alloyed materials which can be used for high performance applications.

EXPERIMENTAL PROCEDURE

Three nickel base superalloys, Rene 80, Hastelloy Alloy X and U-700*, have been evaluated by glass slide deposits, free jet collected powders and consolidated deposits. The powders were sprayed in all cases using a low pressure plasma spraying technique [3].

*Compositions (wt Δ)	Ni	Cr	Co	Fe	Al	Ti	Mo	W	C	other
Hastelloy-X		22	1.5	18.5			9	0.6	0.1	
U-700 LC		14	15		4.3	3.4	4.2		0.02	0.016B, 0.04Zr
Rene 80		14	9.5		3.0	5.0	4.0		0.17	0.015B, 0.03Zr

Glass Slide Deposits

Glass microscope slides were positioned at various distances from the plasma gun exit, low powder feed rates (50 gram/min) were set, and the plasma gun was rapidly (30 cm/sec) traversed over the slides. The glass slides were then viewed via both optical and scanning electron microscopy (SEM).

Free Jet Collected Powders

Another technique to sample the powders from the plasma jets was a powder collector whose aperture could be positioned in the plasma jet. During this time powders deposited onto the cone surface and a small fraction entered the collector aperture and cooled during flight before impacting the collector bottom 95 cm downstream.

Evaluation of these small samples required a modified Field and Fraser technique [4].

Consolidated Deposits

"Sweet-spot" (where the plasma gun is not traversed) deposits on substrates were also made to study the microstructural features (unmelted particles, splat thickness, porosity, etc.) of actual built-up deposits.

Larger deposits of Rene 80 and U-700 6-8 mm thick were also made from which mechanical test specimens were fabricated. These plates were also preheated (800-900°C) and the plasma gun was traversed over the substrate plate to obtain a 80 x 250 mm plate.

RESULTS AND DISCUSSION

Glass Slide Experiments

The results of the glass slide experiments are used to study the morphological changes of individual particles which have interacted with the plasma jet and subsequently impacted the glass substrate. Particle splat morphologies for the Hastelloy-X powder are shown in Figure 1. Figure 1a. shows the splatting of a partially melted particle which was in a "slushy" state when it impacted. Examples of particle fracturing along the melted interdendritic region in a partially melted particle, and the splatting of well melted particles can be seen in Figure 1b. In addition, Figure 1b. also shows solidified streaks caused by the branching of molten splats on impact.

It is apparent from the glass slide morphologies that the nickel-base alloy powders may not be completely molten on impact. The impacted powder particles, if solid, are fractured and deformed; or if liquid are splatted. Liquid fragments splatter across the surface of the substrate. All along more powder particles are continually arriving, covering previously deposited particles. Porosity can also be formed along interdendritic fracture sites, as in Figure 1b., or underneath edges of unmelted powder particles or dendrite fragments or lifted edges of splats.

Free Jet Collected Powder Particles

Figures 2a.-2c. are SEM micrographs of as-received and plasma sprayed powders of Hastelloy-X and Rene 80. Figure 2a. is a SEM photomicrograph of the surface morphology of the as-received Rene 80 powders. It can be seen that the argon gas atomized powders are mostly spherical and quite regular in shape. A few particles which have been impacted or splatted with another particle, during atomization, can be seen. Figures 2b. and 2c. show the most prevalent morphologies of the plasma sprayed Rene 80. When compared to the as-received powders, Figure 2a., the plasma sprayed powders show extensive amounts of fine powders, agglomerated fine particles, fractured powder particles, and various other shapes which have been formed in the plasma jet or during collection. Figure 2b. shows an example of how a coarser screen fraction of powders has been "atomized" into a distribution of powder that has a large volume fraction of particles below 10μm. The probable origin of these fine particles could be i) interdendritic fracturing of powders and their subsequent spheriodization into smaller particles, ii) detachment of the liquid layer from surface melted particles, and iii) secondary atomization of molten droplets into smaller particles.

Consolidated Deposits

As-Deposited Microstructures

The deposits are formed by a rapid succession of powder particle impacts built up on top of one another. The impacting powder particle population has been found to be in various stages of melting and/or solidification. Hence the deposits include features of unmelted particles, partially (interdentritically) melted powder particles, smaller pieces of large dendritic powder particles, or completely melted particles which splat out into thin lamellae. Figures 3a. and 3b. are examples of these microstructures. It has been found that the plasma jet enthalpy level primarily controls the as-deposited microstructures described above, but other variables can also affect microstructure. The plasma jet heat flux and substrate temperature were found to induce directional (epitaxial) grain growth (mostly in the Hastelloy-X), coarsen γ', or precipitate $M_{23}C_6$ carbides in Rene 80 and the U-700. The splat of particles shows up as lamellae in the deposit cross sectional microstructure with splat thicknesses ranging from 1-3 μm. The temperature of deposition and temperature gradients in the deposit from the plasma process heat flux cause cross splat grain growth and re-crystallization. Re-crystallization in Rene 80 plasma deposits was found to be minimal because of the stabilizing effect of the stable precipitating phases (i.e. , carbides, borides). This grain re-crystallization and growth demonstrate the integrity of the intersplat/interparticulate bonds.

U-700 deposits seen in Figure 4 show some similar features to the Rene 80 deposits. This alloy was selected because it represents an alloy with about the same γ' fraction as the Rene 80, but with lower interstitial alloy and carbon content. The microstructure of the U-700 deposit, Figure 4, illustrates that this alloy is less stabilized than was the Rene 80 deposit. The figure shows that substantial grain growth and re-crystallization are occurring during deposition; note the re-crystallization twins.

Hastelloy-X is a solid solution and carbide strengthened alloy with no major precipitating phase such as γ'. Plasma deposits of this alloy shown in Figure 5 indicate that this alloy produces a largely re-crystallized microstructure.

Effect of Post Deposition Heat Treatment

Figures 6 and 7 show the Rene 80 and U-700 microstructures, respectively, after heat treatment. One can see that, in the Rene 80 deposit, there has been substantial grain growth and re-crystallization. However, the grain growth at the solution temperature, 1250°C, was limited to a range of 10-60 μm. The U-700 microstructure, Figure 7, shows considerably more grain growth than the Rene 80. Again, as seen in the as-deposited microstructures, the low interstitial alloy content of this alloy, U-700LC, has resulted in much greater grain growth than in the Rene 80. This plasma deposit now looks quite similar to a wrought P/M structure; however, a potential shortcoming in the large range of grain sizes (50-400 μm) can be seen.

CONCLUSIONS

The low pressure plasma deposition process has been evaluated as an improved plasma deposition and powder consolidation process. The microstructures and properties presented for these selected nickel-base superalloys demonstrate that this new development in plasma deposition can produce consolidated superalloy P/M structures equivalent to other consolidation processes. The plasma deposition parameters, alloy deposited and powder size can all affect the consolidation mechanics by affecting the amount of liquid present during particle impaction. In addition, it has been shown that the alloy can significantly affect both as-deposited and final consolidated (heat treated) microstructure and properties. Alloys which are less stabilized by interstitial elements (i.e. carbon, zirconium, and boron) seem to show more grain growth both during deposition and after final heat treatment. Alloys similar to U-700LC can be expected to show better ductilities and rupture strengths due to the increased grain size.

ACKNOWLEDGEMENT

The work described was sponsored by the Air Force Wright Aeronautical Laboratories/Materials Laboratory, Air Force Systems Command, United States Air Force, Wright-Patterson Air Force Base, Ohio 45433 and the Defense Advanced Research Projects Agency under Contract F33615-81-C-5156.

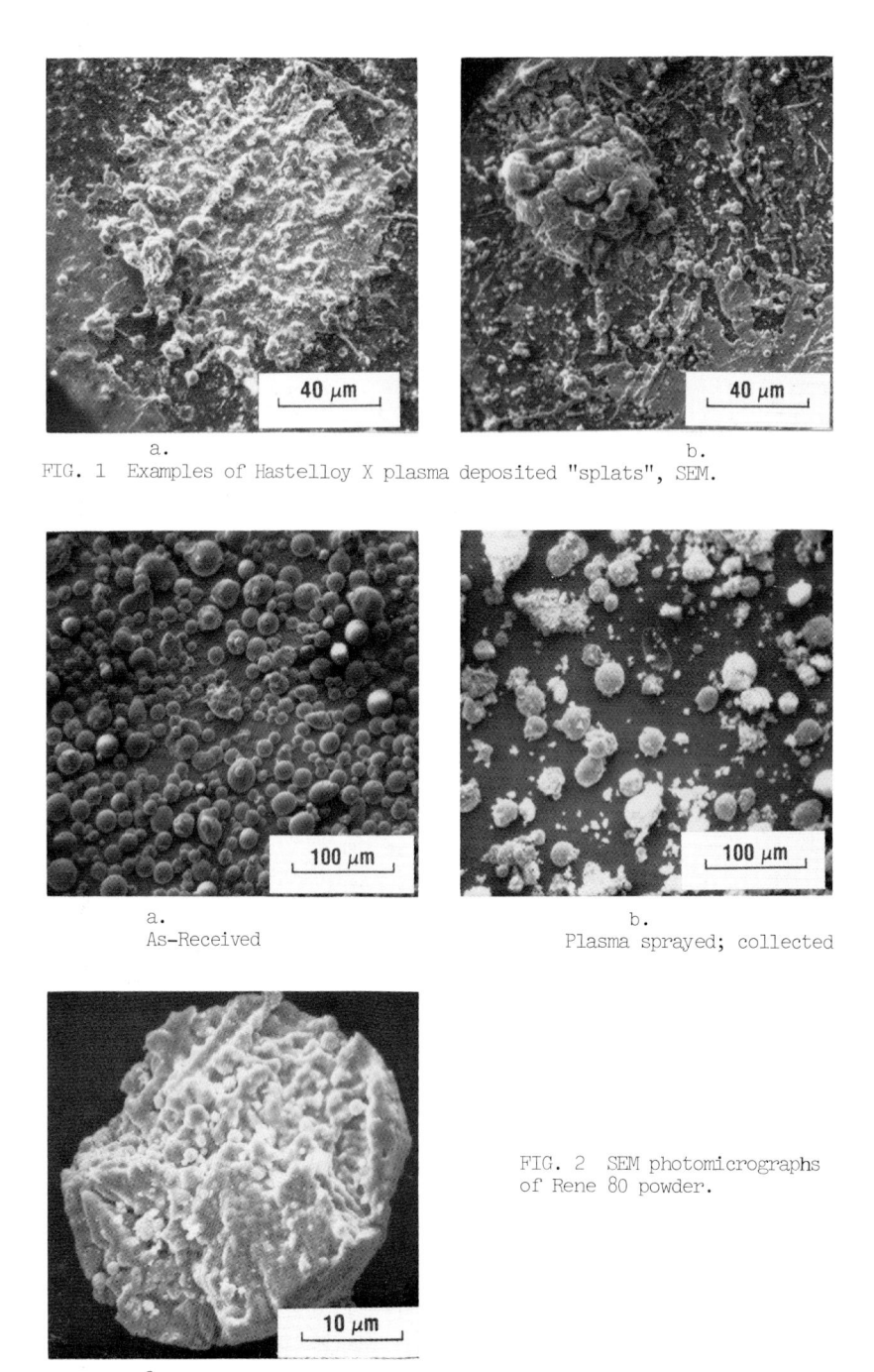

a.   b.
FIG. 1 Examples of Hastelloy X plasma deposited "splats", SEM.

a.   b.
As-Received   Plasma sprayed; collected

FIG. 2 SEM photomicrographs of Rene 80 powder.

c.
Interdendritically fractured powder particle

223

a.   b.
FIG. 3 Rene 80 plasma deposits, etched (HCl/HNO$_3$/H$_2$SO$_4$).

FIG. 4 U-700LC plasma deposit, etched.

FIG. 5 Hastelloy X plasma deposited, etched.

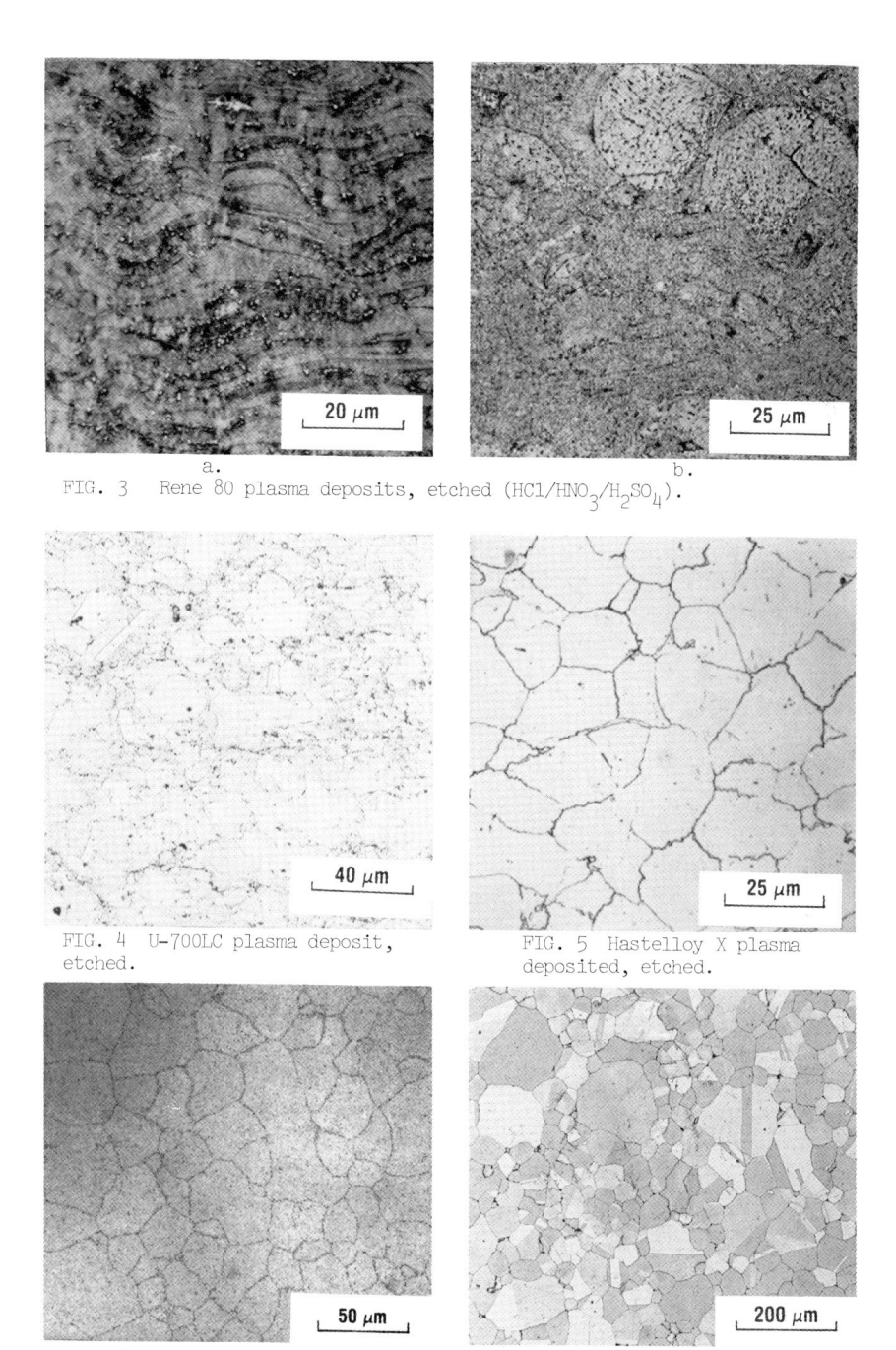

FIG. 6 Heat treated Rene 80 plasma deposit, etched.

FIG. 7 Heat treated U700-LC plasma deposit, etched.

REFERENCES

1. R. Mehrabian, ed., Rapid Solidification Processing, Principles and Technologies III National Bureau of Standards, Dec., 1982, Gaithersburg.
2. M.R. Jackson, J.R. Rairden, J.S. Smith, and R.W. Smith in: Journal of Metals Nov. 1981 p. 23.
3. E. Muehlberger: Seventh International Metal Spraying Conference Proc., London England, 1973, Paper No. 58, Welding Institute.
4. R.D. Field and H.L. Fraser in: Met. Trans. 1978, 9A, No. 1 pp. 131-134.
5. R.H. Bricknell and D.A. Woodford, General Electric Co., Corporate Research and Development TIS Report No. 82CRD222, August 1982.
6. P.A. Siemers, ed. Third Quarterly Report.

METHOD FOR PRODUCING PREALLOYED CHROMIUM CARBIDE/NICKEL/CHROMIUM POWDERS
AND PROPERTIES OF COATINGS CREATED THEREFROM

DAVID L. HOUCK AND RICHARD F. CHENEY
GTE Products Corporation, Chemical and Metallurgical Division, Hawes Street, Towanda, PA 18848

ABSTRACT

A unique process for producing powder whose individual particles contain both chromium carbide and nickel/chromium is discussed. This process involves agglomeration of fine chromium carbide with nickel/chromium-bearing materials using a proprietary binder system. The agglomerates are subsequently sintered, melted in a plasma flame, and sized. This powder does not exhibit the segregation and inhomogeneity problems associated with conventional mechanical blends of chromium carbide and Nichrome powders. Jet Kote* coatings created from the prealloyed powders have uniform microstructures, high hardnesses, and wear characteristics superior to those of plasma-sprayed coatings from the conventional blends.

INTRODUCTION

Chromium carbide (Cr_3C_2) plus Nichrome (Ni-20Cr) powder mixtures have been used extensively for creating coatings for many wear resistance applications. Foremost among these are plasma-sprayed coatings on various turbine components [1]. They are used instead of tungsten carbide/cobalt powders at temperatures over 1000 °F because of their superior oxidation resistance. Several studies have also been conducted on the use of chromium carbide/Nichrome coatings for nuclear reactor applications [2-8]. Some of these studies have included a comparison of plasma-sprayed versus Union Carbide's detonation-gun-applied coatings [5,6]. In the Soviet Union, studies have also been conducted on the wear properties of plasma sprayed [9] and detonation-deposited [10] chromium-carbide-base coatings. These coatings are also used for ambient-temperature wear-resistance applications in which potential chemical attack negates the use of tungsten carbide/cobalt materials.

Taylor [8] studied the phase stability of these coatings in the low-oxygen environment associated with a nuclear reactor. Lai [5] concluded that $Cr_{23}C_6$ performs better in these environments. He also found that detonation-gun coatings outperform their plasma-sprayed counterparts and that a DPH (300g) of at least 650 is required in these coatings. This level of hardness has heretofore been unattainable using conventional powders with plasma spraying.

Typically, two-component mechanical mixtures of approximately 75% chromium carbide (Cr_3C_2) and 25% Nichrome are employed for plasma-spray applications. One of the inherent shortcomings of these mixtures is that they produce an inhomogeneous coating. Mixtures are also inherently difficult to handle due to their tendency to separate during transportation and spraying.

In order to overcome the separation problem, a powder must be produced in which each individual particle contains the desired components of the coating system. Reardon, et al. [11], have described a powder that

*Registered trademark of Cabot Corporation.

has been produced by cladding each of the individual chromium carbide particles with a nickel/chromium alloy. These powders contain 50% Cr_3C_2 and 50% nickel/chromium alloy. The coatings created from them, especially when applied by vacuum plasma spraying, exhibit a high hardness and excellent wear properties. However, these coatings still contain discrete, relatively large islands of chromium carbide and nickel/chromium.

GTE has developed a technique for producing powders which will yield coatings with a uniform, fine dispersion of the chromium-carbide and nickel-rich phases. The purpose of this paper is, (1) to describe the technique used for producing these prealloyed chromium carbide/nickel/chromium powders, and (2) to compare the coatings properties from Jet Kote sprayed prealloyed powders and plasma sprayed mechanical blends.

Additional comparisons of plasma sprayed coatings properties using prealloyed powders and mechanical blends have been performed. Due to text length restrictions, these results will be presented in other publications.

EXPERIMENTAL PROCEDURES

Powder Manufacturing Technique

The technique (patent applied for) used to produce the powders for the test work is as follows:

(1) Fine particles of chromium carbide, nickel and chromium are agglomerated using a proprietary binder.
(2) These agglomerates are subsequently heated in hydrogen to remove the binder and to presinter them, thus instilling enough strength for subsequent processing.
(3) The agglomerates are then prealloyed by passing them through a plasma flame and collecting them in an inert atmosphere [12,13].
(4) The prealloyed powder is classified by screening it through a 270-mesh screen. It is designated SX-195.

Coating Procedures

Plasma Spraying - A conventional mechanical mixture produced to the Aeronautical Materials Specification (AMS) 7875 was sprayed using the conditions shown in Table I. All samples, except those for metallography and microhardness, were grit blasted prior to coating. A 36 grit alumina was used at 60 psi. Grit blasting took place less than half an hour prior to coating and was followed by an alcohol wash.

Jet Kote Spraying - The GTE SX-195 powder was applied using a Jet Kote gun with the conditions shown in Table II. Prior to coating, the samples were grit blasted using 36 grit alumina at 60 psi, followed by an alcohol wash.

Wear Testing

Wear tests were performed using a pin-on-flat device [14]. The coated plates were approximately 2" x 1" x 1/4". The pin was approximately 3/8" diameter. Both the end of the pin and the 1" x 2" surface of the flat were coated with approximately 0.012" of the test materials and then ground flat and smooth prior to wear testing. The as-ground roughnesses were as follows: AMS 7875 - 17μin aa, SX-195 - 17μin aa.

TABLE I. Plasma Spray Conditions for AMS 7875.

Gun	-	Plasmadyne SG-100
Mode	-	80 kw MII
Anode	-	2083-100
Cathode	-	1083A-104
Gas Injector	-	1083A/2083-110
Arc Gas	-	Argon
Orifice	-	56
P_1 (psig)	-	150
Auxiliary Gas	-	Helium
Orifice	-	80
P_1 (psig)	-	200
Powder Gas	-	Argon
Orifice	-	77
P_1 (psig)	-	100
1251 Hopper Setting	-	1.4
Feed Rate	-	1.3 lb/hr.
Volts	-	67
Amps	-	750
Stand Off Distance	-	3 in.
Deposit Efficiency	-	Approximately 60%

TABLE II. Jet Kote spray conditions for GTE SX-195.

PRESSURES (psi)		
Hydrogen	-	95
Oxygen	-	90
POWDER FEED		
Gas	-	Nitrogen
CFH	-	35
WATER		
Pressure (psi)	-	80
Flow (gpm)	-	8
Temperature (°F)	-	125
NOZZLE	-	5/16" x 6"
STANDOFF DISTANCE (in)	-	10
SAMPLE COOLING	-	Air

The wear test consisted of applying a load of 210 psi to the pin while reciprocating (1/2 cycle/second), unlubricated sliding took place between the pin and the flat surface over a 1" distance. Tests were conducted both at room temperature and 1100 °F. The depth of scar and the weight loss of the sample were recorded after a 10-hour period.

RESULTS AND DISCUSSION

Powder Properties

Table III contains data on the chemical and physical properties of the two test materials. The AMS 7875 material contains a Ni-20Cr (Nichrome) alloy which includes some manganese, silicon, and iron. These materials are used as additions in the melting process to improve fluidity during atomization. By comparison, the prealloyed powder does not contain these elements.

TABLE III. Properties of chromium carbide/nickel/chromium test powders.

	AMS 7875			
	Cr_3C_2	Nichrome	Mixture (75/25)*	SX-195
CHEMICAL COMPOSITION (w/o)				
Carbon	12.86	0.034	10.04	10.10
Chromium	85.9	19.20	71	71
Manganese	ND**	0.8	ND	ND
Silicon	0.01	0.9	ND	ND
Iron	0.13	0.5	ND	ND
Nickel	ND	78	19	19
Oxygen	ND	ND	0.1	0.2
PARTICLE SIZE DISTRIBUTION				
Mesh				
+270	0%	0%	0.03%	0.68%
−270+325	0.70%	0%	0.07%	2.88%
−325	99.30%	100.0%	99.90%	96.44%
−20μm***	69.3%	52.3%	43.9%	28.6%
−5μm***	12.9%	2.0%	4.7%	1.2%
Mean (μm)***	ND	ND	27.2	38.3%
Hall Flow (sec/50g)	ND	ND	None	47
Bulk Density (g/cc)	ND	ND	2.56	2.48

*Analysis performed by GTE; components analyzed by manufacturer.
**ND = Not determined.
***GTE Analysis Technique - Leeds & Northrup Microtrac. Manufacturer's technique unknown.

An analysis of the particle size of the mechanical mixture was also performed by GTE using the Leeds & Northrup Microtrac. Note that this technique yields a particle size distribution somewhat different from that of the manufacturer. It was generated to give a comparison between the mechanical blend and the SX-195. Note that the SX-195 is coarser (38.3μm mean) than the AMS 7875 material (27.2μm mean).

Scanning Electron Microscopy and X-ray Energy Spectroscopy

Scanning electron micrographs (SEM) were used to show the particle sizes, shapes, and surface features of the three powders. Individual particle compositions were determined by x-ray energy spectroscopy.

Figures 1 and 2 show the morphologies of the two test powders. Figure 1 shows an SEM of the conventional AMS 7875 material. Note the variety of particle shapes present. Some chromium carbide particles appear to be smooth surfaced plates with irregular fracture surfaces. Others are less distinct and seem to be agglomerates or sponge-like particles. From experience, we find these shapes are related to variations in melting practice. Most of the Nichrome particles are moderately irregular in shape but a few are spherical. This morphology is typical of a water atomized powder.

Figure 2 shows an SEM of SX-195, a prealloyed chromium carbide/nickel/chromium powder. Note that some of the particles within this powder also exhibit a spherical geometry. This geometry is a result of the plasma processing method used in producing these materials [12,13]. As the sintered agglomerates pass through a plasma flame, some of them, especially the smaller ones, melt totally and therefore become spheroidized. X-ray energy spectroscopy analysis performed on the two spherical particles which have been circled revealed that they contain all three components of the prealloyed powder.

Coating Properties

Figures 3 and 4 show the microstructures of the coatings created using the two test materials. Figure 3 shows the nonhomogeneous, complex mixture of lamellar phases which are produced using standard AMS 7875 material. The light grey phase is Nichrome, the dark grey phase is chromium carbide. Also, a certain degree of porosity (black areas), is typically present in these microstructures. Some of this porosity is inherent in plasma sprayed coatings and some is created during sample preparation.

By contrast, Figure 4 shows the microstructure of a coating created using SX-195 with the Jet Kote process. Note that this homogeneous structure does not exhibit massive areas of either metal or chromium carbide. Note, however, that a considerable quantity of oxide phases (dark grey) are present.

Table IV contains data illustrating the properties of the coatings created with the two test materials. Note that the surface of the as-sprayed Jet Kote coating was smoother than that for the plasma sprayed AMS 7875 material, even though the starting particle size was somewhat coarser.

Note also that the hardness of the coating created using prealloyed powder is considerably higher than that of the coating created using a mechanical mixture. The AMS 7875 coating had an average hardness of 594, similar to the average of 600 created by Reardon, et al. [11] using standard Metco plasma spray equipment and AMS 7875 material. The highest hardness (931 DPH) was achieved with SX-195 using the Jet Kote process. This is higher than the hardest coatings created by both vacuum plasma [11] and the Union Carbide detonation gun [5].

The bond strength, as determined by a standard procedure (ASTM C-633-69) was very similar for both materials.

The wear data of Table IV are presented in two forms, the scar depth and the rate of wear. An evaluation of the data reveals that a one-to-one correlation does not exist between the two forms.

For several reasons, the wear rate (weight loss) is the least representative of the two methods for characterization of the wear process: (1) Some transfer of mass may occur during wear testing between the pin and the flat, (2) plastic deformation of the coating and substrate may allow for conformance of the two wear test members to occur during testing, and (3) at elevated temperature, oxidation may also contribute to the weight loss of the sample.

The scar depth can be very accurately determined. It is a true measure of the material which has been removed during the test. Therefore, for the purposes of discussion, the comparison amongst wear data will be limited to the scar depth.

The wear test data of Table IV reveal that the coating created from the prealloyed chromium carbide/nickel/chromium powder is superior to the

230

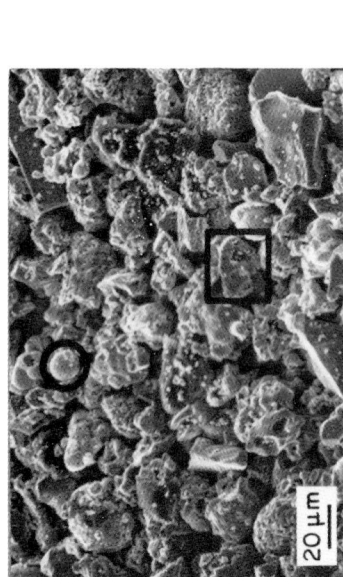

FIG. 1. Scanning electron micrograph of AMS 7875 material 75% chromium carbide (in rectangle) and 25% nickel/chromium alloy (in circle).

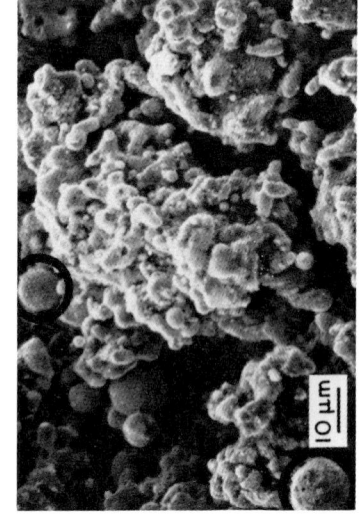

FIG. 2. SEM of GTE SX-195, a "-270 mesh", prealloyed, chromium carbide/nickel/chromium powder. The circled particles also contain all three components.

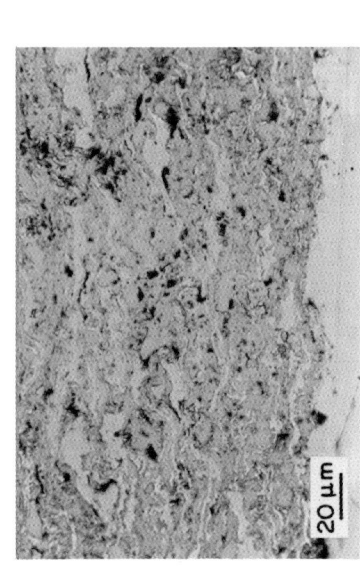

FIG. 3. Micrograph of plasma sprayed AMS 7875 powder.

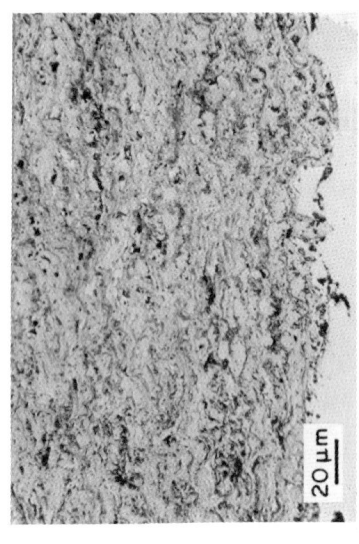

FIG. 4. Micrograph of Jet Kote sprayed GTE SX-195.

TABLE IV. Properties of chromium carbide/nickel/chromium coatings.

Property	Plasma Sprayed AMS 7875	Jet Kote Sprayed GTE SX-195
AS-SPRAYED SURFACE ROUGHNESS (μin. aa)		
Longitudinal	180-240	130-170
Transverse	150-240	130-190
HARDNESS		
Rockwell N (30 kg)	63.1	81.1
DPH (300g)		
Range	454-752	810-1089
Average (8 readings)	594	931
BOND STRENGTH		
[ASTM C-633-69 (psi)]	8990	8673
WEAR (10 hr, 5 ft./min., 210 psi)		
Room Temperature		
Scar Depth (μin)	400	50
Wear Rate (g/hr) x 10^4	4	1
Surface Roughness (μin aa)		
Before	17	18
After	32	8
1100 °F		
Scar Depth (μin)	200	90
Wear Rate (g/hr) x 10^4	2	1.5
Surface Roughness (μin aa)		
Before	17	18
After	100	4
Oxygen (w/o)	ND	7.4

one created with a conventional mechanical mixture. At room temperature, the scar depth in the Jet Kote applied SX-195 coating is 1/8 that of the scar depth in the coating created using a mechanical blend. At 1100 °F, the Jet Kote sprayed SX-195 scar depth is approximately half that of its AMS 7875 counterpart.

Inherent differences exist between the Jet Kote and plasma spray processes. Most notably, the plasma process utilizes inert gases to produce the high temperature plasma, whereas the Jet Kote process relies on the combustion of hydrogen (or other gases) as a source of heat. In addition, the residence time of the particles within the flame may vary for the two processes. These factors may lead to a difference in coating chemistry which is reflected in the difference in the hardness and wear properties.

Quantitative chemical analysis of the SX-195 coatings showed that they contain 7.4% oxygen. Assuming the oxygen to be present as Cr_2O_3, it would contribute 23 weight percent oxide phase to the microstructure. This, of course, could contribute to the differences in hardness and wear characteristics between the plasma and Jet Kote applied coatings. Investigation of the oxygen levels in plasma applied coatings is presently being conducted.

Both the Jet Kote and plasma process can be modified to yield lower oxygen content coatings. Further investigation of the chemistry and properties of modified coatings is also being conducted.

The roughness of the wear-tested surfaces reflects the degree to which damage occurred during testing. An increase in roughness usually indicates that damage has occurred. Unchanged or decreased roughness indicates no damage has occurred or that polishing has taken place. Note that the AMS 7875 coatings were the only samples in which a significant increase in surface roughness occurred. This may be a result of the higher initial level of porosity in these coatings. It could also be a result of other microstructural differences.

CONCLUSIONS

This study shows that:

(1) Uniform, prealloyed powders of chromium carbide/nickel/chromium can be produced by a unique process involving agglomeration and plasma densification.
(2) These powders are sprayable by the Jet Kote* process to produce high-density, homogeneous coatings.
(3) The Jet Kote coatings produced from prealloyed powders are much harder than those plasma sprayed using a mechanical mixture (AMS 7875).
(4) After wear testing, the wear scar depth of prealloyed Jet Kote coatings ranged from 1/2 to 1/8 of the wear scar depth of conventional plasma sprayed coatings.
(5) Jet Kote applied prealloyed coatings contain 7.4% oxygen. These high levels may contribute to their higher hardness and better wear characteristics.

RECOMMENDATIONS

Further investigation of the chemistry and properties of prealloyed chromium carbide/nickel/chromium coatings is required. In particular, the effect of coating chemistry on the ductility, fatigue, and corrosion/oxidation of these materials should be determined. Also, a direct comparison of two coatings techniques using the same prealloyed powder should be conducted. The results of ongoing investigations will be presented in future publications.

ACKNOWLEDGEMENTS

Sincere thanks are due Jet Kote (previously a Division of Browning Engineering, now owned by Cabot Corporation) for preparation of samples for this study. Also, Rensselaer Polytechnic Institute's Tribology Laboratories are to be commended for their performance and evaluation of the wear tests. The efforts of our colleagues at GTE (especially Barb Brown, Bob Burke, Leonid Dorfman, John Gordon, Dave Harrigan, Dave Hunsinger and John Miller) are greatly appreciated.

REFERENCES

1. R. H. Wedge, A. V. Eaves, Paper 19, The Ninth International Thermal Spray Conference, The Hague, May 19-23, 1980.
2. M. Villat, Reprint from Sulzer Research, Number 1974.
3. E. Schwarz, Paper 21, The Ninth International Thermal Spray Conference, The Hague, May 19-23, 1980.
4. G. Y. Lai, Thin Solid Films, 64, 271-280 (1979).
5. G. Y. Lai, Presented at The International Conference on Metallurgical Coatings, San Francisco, California, April 3-7, 1978.
6. C. C. Li, Presented at The International Conference on Metallurgical

*Registered trademark of Cabot Corporation.

Coatings, San Francisco, California, April 21-25, 1980.
7. T. A. Wolfla and R. N. Johnson, Journal of Vacuum Science & Technology, 4, 777-783 (1975).
8. T. A. Taylor, Journal of Vacuum Science & Technology, 4, 790-794 (1975).
9. Y. S. Boriscov, et al., Poroshkovaya Metallurgiya, 10, 81-84 (1978).
10. Y. G. Tkachenko, et al., Poroshkovaya Metallurgiya, 8, 62-65 (1978).
11. J. D. Reardon, et al., Thin Solid Films, 3, 345-351 (1981).
12. U. S. Patent 3,974,245, R. F. Cheney, C. L. Moscatello, and F. J. Mower, Assignee GTE Sylvania Incorporated.
13. U. S. Patent 3,909,241. R. F. Cheney, C. L. Moscatello and F. J. Mower, Assignee GTE Sylvania Incorporated.
14. E. T. Finkin, S. J. Calabrese, M. B. Peterson, Journal of American Society of Lubrication Engineering, 29, 197-204 (1972).

MICROSTRUCTURE AND PHASE COMPOSITION OF SPUTTER-DEPOSITED ZIRCONIA-YTTRIA FILMS

R.W. KNOLL AND E.R. BRADLEY
Pacific Northwest Laboratory, Richland, Washington 99352.

ABSTRACT

Thin ZrO_2-Y_2O_3 coatings ranging in composition from 3 to 15 mole % Y_2O_3 were produced by rf sputter deposition. This composition range spanned the region on the equilibrium ZrO_2-Y_2O_3 phase diagram corresponding to partially stabilized zirconia (a mixture of tetragonal ZrO_2 and cubic solid solution). Microstructural characteristics and crystalline phase composition of as-deposited and heat treated films (1100°C and 1500°C) were determined by transmission electron microscopy (TEM) and by x-ray diffraction (XRD). Effects of substrate bias (0 ~ 250 volts), which induced ion bombardment of the film during growth, were also studied. The as-deposited ZrO_2-Y_2O_3 films were single phase over the composition range studied, and XRD data indicated considerable local atomic disorder in the lattice. Films produced at low bias contained intergranular voids, pronounced columnar growth, and porosity between columns. At high bias, the microstructure was denser, and films contained high compressive stress. After heat treatment, all deposits remained single phase, therefore a microstructure and precipitate distribution characteristic of toughened, partially stabilized zirconia appears to be difficult to achieve in vapor deposited zirconia coatings.

INTRODUCTION

Materials produced by physical vapor deposition (PVD) can differ greatly in microstructure and crystalline phase composition compared to those produced by high temperature bulk processes. PVD materials are formed by direct condensation of atoms from the vapor phase, typically at temperatures below 400°C. Because of the mixing inherent in vaporization, the distribution of elements at the growth surface of the deposit is initially homogeneous. Subsequent atomic ordering, crystallization, and precipitation of second phases within the deposit depend on the atomic mobility at the growth surface,[1] and on self diffusion within the ceramic. Although few studies of multicomponent PVD refractory oxides can be found in the literature, data on refractory oxide crystals[2] indicate very low bulk diffusivities at temperature below ~ 800°C. Therefore, grain growth and the formation of equilibrium phases are restricted, and a homogeneous, nonequilibrium microstructure is expected in a vapor-deposited refractory oxide.

This paper describes the microstructure and crystalline structure of sputter-deposited ZrO_2-Y_2O_3 films containing 3 to 15 mole % Y_2O_3. Changes in the microstructure and phase composition due to heat treatment at 1100°C and 1500°C are also described. The films were characterized primarily by transmission electron microscopy (TEM) and by x-ray diffraction (XRD) in the plane of the deposit. In addition to the effect of Y_2O_3 content, the influence of rf substrate bias on the deposit characteristics was studied. An applied negative bias attracts positive ions from the plasma, causing low energy ion bombardment of the growth surface of the deposit. Previous studies, particularly on metals,[3] have shown that ion bombardment during growth increases atomic mobility and causes resputtering at the growth surface, and therefore can strongly affect deposit composition, microstructure, and defect structure. Substrate bias was varied from 0 to 250 volts rms in this experiment, primarily to determine whether ion bombardment enhances the formation of second phases within the deposit and produces a more equilibrium phase composition.

Characteristics of zirconia films and coatings are of practical interest, and have been studied previously because of applications that relate to the protective[4-9] and electro-chemical[10-13] properties of zirconia. Compared with previous studies of sputter-deposited zirconia[8-14], in this paper the crystal structure and microstructure of the deposits were examined in much greater detail, particularly with regard to effects of alloy composition and annealing. Less attention was paid to electrical and mechanical film properties, and to the influence of various deposition parameters. A more basic motivation for this work was comparison of the known characteristics of bulk-processed zirconia (where high processing temperatures allow diffusion and an approach toward equilibrium) with the characteristics of non-equilibrium vapor deposited zirconia. The wealth of existing data on bulk processed zirconia (e.g. Ref [15]), including the equilibrium phase diagrams, serve as a frame of reference for the present study.

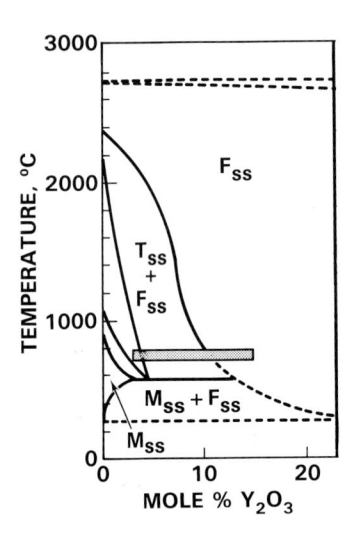

Fig. 1. ZrO_2-rich end of the equilibrium phase diagram for the zirconia-yttria system, redrawn from Stubican, et. al. [17]. The shaded bar indicates the composition range studied in this paper.

The Zirconia-Yttria System

Oxides such as Y_2O_3, CaO, or MgO are alloyed with ZrO_2 to inhibit the destructive phase transformation that occurs in pure ZrO_2 near 1100°C (tetragonal to monoclinic, accompanied by a 10% volume change), and to alter the electrical and chemical properties of ZrO_2 [16]. Studies of the equilibrium phases present in bulk-processed ZrO_2-Y_2O_3 [17] have shown that addition of Y_2O_3 to ZrO_2 produces a high-temperature cubic solid solution (stabilized zirconia), or a mixture of cubic + tetragonal phases (partially stabilized zirconia), depending on Y_2O_3 content (Fig. 1). Alloying with Y_2O_3 reduces the transition temperature of the high temperature phases to below 600°C, where atomic mobility is low, therefore these phases can be easily retained at low temperatures. Extensive research on bulk zirconia has shown that after heat treatment to form an appropriate microstructure, partially-stabilized zirconia is strengthened and toughened beyond levels typically associated with a ceramic.[15, 18, 19] A fine dispersion of metastable tetragonal precipitates in the cubic matrix can toughen the zirconia by inhibiting the growth of microcracks. Cracks impinging upon the precipitate transform it to a monoclinic structure, absorbing the energy of fracture. Because of the importance of microstructure on mechanical properties, a point of particular interest in this study was whether second phase particles could be easily precipitated in the vapor-deposited zirconia.

EXPERIMENTAL PROCEDURE

Deposition

The zirconia coatings were produced by rf diode sputter-deposition in a system that has been described in detail elsewhere.[14, 20] Sputtering targets consisted of 15 cm-diameter hot-pressed zirconia discs nominally containing 3, 6, 8, or 15 mole % Y_2O_3. The plasma was composed of 80% high purity Ar and 20% O_2 at a total pressure of 2.7 Pa (20 mTorr). Coatings were deposited at a rate of ~ 0.6 μm/hr using a target power density of 3W/cm^2, and a substrate-target spacing of 5 cm. To bias the substrates during deposition, power was applied to the substrate holder using a power splitting and matching network. Coatings of each composition were deposited with rf substrate bias of -50, -100, -200, and -250 volts, and also with the substrate grounded.

Several types of substrates and coating thicknesses were used, depending on the intended heat treatment and method of analysis. Coatings that were analyzed by x-ray diffraction in the as-deposited condition or following heat treatment at 1100°C, were 1 ~ 2 μm thick and were deposited on polished fused-SiO_2, sapphire, and Ni discs. All of the metal substrates were polished with alumina abrasive to a surface roughness of ~ 100 nm, while the SiO_2 substrates had a surface finish of better than 3 nm. Coatings heat treated at 1500°C were 2-3 μm thick, deposited on polished sapphire substrates. Finally, films for TEM analysis were 0.1-0.2 μm thick, on polished 3 mm-diameter Ni or 304 stainless steel substrates. Thin films were used here to simplify TEM specimen preparation, and these films were deposited simultaneously with the thicker coatings by rotating a shield over the TEM substrates after the required thickness was attained. The substrates rested on a water-cooled Zr metal base during deposition, but were in poor thermal contact with the base. Under the deposition conditions used, substrate temperatures were 250°C ~ 300°C due to heating by radiation and secondary electrons from the target.

Characterization

Data on crystal structure and crystallographic texture were obtained with a standard Norelco diffractometer using CuKα x-rays. The diffractometer was stepped over a 2θ range of 20° to 80° with 0.05 degree increments and 40 sec counting periods. Images of the microstructure were obtained with a Phillips 400 transmission electron microscope (TEM) operating at 120 KeV. Electron diffraction and dark field analysis were used to search for the presence of second phases. Specimens were prepared for TEM analysis by electrolytically thinning the metal substrate from the uncoated side until perforation occurred, in a twin-jet electropolisher. The electrolyte (5% perchloric acid in methanol) attacked the metal but not the zirconia. In a successfully thinned specimen, the zirconia film remained intact over the perforation in the substrate. Final thinning of the film, if any, was done in an ion mill with a 6 KeV Ar$^+$ beam incident at 20° to the plane of the film.

Surface topography of some coatings was examined in a scanning electron microscope (SEM), with a thin C layer evaporated on the surface to prevent charging. Deposit composition was measured by energy dispersive x-ray spectroscopy in the SEM, and an impurity content of about 1 mole % (mainly CaO) was found.

Selected specimens of each composition were heat treated in vacuum (~ 6 X 10^{-4} Pa) at 1100°C (for 20 hrs) and at 1500°C (for 2½ hrs). All heat treated deposits were examined by x-ray diffraction, and the 1100°C specimens were examined by TEM also. An attempt was made to heat treat TEM specimens at 1500°C, however the coatings spalled from their Mo substrates during annealing. All coatings as-deposited were highly transparent in the visible spectral range (thus they were stoichiometric or nearly so), and they remained transparent after heat treatment.

RESULTS AND DISCUSSION

Crystal Structure

The zirconia-yttria deposits were single phase and of the same crystal structure, regardless of yttria content, substrate bias (V_B), or heat treatment. However, the crystallographic texture of the films did depend on those variables. X-ray diffraction plots typical of as-deposited, unbiased films are presented in Fig. 2 and those of annealed films are given in Fig. 3. With the exception of several weak, extra lines in the XRD plots, both the XRD and electron diffraction data fit an fcc crystal structure with a lattice parameter (a_o) of about 0.515 nm for the as-deposited (AD) films, and 0.512 nm for the annealed films. Interplanar spacings (d_{hkl}) corresponding to the principal x-ray diffraction peaks are listed in Table I.

The observed structure is that of the high temperature cubic phase on the equilibrium phase diagram in Fig. 1. This phase, rather than the low temperature equilibrium phases near the deposition temperature, was quenched in during deposition. The observed cubic structure of the AD films agrees with previous studies[11-14] of sputter-deposited zirconia films alloyed with yttria or calcia in the range of 8-16 mole %. However, these results differ from those of Scott[21], who determined the structure of zirconia-yttria alloys that were arc melted and quenched. Scott observed a metastable tetragonal structure in the composition range of 3 ~ 8 mole % Y_2O_3, and the cubic structure otherwise.

TABLE I. Room temperature lattice spacings (nanometers) corresponding to the principal observed x-ray diffraction peaks in as-deposited (AD) and heat treated (1100°C and 1500°C) zirconia-yttria films. Interplanar spacings calculated for an fcc structure with lattice parameter 0.512 nm are given in parentheses for reference.

hkl	111 (.2956)			200 (.2560)			202 (.1810)			113 (.1544)			400 (.1280)		
%Y_2O_3	AD	1100	1500	AD	1100	1500	AD	1100	1500	AD	1100	1500	AD	1100	1500
3	.299	.294	a	.258	.256	a	.182	.181	a	.156	.155	a	n.o.	.1286	a
6	.300	.295*	.296	.258	.254*	.256	.183	.181*	.182	.156	.154*	.155	n.o.	.1277*	.1286
8	.298	.295	n.o.	.258	.256	.256	.183	.181	.181	.156	.155	.155	.130	.1286	.1286
15	.296	.296	.296	.257	.256	.256	.182	.181	.181	.156	n.o.	.155	.129	.1286	.1286

a) No specimen of this composition was heat treated in 1500°C.
*) This film blistered and partially spalled from the substrate during heat treatment.

239

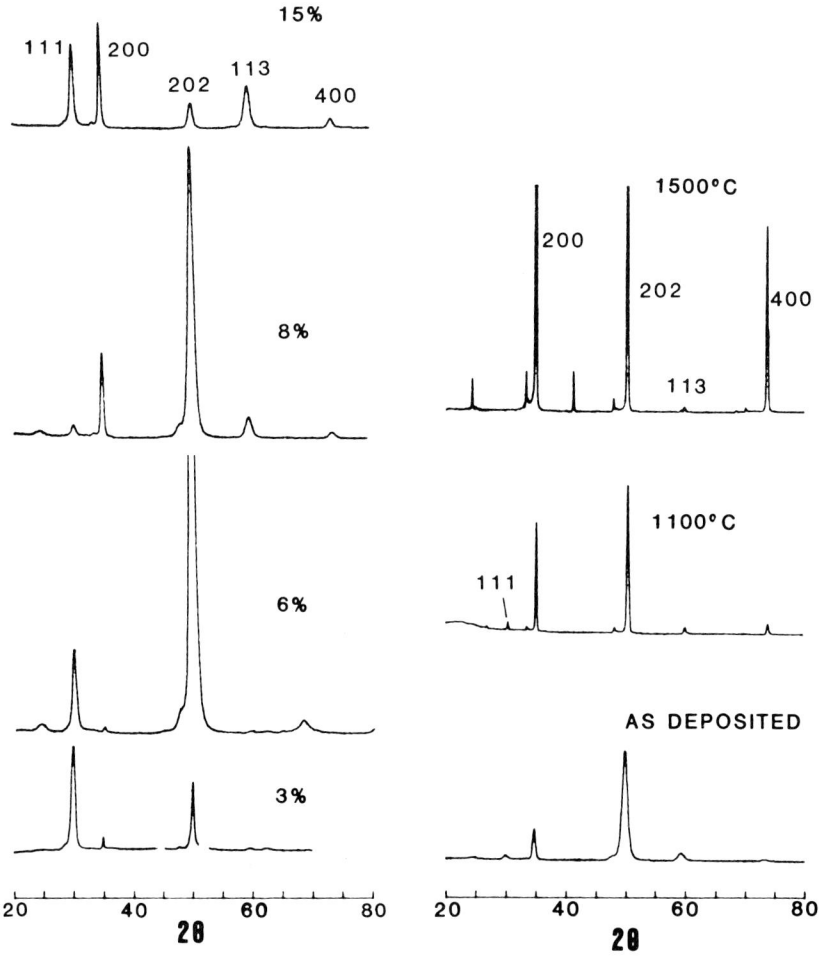

Fig. 2. X-ray diffractometer plots for as-deposited, unbiased zirconia films containing 3,6,8 and 15 mole % Y_2O_3, plotted using identical vertical scales. The 6%-15% films were 2 μm thick, on sapphire substrates. The 3% film, about 1 μm thick, was deposited on Ni, and the Ni peaks have been removed from the plot. Labeled peaks correspond to cubic structure with lattice parameter of about .516 nm.

Fig. 3. Comparison between the x-ray diffractometer plots of as-deposited and annealed zirconia - 8 mole % yttria deposits. Labeled peaks correspond to the cubic zirconia structure. With the exception of the peak at 24.5°, the extra peaks particularly evident in the 1500°C specimen correspond to an unidentified bcc phase.

Extra lines indicative of the tetragonal solid solution (T_{ss})[22] or monoclinic phase were not present in the diffraction data from annealed (1100 or 1500°C) or AD specimens. Careful TEM analysis of the AD films and of those annealed at 1100°C revealed no T_{ss} particles, although these precipitates should have been resolvable in the cubic matrix[19,23] if present in the 3 to 8 mole % Y_2O_3 deposits. The absence of tetragonal phase in the annealed deposits was somewhat surprising in view of a recent study by Marder, et. al.[23] on precipitation in bulk ZrO_2-CaO specimens (the phase diagrams of the ZrO_2-CaO and ZrO_2-Y_2O_3 systems are similar). These workers found a dispersion of small T_{ss} particles in ZrO_2-8 mole % CaO that had been solution-annealed at 1850°C and then cooled rapidly to 1000°C.

Annealing at 1500°C markedly increased the XRD line intensities (Fig. 4), but did not significantly alter the interplanar spacings of the cubic phase, relative to the 1100° anneal. Several extra lines that were weak or nonexistent in the XRD data from the as-deposited films became distinct during the 1100° and 1500° anneals (some extra lines on the 1500°C plot in Fig. 4 result from the Al_2O_3 substrate). One peak represented the (110) planes of the cubic lattice, however this peak is forbidden for an fcc lattice. A weak set of lines represented a bcc structure with $a_o \sim 0.536$ nm. The origin of these lines - either impurity or nonequilibrium phase - was not determined.

Diffraction lines from the AD films were relatively broad and of low intensity indicating considerable imperfection in the crystalline structure of the AD films. There was some variability in the interplanar spacings, particularly in d_{111} and d_{200}, which decreased somewhat as Y_2O_3 content increased. Following the 1100°C anneal, the diffraction lines narrowed, and increased moderately in intensity, as shown in Fig. 3 for ZrO_2-8 mole % Y_2O_3. Based on interplanar spacings other than d_{111}, the lattice parameters of the films decreased to 0.512 nm for all compositions after annealing. The relaxation in a_o upon annealing, and the decrease in the width of the diffraction lines (beyond that expected from the observed grain growth) suggest a high defect density and lattice distortion in the as-deposited material. Using the observed grain size (from TEM data) in the Scherrer equation, which relates the width of the diffraction peak to the mean grain size in the absence of lattice imperfections, the observed line width is twice that predicted for a perfect lattice.

Crystallographic texture of the films varied with type of substrate, yttria content, heat treatment, and bias. On SiO_2 and sapphire substrates, the relative intensity of the 202 peak decreased as yttria increased, as shown in Fig. 2 for the 6-15% specimens. Preferred orientation of films on Ni substrates followed no pattern, and must have been influenced by the texture of the Ni sheet itself.

Ion bombardment of biased films during deposition did not affect crystal structure, although film microstructure, stress, and crystallographic texture changed as V_B increased. Preferred crystallographic orientation shifted from (202) at zero bias to (111) or (200) at 50-100 V bias, then returned to (202) at higher bias voltages. At 150 V and above, the total intensity of the diffraction lines began to decrease and the lines broadened further, suggesting a disordering effect of ion bombardment. On the other hand, ion bombardment during growth has previously been found to increase the crystallinity and the formation of second phases in some sputter-deposited metal alloys[24]. Substrate bias did not alter Y_2O_3 content of the film, in contrast to a previous study of sputter-deposited ZrO_2-CaO films[20], where Ca content decreased as V_B increased.

Microstructure

Microstructural characteristics such as grain size or the presence of growth defects were more sensitive than was crystal structure to the experimental variables. Examples of the microstructure of unannealed films

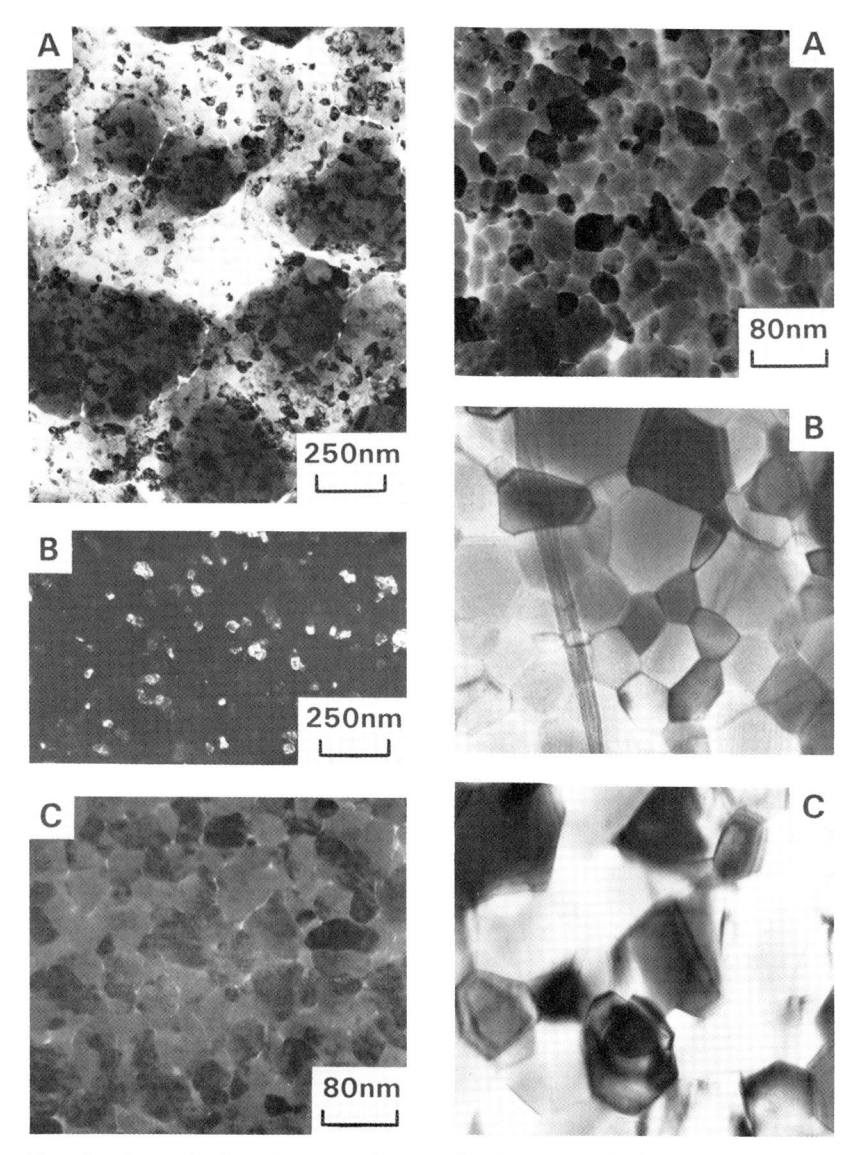

Fig. 4. Transmission electron micrographs of as-deposited, unbiased films. (A) 8% yttria, showing columnar growth and cracks between columns, (B) dark field micrograph of same film, showing individual grains, and (C) 3% yttria films, with typical intergranular voids.

Fig. 5. Transmission electron micrographs typical of films annealed at 1100° for 20 hrs. (A) 15% yttria film, still containing intergranular voids, (B) 8% yttria, and (C) 3% yttria. Note the increase in grain size with decreasing yttria content. (All micrographs at same mag.).

deposited without bias are shown in TEM micrographs in Fig. 4. Both individual grains and columnar growth structures approximately an order of magnitude larger than the grains are evident in the micrograph of ZrO_2-8 mole % Y_2O_3 in Fig. 4(a). Grain size is clearly shown in the dark field micrograph of the same film in Fig. 4(b). Grain size in the AD films was ~ 40 nm, and was apparently not sensitive to yttria content. The columnar growth structures are well known features of vapor deposited materials[9, 12], but are noted here because of the porosity at the intercolumnar boundaries. SEM examination showed that columnar growth structures were more pronounced on the metal substrates, which had a rougher surface finish. Intergranular voids, evident in Fig. 4(c), also formed where grain corners failed to meet in the unbiased deposits. Film porosity is an important factor in protective and electrochemical applications. Croset, et. al.[11] detected porosity in ZrO_2-Y_2O_3 thin-film sensors using an electro-chemical technique, but did not characterize the porosity.

In biased films, grains were more densely packed and appeared more strained in the TEM images. On a macroscopic scale, noticeable bowing of the Ni substrates occurred, indicating a high compressive stress in the bias-sputtered films. Quantitative measurements of film stress are in progress. Annealing of the unbiased films produced grain growth and healing of the intergranular voids. TEM micrographs of 3, 8, and 15% Y_2O_3 specimens annealed at 1100°C for 20 hrs. are presented in Fig. 5. The increase in grain size during annealing was greater in deposits of lower yttria content. Grain size was ~ 100 nm in the 3% specimens and only ~ 45 nm in the 15% specimens. Intergranular voids were still present in the 15% yttria films, and the grain morphology resembled that in as-deposited films. This result suggests that increased yttria content inhibited atomic mobility.

SUMMARY AND CONCLUSIONS

Characteristics of the crystalline structure and microstructure of sputter-deposited ZrO_2-Y_2O_3 films containing 3 to 15 mole % Y_2O_3 were described. The high temperature cubic structure was quenched into the films during deposition, and neither yttria content nor substrate bias voltage affected the structure. During annealing for moderate times at temperatures within the tetragonal + cubic field on the phase diagram, tetragonal particles did not precipitate from the cubic phase and thus the zirconia is not "partially stabilized." Because of the stability of the as-deposited phase, improved mechanical properties resulting from the presence of tetragonal precipitates in zirconia appear to be difficult to attain in the vapor deposited zirconia, except by prolonged heat treatment. Generalization of these results, i.e. correlation between the equilibrium phase diagram of a multicomponent ceramic and the phases expected in a vapor deposited material, would require further study of other ceramic systems. In the microstructural studies, intergranular and intercolumnar porosity was observed. Increased substrate bias reduced this porosity, but also introduced high compressive stress into the films. This trade-off may be important in practical applications of zirconia deposits.

ACKNOWLEDGEMENTS

The authors acknowledge H.E. Kjarmo for obtaining the x-ray diffraction data and for performing SEM analysis; W.T. Pawlewicz and J.T. Prater for valuable discussions; and I.B. Mann, R.F. Stratton, and D.D. Hays for other contributions. This work was supported by the Materials Sciences Division of the Office of Basic Energy Sciences, U.S. Department of Energy.

REFERENCES

1. K.L. Chopra, *Thin Film Phenomena* (McGraw-Hill, 1969) Ch. IV.
2. W.D. Kingery, H.K. Bowen and D.R. Uhlmann, *Introduction to Ceramics* (John Wiley and Sons, 1976, second edition) Ch. 6.
3. D.M. Mattox and F.J. Kominiak, J. Vac. Sci. Tech. 9 (1972) 528-532.
4. R.A. Miller, J.L. Smialek, and R.G. Garlick in Ref. 15, 241-253.
5. R.J. Bratton and S.K. Lau in Ref. 15, 226-240.
6. R.C. Bill, J. Sovey and G.P. Allen, Thin Solid Films 84 (1981) 95-104.
7. R.E. Benner and A.S. Nagelberg, Thin Solid Films 84 (1981) 89-94.
8. P.G. Valentine and R.D. Meier, NASA-CR-165126, (1980).
9. J.W. Patten, M.A. Bayne, D.D. Hays, R.W. Moss, E.D. McClanahan, and J.W. Fairbanks, Thin Solid Films 64 (1979) 337-343.
10. T.L. Barr, L.B. Welsh, F.R. Szofran, J.E. Greene and R.E. Klinger, J. Vac. Sci. Tech. 15 (1978) 341.
11. M. Croset, P. Schnell, G. Velasco and J. Siejka, J. Vac. Sci. Tech. 14 (1977) 777-781 and J. Appl. Phys. 48 (1977) 775-780.
12. J.E. Greene, R.E. Klinger, L.B. Welsh and F.R. Szofran, J. Vac. Sci. Tech. 14 (1977) 177-180.
13. J.E. Greene, C.E. Wickersham, J.L. Zilko, L.B. Welsh, and F.R. Szofran, J. Vac. Sci. Tech. 13 (1976) 72-75.
14. W.T. Pawlewicz and D.D. Hays, Thin Solid Films 94 (1982) 31.
15. A.H. Heuer and L.W. Hobbs (editors), Advances in Ceramics Vol 3, The Science and Technology of Zirconia (American Ceramic Society, 1981).
16. E.C. Subbarao in Ref. 15, 1-24.
17. V.S. Stubican, R.C. Hink, and S.P. Ray, J. Amer. Ceram. Soc. 61 (1978) 17-21.
18. R.C. Garvie, R.H. Hannink, and R.T. Pascoe, Nature 258 (1975) 703-704.
19. G.K. Bansal and A.H. Heuer, J. Amer. Ceram. Soc. 58 (1975) 235-238.
20. W.T. Pawlewicz, J. Appl. Phys. 49 (1978) 5595-5601.
21. H.G. Scott, J. Mater. Sci. 10 (1975) 1527-1535.
22. T.K. Gupta, J.H. Bechtold, R.C. Kuznicki, L.H. Cadoff, and B.R. Rossing, J. Mater. Sci. 12 (1977) 2421-2426.
23. J.M. Marder, T.E. Mitchell and A.H. Heuer, Acta. Metall. 31 (1983) 387-395.
24. R.W. Knoll and E.D. McClanahan, J. Vac. Sci. Tech. A-1 (1983) 271-274.

MULTIPLE ARC DISCHARGES FOR METALLURGICAL REDUCTION OR METAL MELTING

J.E.HARRY, R.KNIGHT
Department of Electronic and Electrical Engineering, University of Technology, Loughborough, Leicestershire, LE11 3TU., U.K.

ABSTRACT

Multiple arc discharges enable more than one arc to be operated in close proximity supplied from the same supply. This enables reduced individual electrode current to be used. Multiple electrodes may be used in d.c. arc furnaces and d.c. plasma furnaces for melting, liquid-phase smelting and particulate phase processes. Non-consumable multiple electrodes may be used to replace graphite electrodes in arc furnaces. Other applications include high temperature gas heating and in flight plasma reactions.

INTRODUCTION

Multiple discharges are a number of electric discharges (glow or arc) operated from the same power source in close proximity to each other so that they interact.

The use of multiple electric discharges allows the constraints on small volume and the high current loadings of conventional arc and plasma processes to be overcome since several discharges can be operated in close proximity to provide a large volume of ionised gas suitable for gas heating or for particulate phase reactions.

Multiple arcs also enable; (i) a higher total current or (ii) reduced electrode loadings to be used and are applicable to plasma reactors, operation of multiple plasma torches from a single supply and high current non-consumable electrodes for use in metal melting and submerged arc furnaces.

OPERATION OF MULTIPLE DISCHARGES

Normally when two or more electric discharges are operated from the same supply separate transformers and rectifiers are used for each discharge for instance in arc welding. In other applications separate stabilising impedances are used, for example in fluorescent and high pressure discharge lamps and welding while in the case of the a.c. arc furnace each electrode is supplied from a separate winding of the supply transformer, the current flowing from one electrode to the common bath and returning through one or both of the other two electrodes. The electric arcs are normally separated by an appreciable distance (approximately 1m) compared with the distance between the electrodes so that no significant electromagnetic interaction occurs between arc columns.

A simple electric circuit for the operation of two electric arcs is shown in Fig.1 in which each arc is supplied from a different power source and separately stabilised. Fig. 2(i) and (ii) show two arcs operated from the same supply with separately stabilising resistors on one side of the

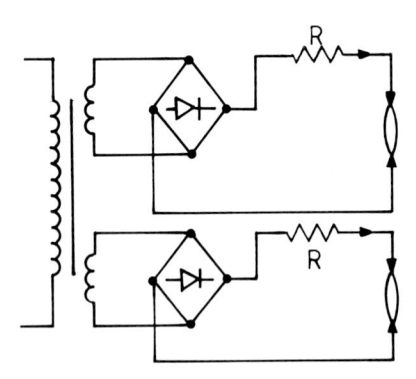

Fig 1. Operation of two arcs from the separate power supplies

discharge and a common electrode on the other. If the arcs are operated sufficiently far apart so that the electromagnetic forces between them are small (Fig.2(i)) both discharges will be stable and exist separately, however, if they are brought together so that the separation is comparable with the arc length then the arc columns will coalesce to form a three-root discharge with a common root (Fig.2(ii)) If now four individually stabilised electrodes are used then the discharges coalesce between the electrodes as the arcs are moved together to form a stable four root discharge (Fig.2(iii)). The direction of the current flow in adjacent electrodes is the same (parallel discharges); it is also possible to operate with the current flow in adjacent discharges in opposite directions (Fig.2(iv)) (anti-parallel discharges). In the anti-parallel configuration as the discharges are brought together they repel each other so that despite the small distance between adjacent electrodes the discharges never coalesce, however in some circumstances breakdown occurs between adjacent electrodes of opposite polarity completing the series circuits through the power supplies.

The equivalent circuit of two parallel coalesced discharges is shown in Fig. 3. The individual discharges may be considered in terms of the resistance of the cathode regions corresponding to r_1, r_2 and the anode regions corresponding to r_3, r_4 with r_o corresponding to the coalesced region between the arcs. R_1, R_2, R_3, R_4 are the stabilising resistors. This circuit can be analysed in the same way as an unbalanced bridge[1].

MULTIPLE ARC PROCESSES

Multiple electric discharges can be used in a number of different configurations to carry out various processes ranging from particulate phase reactions to gas heating.

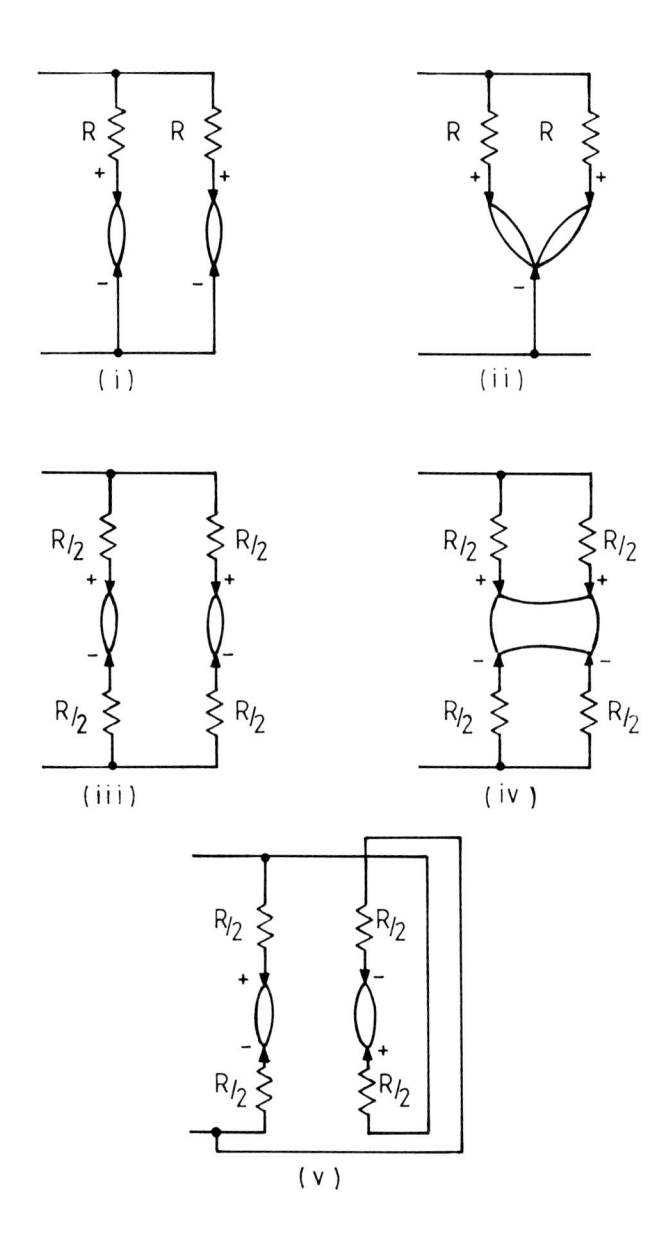

Fig. 2 Modes of operation of two arcs from a single power supply
    (i)   four-root discharge with common electrode
    (ii) three-root discharge
    (iii) four-root separate discharges
    (iv) four-root parallel coalesced discharge
    (v)   four-root anti-parallel discharge

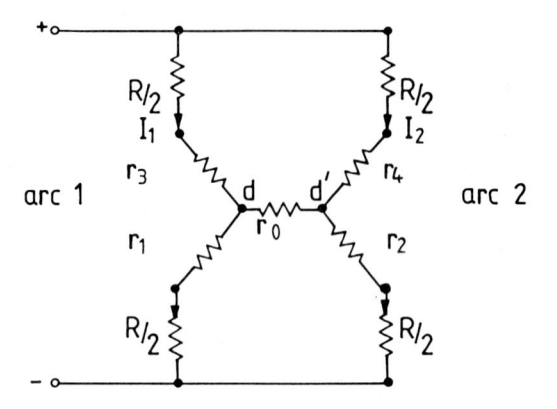

Fig. 3. Equivalent circuit of coalescing arcs with individual stabilising resistors

Multiple Electrode Plasma Reactor

The radial arc reactor is ideally suited to particulate phase reactions where a large volume of uniformly heated ionised gas is required or for heating gases. A multiple electrode plasma reactor using six pairs of radially mounted graphite electrodes is shown in Fig.4(1)[2], [3]. A stable discharge with a high degree of uniformity of intensity distribution (± 10%) over a region up to 50mm diameter can be achieved with the configuration shown in Fig. 4(i)[4].

Multiple Electrodes in d.c. and a.c. Arc Melting and Smelting Furnaces

The use of d.c. for arc furnaces offers a number of advantages including reduced electrode wear, and greater arc stability. A number of d.c. arc furnaces have been developed and more recently d.c. has been used in electric smelting furnaces. Two different approaches have been used, (i) based on conversion of existing furnaces to d.c. using a single central electrode operated in the same way as a conventional a.c. furnace and (ii) the replacement of the arc furnace electrodes by one or more d.c. plasma torches mounted in the roof or side walls of the furnace.

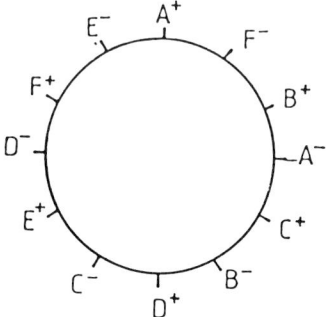

Fig. 4. Radial arc reactor and electrode configuration
(i) Reactor
(ii) Electrode configuration

The application of multiple discharges to d.c. arc furnaces using graphite electrodes enables more than one electrode to be used resulting in increased current carrying capacity or reduced electrode loading as well as more uniform heating of the molten bath. Operation is achieved by separately stabilising each electrode with inductive reactance in the secondary circuit of the transformer with individual rectifiers for each electrode. (Fig.2 (i)).

In the case of the a.c. furnace further advantages are possible resulting from better utilisation of the electrode material and reduced inductance, however the additional mechanical complexity is a major disadvantage.

Non-Consumable Multiple Electrode for Metal Melting

The cost of graphite electrodes used in arc furnaces for the manufacture of steel from scrap contributes about 10% to the overall cost of the steel. Existing water cooled electrodes are not capable of carrying the high currents required however the use of separately stabilised multiple electrodes enables very high currents to be used.

A compact multiple electrode assembly in which three tungsten cathodes are mounted in a common water cooled copper body is shown in Fig. 5 [5]. The insulated nozzle enables a pilot arc to be ignited which is subsequently used to transfer the main arc to the bath without contact with the anode. Operation with a common nozzle enables the design of the electrode assembly to be simplified and is made possible by separately supplying each pilot arc since only a low power is required in the pilot arc circuit. The arcs are spatially stabilised on the axes of the electrodes by vortex flows of argon and coalesce in front of the nozzle to produce a common anode root on the hearth of the furnace. The arcs are ignited simultaneously using a single hf ignition unit with separately stabilised outputs. The device is only 100 mm in diameter and has been operated at currents up to 800A per electrode, 2400A total current with arc voltages up to 60V at arc lengths up to 125mm dissipating powers of the order of 140kW (Fig. 6). This demonstrates the feasibility of a compact high current (30kA) non-consumable electrode since conventional plasma torches incorporating single tungsten cathodes can be operated at currents up to 10kA.

High current d.c. non-consumable electrodes incorporating multiple cathodes may be used in a number of existing processes. In d.c. arc furnaces their use offers advantages in terms of reduced operating costs, operation in an inert atmosphere and increased stability.

Submerged arc smelting furnaces with typical shell diameters of 9m use Söderberg electrodes up to 1.5m diameter. The furnace diameter is determined by the size of the electrodes and the spacing required between them to prevent interaction between the arcs. Replacement of these electrodes by multiple cathode assemblies only 100mm in diameter would enable the diameter of the furnace to be reduced to approximately 5m resulting in increased power density and throughput and reduced capital costs.

Plasma Furnace with Multiple Torches

Multiple plasma torches operating from a single supply can be used in plasma furnaces to provide a larger reaction zone and a more uniform energy distribution in the furnace together with a higher total power input. A pilot-plant plasma reactor rated at 0.1mW, incorporating three d.c. plasma torches has been developed for high temperature slag-melt phase reduction processes.[6]. The principle of operation is shown in Fig. 7. The power supply (Fig. 8) features a single transformer with low leakage reactance, to minimise the shared impedance, supplying separate rectifiiers for each torch. Individual stabilising impedances are connected at the input to the rectifiers. The arc current is controlled by saturable reactors. A single high frequency high voltage ignition unit is connected in parallel with the power supply and separately coupled to each torch.

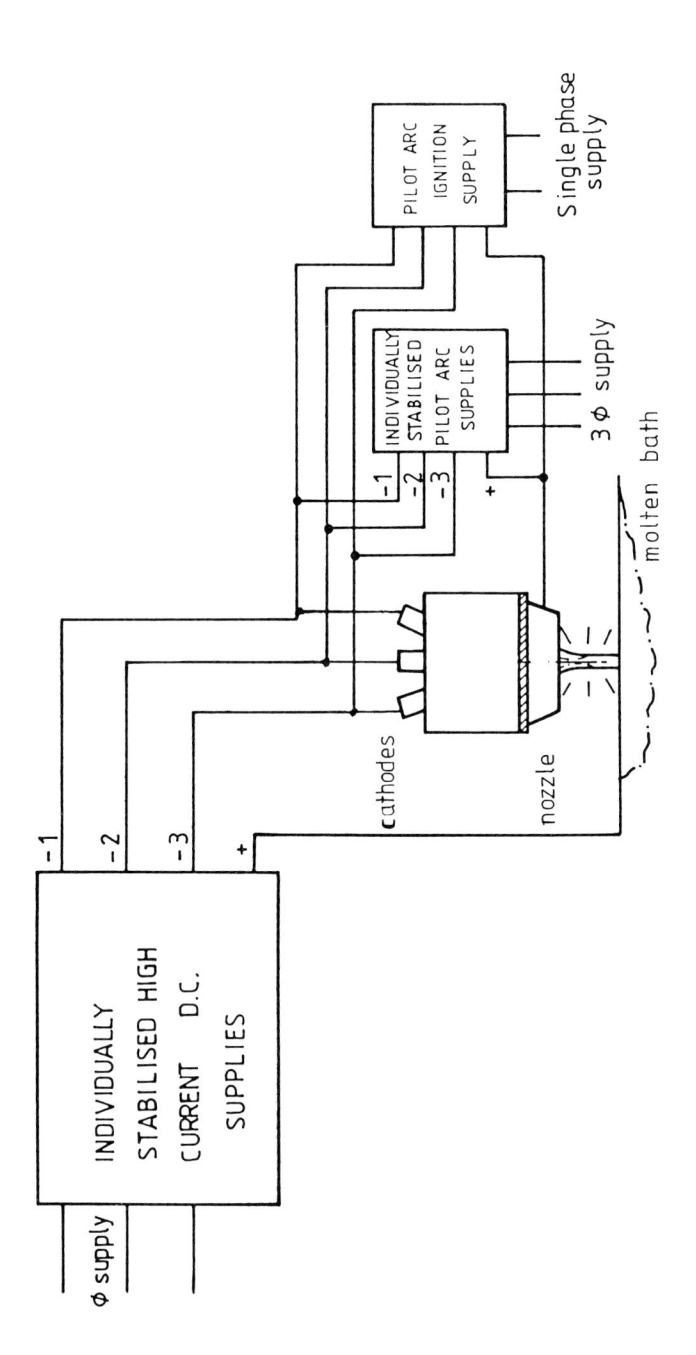

Fig. 5 Non-consumable multiple electrode assembly and power supply

Fig. 6 Multiple electrode assembly in operation

The plasma torches operate in the transferred mode with long arc columns and a coalesced anode root on the crucible. The transferred arcs are established using an auxiliary starting electrode which enables the torches to be fixed within the reactor allowing a simple furnace construction to be used. Arc lengths in excess of 0.5m at currents of the order of 200A per torch can be maintained and the furnace has been used for processing material at temperatures above 1200°C at feed rates up to 50kg/hr.

CONCLUSIONS

The principle of operating separate and coalesced multiple arcs from a common power supply can be applied to a number of arc processes including gas heating, plasma reactors for processing particulate materials, and arc furnaces. A larger volume of ionised gas may be produced or a more uniform distribution of heat obtained. The technique offers advantages over conventional equipment in terms of the capital cost of the power supply and can be readily scaled up to higher currents by the addition of more plasma torches or electrodes.

REFERENCES

1. Harry J E, Knight R, Simultaneous operation of electric arcs from the same supply. IEEE Trans Plasma Science, PS. 9 (4) pp.248-254. (1981)
2. Harry J E, Hobson L, Production of a large volume discharge using a multiple arc system. IEEE Trans on Plasma Science PS. 7 (3) pp.157-162. (1979)
3. Harry J E, Hobson, A multiple arc system. J. Phys. E. Sci. Instrument Vol 12 pp.357-358. (1979)
4. Harry J E, Knight R, Investigation of the intensity distribution of large volume multiple discharges. J. Phys. D. (To be published)
5. Harry J E, Knight R, Nonconsumable electrodes for arc furnaces. Paper 2.2.2 10th International Congress on Electroheat, Stockholm. (1984)
6. Harry J E, Knight R, Power supply design for multiple discharge arc processes. 6th International Symposium on Plasma Chemistry, International Union of Pure and Applied Chemistry, Montreal. pp 150-155. (1983)

ON THE ALLOWANCE FOR THE TEMPERATURE DEPENDENCE OF PLASMA PROPERTIES FOR SELECTION OF DIMENSIONLESS NUMBERS TO CORRELATE CHARACTERISTICS OF ELECTRIC ARCS

V.A.VASHKEVICH, S.K.KRAVCHENKO, T.V.LAKTYUSHINA AND O.I.YASKO
Luikov Heat and Mass Transfer Institute, Minsk 22o728, USSR

ABSTRACT

Plasma properties as a function of heat conduction are approximated by the power-law relation, for instance, $\sigma = \sigma_0 (S*/S_0)^{n_\sigma}$. Then, the temperature effect on the dimensionless number is allowed for by a factor, $(S*/S_0)^\gamma$, the value of which is estimated by equating the dominant number or the whole set of numbers to unity. Substituting this value into the generalized equation gives the form of the correlation function. A relative contribution of each number, κ_i, is determined from the comparison of the exponents of the dimensionless numbers with the experimental ones.

INTRODUCTION

The electric arc is used or appears as an undesirable phenomenon in many devices such as circuit breakers, arc lamps, electric welding apparatus, plasmatrons, etc. Estimation of electric arc characteristics is a difficult problem. Theoretical calculation of geometrically nonstabilized arcs is practically impossible because of complex and unstable configuration, chaotic movement, strong temperature dependence on gas properties, lack of the required data and so on. The approximate similarity methods have been developed to generalize arc dis - charge characteristics [1-5] but these as well face a lot of difficulties, namely, the dependence of dominant processes on arc-to-gas heat transfer conditions, a difficult choice of dominant criteria, temperature dependence of plasma properties being different for particular gases.
The analysis, however, has shown that under certain conditions for a particular heated gas, one or two dominant numbers can be chosen to fairly generalize current-voltage characteristics. The "energy" criterion for convective heating dominates almost in all cases but it changes its form, however, depending on arc discharge conditions [3]. This method has proved to be most effective when generalizing the characteristics of highly unstable longitudinally and transversely blown arcs [4].
The attempts to describe the characteristics of arc discharges in different gases by a single formula have yielded some favourable results [5] but the suggested approach is not widely used because of insufficient substantiation of the method of choosing the characteristic temperature.
The r.m.s. deviations for different numbers are compared to choose the dominant ones [4]. The method is rather effective for choosing one or two dimensionless numbers but it does not estimate the quantitative contribution of individual phenomena described by these numbers to the whole process.

Mat. Res. Soc. Symp. Proc. Vol. 30 (1984) Published by Elsevier Science Publishing Co., Inc.

SPECIFICITY OF THE APPROACH

In the present contribution an attempt has been made to take into account specific features of the temperature dependence of plasma properties for definition of the characteristic arc temperature. This enables one, when the single number is used, to theoretically calculate its exponent and, in case of two or more numbers, to estimate their relative contribution to the electric arc characteristic.

The choice of the scale values of plasma properties using the suggested approach allows unification, by a single formula, of the characteristics of an arc burning in different gases under similar heat transfer conditions.

It is distinctive of the method to turn to its own advantage the disadvantages of using for the electric arcs the traditional generalized method developed for prescribed boundary temperatures. As the arc temperature itself depends on the discharge conditions, the application of the traditional method encounters the insurmountable barrier for choosing the characteristic temperature.

In the proposed method, the scale temperature is estimated in terms of the prescribed values by equating the dimensionless number for the dominating process to unity. Accordingly, the scale temperature value is not constant and depends on discharge parameters. When generalizing, however, it is desirable to use some constant "characteristic" temperature. It is required in this case to introduce a temperature dependence of arc plasma properties.

The main contribution to the current-voltage arc characteristic is made by the core of the arc column where temperature changes are less appreciable than in the periphery, the contribution of which to electric conductivity can be neglected to a first approximation. Therefore, the power-law approximation can be used for the arc core despite rather a complex temperature dependence of plasma properties. The temperature range for the power-law approximation can be extended by using the heat flux potential rather than a temperature. As an example, for electric conductivity:

$$\frac{\sigma^*}{\sigma_0} = \left(\frac{S^*}{S_0}\right)^{n_\sigma}. \tag{1}$$

Similar dependences are used for other properties.

ESTIMATION OF THE EXPONENT USING ONE NUMBER

S^*/S_0 vs discharge parameters is estimated from the dominant number. For example, in the case of a longitudinally blown arc, the dominant number is $\sigma^* h^* G d / I^2$. By using the expression of type (1) and equating this dimensionless number to unity, we obtain:

$$\frac{\sigma_0 h_0 G d}{I^2} \cdot \left(\frac{S^*}{S_0}\right)^{n_\sigma + n_h} = 1 \tag{2}$$

whence

$$\frac{S^*}{S_0} = \left(\frac{\sigma_0 h_0 G d}{I^2}\right)^{-\frac{1}{n_\sigma + n_h}} . \quad (3)$$

The arc current-voltage characteristic may be represented as generalized resistance:

$$\frac{U d \sigma^*}{I} = c . \quad (4)$$

Then, using S^*/S_0 from (3) gives:

$$\frac{U d \sigma_0}{I} = c \left(\frac{\sigma_0 h_0 G d}{I^2}\right)^\delta \quad (5)$$

where

$$\sigma = \frac{n_\sigma}{n_\sigma + n_h} . \quad (6)$$

Similarly, the exponent may be found for the numbers representing other dominating processes.

The value of S_0 and the temperature dependence of plasma properties being known, the exponent, δ, may be calculated theoretically. Then, the arc current-voltage characteristic is obtained within the constant factor.

Expression (5) may be used in two ways:
(1). In weakly stabilized arcs, the discharge is unstable both in time and space. In this case, the arc column temperature depends on physical gas properties rather than on discharge parameters and is retained at a certain level, the deviation from which to any side causes a voltage increase. This phenomenon underlies the known "voltage minimum" principle. Then, plotting $\delta = \delta(S)$, S_0 may be estimated and the exponent, δ, be found theoretically.

(2). In the general case applicable for arcs of any kind, the experimental value of the exponent, α, is considered to be known. Then, such a temperature (heat flux potential) can be found, at which $\alpha = \delta$. The latter will be the very true value of T_0 that must be used as a characteristic one for correlation of arc characteristics if different-kind gases are used. Both a minimum value of the r.m.s. deviation and an equality of the exponents α and δ will characterize the dominant number in this case. If these two characteristics do not coincide, this means the change of the dominating process with varying the kind of a gas.

Figures 1 and 2 present the generalized current-voltage characteristics for a weakly stabilized arc in a longitudinally vortex flow. Correlations are made both in the conventional way (fig. 1) [5] and using the present method (Fig. 2).

Air, nitrogen, argon, hydrogen and helium were heated. Despite the great difference in the properties causing the difference in the specific contribution, for different gases, of the $\frac{\sigma_0 h_0 G d}{I^2}$ number to the generalized current-voltage characteristic, the single characteristic obtained by the given method proves to be fairly satisfactory. In any event, the r.m.s. scatter reduces from 0.86 to 0.44 against the previous approach. Such a generalized characteristic may be used for tentative calculation of the current-voltage characteristic of the gases not yet studied. This is especially useful when designing powerful plasmatrons requiring high expenses.

The method may be extended to the case when arc characteristics are determined by more than one dominating process.

Let κ_i be the characteristic of a quantitative contribution of some process i to the formation of the current-voltage characteristic. Then,

$$\sum_{i=1}^{m} \kappa_i = 1. \qquad (7)$$

Similarly to expression (2)

$$\prod_i A_{io}^{\kappa_i} \left(\frac{S*}{S_0}\right)^{\gamma_i \kappa_i} = 1 \qquad (8)$$

and

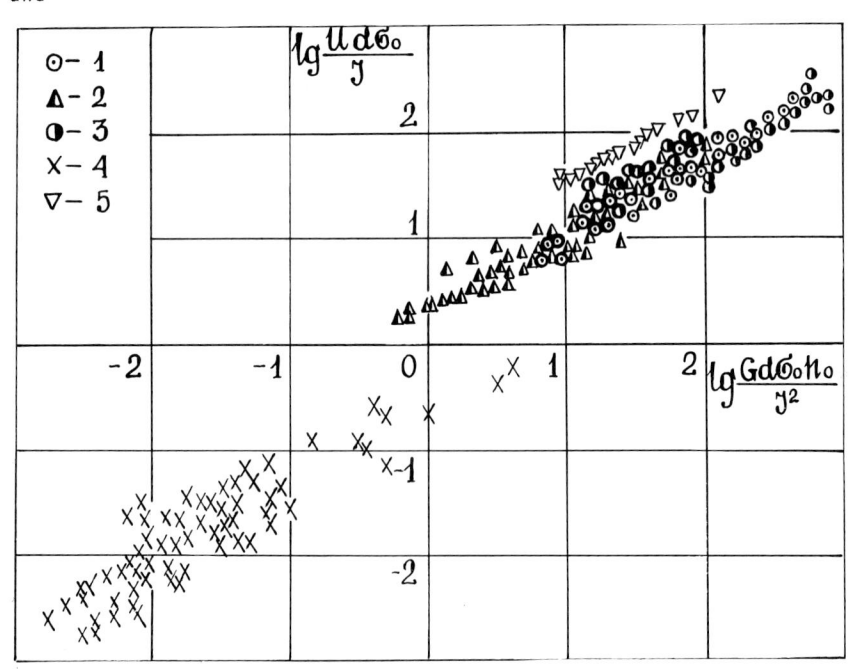

FIG. 1. Generalized current-voltage plasmatron characteristic obtained by the conventional approach.

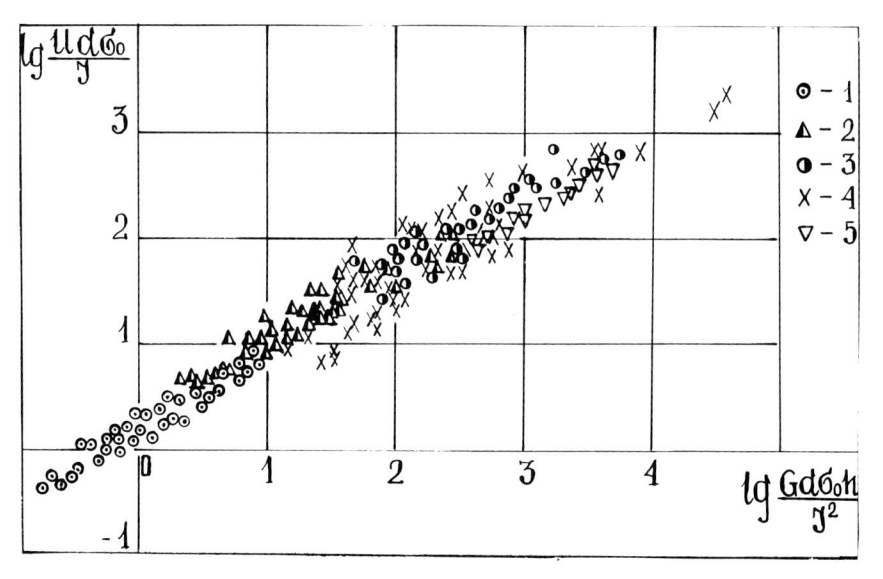

FIG. 2. Generalized current-voltage plasmatron charactetistic allowing for gas properties vs temperature.

$$\frac{S^*}{S_o} = \prod_i A_{io}^{-\frac{K_i}{\Sigma \gamma_i K_i}} \tag{9}$$

Substituting this value into the generalized resistance, we get:

$$\frac{U d \sigma_o}{I} = c \prod_i A_{io}^{\alpha_i} \tag{10}$$

where

$$\alpha_i = \frac{n_\sigma K_i}{\Sigma \gamma_i K_i} \quad . \tag{11}$$

If

$$\delta_i = \frac{n_\sigma}{\gamma_i} \quad , \tag{12}$$

$$\phi_i = \frac{\gamma_i K_i}{\sum_i \gamma_i K_i} \quad , \tag{13}$$

then

$$\alpha_i = \delta_i \phi_i \quad . \tag{14}$$

It is evident

$$\sum_i \phi_i = 1. \tag{15}$$

Hence,

$$\kappa_i = \frac{\alpha_i}{\sum_i \alpha_i}. \tag{16}$$

Thus, a quantitative contribution of the numbers may be determined by the exponents of the generalized characteristic, and the most essential numbers can be chosen for correlation.

If the exponents of all essential (for example, two) numbers for different gases are approximately equal, the experimental data may be generalized by the single current-voltage characteristic. As the difference between exponents grows, the generalization accuracy decreases.

The following correlation procedure may be suggested:

(1). Theoretical exponents $\delta_i = \delta_i(T)$ are found from the temperature dependences of plasma properties. This procedure is similar to δ_i estimation for one dominant number, when each number is considered to be dominant.

(2). Comparison of the experimental and theoretical values gives $\phi_i = \phi_i(T)$.

(3). Characteristic temperature is found for $\sum_i \phi_i(T_0) = 1$.

(4). The values of A_0 numbers are found for T_0 and experimental data are generalized by a single current-voltage characteristic for different gases.

(5). Exponents of α_i are compared for different gases. In case the deviation from the generalized value is in excess of the admissible one, new T_0 and A_0 values are sought and the generalization procedure is repeated. Several iterations may be required.

In case of rather appreciable difference of gas properties, it may appear that even the first most essential numbers are different. To be specific, it will be the "blowing through" number $\sigma_0 h_0 G d / I^2$ for air and the turbulent heat transfer number $d^3 \sigma_0 \rho_0 h_0^{1.5} / I^2$ for helium. This is especially true for the less important numbers. The superimposition of arc characteristics for such gases is unreasonable. However, the generalized formulae may be found for some groups of gases similar in their properties.

NOMENCLATURE

A, dimensionless number; c, constant; d, electrode diameter; G, gas flow rate; h, enthalpy; I, current; m, number of criteria; n, exponent in the expressions approximating plasma properties; S, heat flux potential; T, temperature; U, voltage; α, exponent at experimental numbers; γ_i, exponent at (S^*/S_0) in the A_i number; δ, theoretical exponent of the number calculated by the temperature dependence of gas properties; κ, specific contribution of a process to the generalized characteristic; ρ, plasma density; $\phi=\alpha/\delta$, experimental to theoretical exponent ratio.

REFERENCES

1. S.S. Kutateladze, O.I. Yasko, Journ. Engng Phys. $\underline{7}$, 25-27 (1964).
2. G.Yu. Dautov and M.F. Zhukov, Prikl. Mat. i Teor. Fiz., No. 6, 111 (1965).
3. O.I. Yasko, Brit. J. Appl. Phys. (J. Phys. D), ser. 2, $\underline{2}$ 733-751 (1969).
4. A.G. Shashkov and O.I. Yasko, IEEE Trans. of Plasma Science, Ps-1, No. 3, 21-35 (1973).
5. A.S. Koroteev and O.I. Yasko, Journ. Engng Phys., $\underline{10}$, No.1, 26-31 (1966).

NON-EQUILIBRIUM MODELING AND DISSIPATIVE STRUCTURES IN SOLID MATERIAL - PLASMA INTERACTIONS

Yu.L. Khait
Department of Physics, Ben Gurion University of the Negev, Beer Sheva, Israel.

ABSTRACT

Topics discussed: (a) The dissipative structure (DS) composed of the plasma bulk (PB), near-to-surface plasma layer (PL), surface and the adjacent material layer (ML) coupled by mass, electric charge, etc. fluxes and applications to plasma deposition. (b) The transient local dissipative structure (TLDS) formed by a single plasma ion impinging on the surface and associated with sputtering, etc.

INTRODUCTION

Interactions of plasmas of different gas discharges with solid materials are presently of great interest due to a large variety of industrial and laboratory applications in plasma coating and deposition, in plasma etching, in processing of various materials, etc. [1-13]. At the same time, theoretical foundations of the plasma-solid interaction (PSI) have not yet been constructed and many related phenomena are not quite understood. The PSI kinetics differs significantly from that of the conventional interaction of gases with solid surfaces, which has been studied for decades in physics, physical chemistry, etc. Specific properties of the PSI kinetics are determined mainly by the following facts [10-14 19]:

(i) Plasma contains electrically charged particles (electrons and ions) which can be affected by electric, electromagnetic and magnetic fields. As a result energy distributions of plasma electrons, $f_e(\varepsilon_e)$, and ions, $f_i(\varepsilon_i)$, can differ significantly from the Maxwell distribution. The mean energy of charged particles (especially of plasma electrons)

$$\bar{\varepsilon}_y = \int \varepsilon_y f_y(\varepsilon_y) d\varepsilon_y \; , \; y = e,i \qquad (1.1)$$

can be much larger than kT_g, where T_g is the gas temperature. $f_y(\varepsilon_y,t)$ and $\bar{\varepsilon}_y(t)$ can depend on time t in non-stationary conditions, e.g. in pulse discharges [15-19] or when gas portions pass spatially non-homogeneous fields [15,16,19,20]. For example, $\bar{\varepsilon}_e \simeq (1-5)$ev and $kT_g \simeq (0.04-0.1)$ev in many cases.

(ii) Atoms, molecules, free radicals and ions in plasmas and on the surface can have non-equilibrium populations of their rotational, vibrational and electonic levels due to the influence of charged particles with non-equilibrium $f_e(\varepsilon_e)$ and $f_i(\varepsilon_i)$. This can lead to new reaction routes.

(iii) Electrically charged particles can participate directly in reactions and initiate new reaction routes (in ion-molecular reactions, etc.).

(iv) The processes in plasma can have a non-linear character since: (a) electric, electromagnetic and magnetic fields can change electron, n_e, and ion, n_i, concentrations, $f_e(\varepsilon_e)$ and $f_i(\varepsilon_i)$ and rate coefficients, and (b) these changes, in turn, affect plasma parameters and field strengths [15-19].

(v) Various collective phenomena can occur in plasmas due to the Culomb interaction.

(vi) Mass, energy, momentum and electic charge exchanges between the plasma bulk (PB), near-to-surface plasma layer (PL), the solid surface and the near-to-surface material layer (ML) of thickness $\Delta\ell$ of a few $d\simeq(2-4)\ 10^{-8}$cm. take place. The directed fluxes performing these exchanges create a kinetic coupling of the PB, PL, solid surface and ML which form a non-equilibrium system involved in the PSI kinetics [10,11-13]. This system can be affected and controlled by electric, electromagnetic and magnetic fields.

Such situation can be described in terms of the fundamental concept of "dissipative structure" (DS) which has been effectively used in physics, physical chemistry, chemical kinetics, hydrodynamics, etc. [21-26]. In the above particular case one has the DS composed of four kinetically coupled "boxes", the PB, PL, solid surface and ML. Each of these boxes has certain specific properties and takes a specific role in the PSI kinetics, which are discussed in the next sections [10-13].

In this paper we suggest to use the concept of the near-to-surface DS composed of the above four boxes (in some cases one should include the 5th box - the material bulk) as one of the basic concepts of the PSI kinetics. The macroscopic DS composed of macroscopic boxes can be stationary and non-stationary (e.g. in pulsed discharges).

Another kind of DS, namely, semi-microscopic transient local dissipative structures (TLDS's) with life time $\Delta\tau \simeq 10^{-13}$-$10^{-11}$s in macroscopically small volumes $V_f \simeq N_f d^3 \sim R_f^3$ of radius $R_f \sim c_o \Delta\tau$ containing $N_f \simeq 10$ to 10^3 particles are formed by ions with energies

$$\varepsilon_i \simeq (10\ \text{to}\ 10^3)\text{ev} \simeq (10^2\ \text{to}\ 10^4) kT_s \qquad (1.2)$$

impinging on the surface [12,19,29]. Here c_o is the finite energy transfer velocity in the material. When energy ε_i is released during a short "stoppage time" $t_s \simeq 10^{-14}$-10^{-13}s in a small volume $V'_f \sim (R'_f)^3 < V_f$ of radius $R'_f \simeq c_o t_s$, it creates the initial fireball (with the large energy $\varepsilon'_f \simeq \varepsilon_i / N'_f$ per particle) for the development of the short-lived hot spot (SLHS) and the TLDS [12,13,29]. The SLHS producessputtering, electron emission, changes of local material properties, etc. and takes an important role in the PSI kinetics [10-13,29]. The SLHS and TLDS are associated with many-body non-equilibrium transient phenomena. The observed macroscopic effects are composed of tremendous numbers of such semi-microscopic transient many-body SLHS phenomena associated with the TLDS. The TLDS has been introduced by the author [27,28] in connection with the kinetic many-body theory of short-lived large energy fluctuations (SLEF's) of small numbers, $N_o \geq 1$, of particles in and on solids up to $\varepsilon_o \gg N_o kT_s$ and with its applications to SLEF-assisted rate processes (diffusion, desorption, etc.) and melting. In this case TLDS also has a short duration $\Delta\tau \simeq 10^{-13}$ to 10^{-12}s and is localized in volume $V_1 \simeq N_1 d^3$ containing $N_1 \simeq 30$ to 10^2 particles. The SLEF theory and its applications lead to a significant narrowing of the theory-observations gap [27,28]. This theory: (a) can be used directly in some thermally activated rate processes on the surface and in the adjacent bulk involved in the PSI kinetics, and (b) certain modifications of the concepts and methods used in the SLEF theory [27,28] can be and partly have been used in the PSI kinetics, sputtering, etc. [10-13,29]. In this paper we present a further development and applications to the PSI kinetics of the DS and TLDS concepts and the ideas presented in [10-13 27-29]. We use dimension and similititude theories to obtain qualitative and semi-quantitative results for the main features of the problems under consideration, since at present one can hardly expect to solve them exactly.

2. NON-EQUILIBRIUM NEAR-TO-SURFACE DISSIPATIVE STRUCTURE IN PLASMA DEPOSITION

The PB, PL, surface and ML taking important roles in the PSI kinetics, e.g. in plasma deposition [10,11,13], are combined into a single dissipative structure by energy, mass, momentum and electic charge fluxes transferring different substances between the DS parts during plasma deposition, etching, etc. The DS has a specific non-homogeneous character and different parts of the Ds take different roles in the PSI kinetics. The near-to-surface PL of thickness [10,11,13]

$$\Delta L = b \cdot \lambda_{es} = b \ kT_g (P \cdot \sigma_{es})^{-1} \quad (2.1)$$

plays a key role of plasma-tosolid "bidge" in the DS associated with plasma deposition, since:
(a) The PL properties differ from those of both the PB and solid.
(b) The PL coupled with the plasma bulk and the surface by the corresponding direct fluxes, combine them into a single DS.
(c) Specific processes taking place in the PL are closely connected with those occurring in the PB and the solid and lead to experimentally detectable kinetic and spectroscopic effects in plasma coating, ect. [10,11].
(d) Value

$$b = \frac{\Delta L}{\lambda_{es}} = \frac{\Delta L \cdot P \cdot \sigma_{es}}{kT_g} \quad (2.2)$$

can be treated as the dimensionless PL thickness, where $\lambda_{es}=kT_g(P\cdot\sigma_{es})^{-1}$ is the mean free path length of electrons emitted from the surface, accelerated (towards the surrounding plasma) in the near-to-surface electric field E_s up to energy $\varepsilon_{es} \simeq 10^2$ev and forming the high energy electron beam (HEEB) [10,11]; σ_{es} is the total cross section of collisions of HEEB electrons with gas particles; P and T_g are the gas pressure and temperature.

Kinetic coupling of the PL with the surface is associated mainly with the following phenomena [10,11,13]: (i) The flux of positive ions with density

$$J_i = 0.25 n_i \cdot \bar{u}_i = 0.25 a_i \cdot \bar{u}_i \cdot N \quad (2.3)$$

which are directed towards the surface and accelerated by the near-to-surface electric field E_s up to large energies $\varepsilon_i >> kT$, e.g.

$$\varepsilon_i \simeq 50\text{-}100\text{ev} \ \text{or} \ \varepsilon_i \simeq (500\text{-}10^3)kT_s \ \text{at} \ T_s \simeq 10^3 K \ , \quad (2.4)$$

impinges on the surface. Here $N=P(kT_g)^{-1}$ is the total number of gas particles, $n_i=a_i \cdot N$ is the number of ions in 1cm^3 corresponding to ionization degree a_i, $\bar{u}_i$ is the mean velocity of ions before the acceleration starts. (ii) An ion with energy $\varepsilon_i >> kT_s$ impinging on the surface forms the SLHS and TLDS, discussed in the previous section. The SLHS produces electron emission, sputtering, desorption, etc., which form corresponding fluxes of particles directed towards plasma [10-13,29]. (iii) The flux $J_{es}=\gamma_s^* J$ of HEEB electrons is emitted from the surface and accelerated towards the surrounding plasma by the electic field E_s up to high energy

$$\bar{\varepsilon}_{es}(z) = \int \varepsilon_{es}(\varepsilon_{es};z) \ d\varepsilon_{es} >> \bar{\varepsilon}_{ep} = \int \varepsilon_{ep} \cdot f_{ep}(\varepsilon_{ep}) \cdot d\varepsilon_{ep} \quad (2.5)$$

at $z<\Delta L$. Here γ_s is the electron emission coefficient of the solid material. $f_{es}(\varepsilon_{es};z)$ and $\bar{\varepsilon}_{es}(z)$ are the HEEB energy distribution and mean energy which depend on distance z from the surface, since HEEB electrons lose their energy in ionizing, exciting and elastic collisions with gas particles within the PL. The averaged PL thickness ΔL is determined by condition $\bar{\varepsilon}_{es}(\Delta L) \simeq \bar{\varepsilon}_{ep}$, where $\bar{\varepsilon}_{ep}$ is the mean energy of electrons in the PB which have energy distribution $f_{ep}(\varepsilon_{ep})$ [10,11]. Partial PL thickness with respect to particular components populating PL and HEEB-induced processes can also be determined [11]. Therefore, electrons in the PL have bimodal z-dependent energy distribution

$$\phi_L(\varepsilon;z) = f(\varepsilon_{es};z) + f_{ep}(\varepsilon_{ep}) \quad \text{at} \quad z < \Delta L \tag{2.6}$$

which produces the corresponding large PL rate coefficients K_L of excitations and ionizations associated with cross sections $\sigma_{ex}(\varepsilon)$ and $\sigma_{ion}(\varepsilon)$

$$K_L(z) = K_{es}(z) + K_{ep} \sqrt{\int \sigma_y(\varepsilon) \cdot \varepsilon^{\frac{1}{2}} \cdot \phi_L(\varepsilon;z) d\varepsilon} \; ; \; y=ex,ion \tag{2.7}$$

here $K_{es}(z)$ and K_{ep} are partial rate coefficients associated with excitations caused by HEEB and plasma electrons respectively. Hence one can see that $K_{es}(z<\Delta L) >> K_{ep}$ for excitation energies $\Delta e_{jk} >> \bar{\varepsilon}_{ep}$, since $\varepsilon_{es}(z<\Delta L) > \Delta e_{jk}$, e.g. $\varepsilon_{es}(z<\Delta L) \simeq 50\text{ev}$ and $\varepsilon_{ep} \simeq 2\text{ev}$ whereas $\Delta e_{jk} \simeq 10-15\text{ev}$. The effectiveness of the HEEB with respect to various excitations in the PL is confirmed experimentally by spectroscopic and kinetic data [10,11]. The PL itself has a fine structure which is, probably, similar to some degree to that of the cathod region in d.c.electic discharges. This question deserves a special study. Here we only want to note that Eqs(2.1) and (2.2) can be presented in the form

$$\Delta L \cdot P = b \cdot k T_g \cdot \sigma_{es}^{-\frac{1}{2}} \quad \text{with} \quad b \simeq \alpha^{\frac{1}{2}} \tag{2.8}$$

which reminds the well-known Paschen law for electic discharges. Here value $\Delta L \cdot P$ is approximately constant for a given gas composition and T_g. Parameter $b \simeq \alpha^{\frac{1}{2}}$ has been estimated in [11] through the consideration of the Brownian motion of an averaged HEEB electron which is scattered in $\alpha >> 1$ (e.g. $\alpha \simeq 50-100$) collisions with gas particles during its motion within the PL. The electron anizotropic Brownian motion in the PL can be described by the Fokker-Plank equation with the "drift" velocity $u_d(z)$ directed along the z-axis perpendicular to the surface which depends on $\bar{\varepsilon}_{es}(z)$ and on a degree of anizotropy of HEEB electrons scattering in their collisions with gas particles. The ratio $\xi(z) = u_d(z) \cdot [2m \varepsilon(z)]^{-\frac{1}{2}}$ can serve as a measure of this anizotropy of the HEEB, where $\xi(z) \longrightarrow 0$, if $z \longrightarrow \Delta L$. (iv) The plasma ions bombarding the surface produce fluxes $J_{sk} = \alpha_{sk} \cdot J_i$ of sputtered particles of various types k=1,2 directed towards the PL. The existence of sputtered particles in the PL conformed spectroscopically [11] presents an additional evidence of the fact that the ions impinging on the surface have rather high energies, at least substantially higher than sputtering thresholds $\varepsilon_{th} \simeq 20-30\text{ev}$. (v) Processes on the surface also produce fluxes of desorbed particles towards the PL. (vi) Fluxes of various neutral particles impinging on the surface and coming from the PL contribute to the PL-solid interaction too.

All these processes create a rather strong kinetic PL-solid coupling and influence substantially the PSI kinetics.

On the other hand, the PL is coupled kinetically with the plasma bulk of the characteristic length $L >> \Delta L$. The coupling is performed by fluxes of neutral and charged particles. The mass PB-PL exchange supplies the PL with particles which are "treated" in the PL by HEEB and participate in plasma deposition on the surface [10,11,13]. Different parts of the near-to-surface DS require different experimental and theoretical means to be studied. However in spite of these differences a better understanding of the PSI kinetics can be achieved, in our opinion, only through coherent studies of interrelated phenomena occurring in various parts of the DS and by combining all these data to construct the entire picture of the PSI kenetics. Some related questions associated with phenomena on the surface and in the adjacent ML will be discussed in the next section [11-13,27-29]. Here we consider some non-equilibrium phenomena occurring mainly in the PL and involving PL-PB and PL-surface kinetic coupling to explain one interesting but puzzling question: why do changes of experimental low velocities $v_g \approx 0.3$-$1 cm/s$ of the gas passing through the reactor in plasma deposition experiments markedly affect the deposition kinetics which involves electrons, ions and neutrals whose velocities, $u_e \approx 10^8$-$10^9 cm/s$, $u_i \approx 10^5$-$10^6 cm/s$ and $u_n \approx 10^5 cm/s$, are much larger than v_g? The answer to this question can be obtained from the model presented in [10,11,13] and the Ds concept: the optimum gas velocity for plasma deposition should satisfy the condition

$$v_g \approx \bar{v} = 0.25 \cdot q_{es}(\varepsilon_{es}) \cdot a_i (T_g,P,E) \cdot \gamma_s(\varepsilon_i) \bar{u}_i \qquad (2.9)$$

taking into account the PL-PB and PL-surface kinetic coupling. Here $\bar{v}$ is the characteristic kinetic parameter with velocity dimension introduced in [10], which is $\bar{v} \approx 0.3$-1 cm/s for many experimental conditions [10,11,13]; q_{es} is the dimensionless effectiveness of excitations of gas particles in the PL by HEEB electrons which is determined by the corresponding cross sections. The condition (2.9) being in agreement with experimental data means that the optimum residence time $\Delta \tau_g \approx L/v_g$ of every gas portion in the reactor is equal to the characteristic time $\Delta \tau \approx L/\bar{v}$ the HEEB (of a given intensity J_{es} and energy distribution $f_{es}(\varepsilon_{es};z)$) needs to "treat" particles of the gas portion passing through the reactor, taking into account the influx of particles from the PB into the PL. Therefore Eq(2.9) leads to the "resonance" condition $\Delta \tau_g \approx \Delta \tau$. This condition can be expressed in the form $h_g = \bar{v}/v_g \approx 1$, if one uses dimensionless criterion $h_g = \bar{v}/v_g$. On the other hand, the calculated efficiency of plasma deposition g which can be measured experimentally [10], is a function $g=g(h_g)$ of h_g. Hence one obtains an important practical conclusion: one can increase gas velocity v_g without breaking down of the optimum condition $h_g \approx 1$ and decreasing of $g(h_g)$, if one is able to enhance $\bar{v}$ by the corresponding changes of parameters which determine $\bar{v}$. Such increase in v_g leads to the corresponding enhancement of the technological efficiency of plasma deposition. At the same time small values of dimensionless criteria

$$h_x = \bar{v} \cdot \Delta L / D_x << h_x = \bar{v} \, L/D_x << 1 \quad \text{at} \quad x=n,i$$

for many experimental conditions show that diffusion velocities $\Delta \bar{v}_x = D_x / \Delta L$ and $v_x = D_x / L$ for neutrals and ions in the PL and PB are large compared to $\bar{v}$. This ensures a rapid diffusion transport and the HEEB-controlled kinetics.

3. TRANSIENT LOCAL DISSIPATIVE STRUCTURE (TLDS) ON THE SURFACE AND IN THE ADJACENT MATERIAL LAYER.

Every energetic ion impinging on the surface forms the SLHS and TLDS on the surface and in the ML, as discussed in section 1. The TLDS (or its parts) can have the "inward" or "outward" character associated with the "inward", $p_{\perp}^{(in)}$ and $j_{y\perp}^{(in)}$, or "outward", $p_{\perp}^{(out)}$ and $j_{y\perp}^{(out)}$, components of substrate particles momenta $p=\{p_\perp, p_\parallel\}$ and of microfluxes $j_y = \{j_\perp, j_\parallel\}$ within the SLHS, where subscripts $\perp$ and $\parallel$ point out the directions perpendicular and parallel to the surface and $y=E,m,p$ and q are connected with the energy, mass, momentum and electric charges respectively. The initial fireball in volume V_f^i containing a macroscopically very small number of particles N_f^i (e.g. $N_f^i \approx 10\text{-}100$) [12,13,29] is characterized by: (a) high evergy per particle $\varepsilon_f^i \approx \varepsilon_i / N_i^i$ (e.g. $\varepsilon_i \approx 1\text{-}5\text{ev}$), according to the energy conservation law, the finiteness of the energy transfer velocity c_o and the shortness of the ion "stoppage time" $t_s \approx 10^{-14}\text{-}10^{-13}\text{s}$ [12,13,29]. (b) "Inward" momentum components $p_\perp^{(in)}$, according to the momentum conservation law, and (c) The absence of fluxes J_x outside the fireball. It is worthwhile introducing the local "rapid" time scale τ with the reference point $\tau_p = 0$ at the instant when the projectile energy $\varepsilon_i(\tau)$ becomes equal to the threshold energy E the projectile needs to propagate in the solid, i.e. $\varepsilon_i(\tau_p = 0) \approx E$ (e.g. $E \approx 5\text{-}10\text{ev}$). Therefore the fireball formation is associated with advanced processes at $\tau_f^A < 0$. This approach presents a modification of the one used in the kinetic many-body SLEF theory [27,28]. Then the fireball decays and initiates various local transient phenomena associated with the TLDS [12.13.29]: (a) At $\tau_f^R > 0$ a part of the fireball energy forms inward microshock waves and fluxed $j_x^{(in)}$ associated with the inward TLDS and formations of the corresponding short-term particle coopetative motion. (b) Another part of the fireball energy is linked with explosion-like outward microshock waves directed to the surface and formed due to very high fireball effective kinetic temperature $T_f \approx \varepsilon_f / 3k$ and pressure P_f. These waves can cause material microjets towards the PL, which can produce sputtering (of particles able to overcome surface energy barriers), and microcraters (or microfunnels) and protrusions on the surface. (c) The inward microshock waves can form a quasi-gas microbulb in the small near-to-surface volume V_g. This shockwaves can be partially reflected from the "cold" surrounding substrate medium. Such reflection can initiate the cumulative shockwave directed to the surface which can be made strong enough due to rapid explosion-like closing of the microbulb when the cumulative microwave comes to the surface, especially if it occurs during the microfunnel formation. Then this cumulative wave can be transformed into a cumulative microjet directed towards the PL. Such cumulative microjet can produce high-energy volcano-like microsplash of material particles into the PL, similar to that considered in hydrodynamics during the formation of liquid or gas "plumes". The total number of material particles coming to the surface, the number of sputtered particles, etc. during the TLDS and SLHS evolution depends on ε_i, fireball and material parameters, etc. and repuire a special consideration. Some related estimates have been done in [12,13,29]. Some ideas and results of the kinetic many-body theory [27,28] and their modifications can be used in this case. Here we want to add the following: (i) The above short-term phenomena associated with the SLHS and TLDS can cause drastic transient changes of material parameters in the SLHS volume V_f and on the surface of areas $\Delta s_f \approx V_f^{2/3}$. (ii) The SLHS phenomena and TLDS cause local transient breakdowns of symmetry and stability in the solid (see also [27]). (iii) The above SLHS phenomena can perturb strongly electronic states in and on the solid (see pertinent discussions in [27,28] and in [23.13,29]), since: (a) the Bloch theorem and the Born-Oppenheimer principle (which enables one to separate atomic and electronic wave functions) lose their validities during the SLHS and TLDS lifetime $\Delta\tau \approx 10^{-13}\text{-}10^{-11}\text{s}$. Electons in the SLHS volume V_f can experience various transitions consuming the fireball energy which is the energy source for the TLDS. (b) The same SLHS volume V_f and the surface areas $\Delta s_f \approx V_f^{2/3}$ undergo a sequence of many ion strikes with the mean frequency

$$\nu_f \approx j_i \cdot V_f^{2/3} \approx 0.25 a_i \bar{u}_i \cdot V_f^{2/3} \cdot N \qquad (3.1)$$

which can be rather high, e.g. $\nu_f \approx 10^6 s^{-1}$ at $V_f^{2/3} \approx 10^2 \cdot d^2$, $a_i \approx 10^{-3}$, $N \approx 10^{17} cm^{-3}$ and $d^2 \approx 10^{-15} cm^2$. If time $\Delta t_f \approx \nu_f^{-1}$ between the two subsequent ion hits on the same surface area Δs_f satisfies conditions

$$\Delta t_f \approx \nu_f^{-1} << \theta_R = \theta_{OR} \cdot \exp(\frac{\Delta E}{kT_s}) \quad \text{or} \quad \Delta t_f \leq \theta_R , \qquad (3.2)$$

the material in volume V_f has not enough time for recovering after changes caused in this volume by the previous SLHS. Here θ_R is the recovery time controlled mainly by thermally activated rate processes in and on the solid. In this case the substrate surface and ML can have a "memory" associated with residual changes of material properties in every V_f. This memory can influence the kinetics of SLHS phenomena and of plasma-assisted surface processes. Thermally activated rate processes occurring in and on the solid during Δt_f between the two subsequent ion hits can be treated with the help of the kinetic many-body theory of SLEF's and SLEF-assisted rate processes in and on solids and melting [27,28]. The kinetic many-body SLHS phenomena during $\Delta \tau \approx 10^{-13} - 10^{-11} s$ can be considered with the help of the approach suggested in [12,13,29].

REFERENCES

1. Proc. 7th Int. Vac. Congr. and 3rd Int. Conf. on Solid Surfaces (Vienna, Austria 1977).
2. Proc. 4th Int. Symp. on Plasma Chemistry (Zurich, Switzerland 1979).
3. Proc. 5th Int. Symp. on Plasma Chemistry (Edinburgh, U.K. 1981).
4. Abstracts Int. Workshop on Plasma Chemistry in Technology (Ashkelon, Israel 1981).
5. Proc. Int. Coll. on Plasma and Sputtering (Nice, France 1982).
6. Proc. 6th Int. Symp. on Plasma Chemistry (Montreal, Canada 1983).
7. H.F. Winters, J.W. Coburn, T.J. Chuong, Research Report R.J. 3617 (42243) 9/28/82 IBM Research Lab. (San Jose, California, U.S.A.).
8. D.L. Flamm and V.M. Donnelly, Plasma Chemistry and Plasma Processing 1, 317 (1981).
9. A.T. Bell and J.R. Holluhan, Techniques and Applications of Plasma Chemistry (Wiley, N.Y. 1974).
10. Yu.L. Khait, A. Inspektor and R. Avni, Proc. 4th Int. Symp. Plasma Chemistry (Zurich 1978); Thin Solid Films 72, 420 (1980); Invited Paper at the Int. Conf. on Metallurgical Coating (San Diego, U.S.A. 1980).
11. Yu.L. Khait, U.Carmi and R. Avni, Proc. 6th Int. Symp. Plasma Chemistry (Montreal 1983).
12. Yu.L. Khait, Proc. 4th Int. Symp. Plasma Chemistry (Zurich 1979); Proc. 5th Int. Symp. on Plasma Chemistry (Edinburgh 1981); Proc. 4th Int. Coll. on Plasma and Sputtering (Nice 1982).
13. Yu.L.Khait, Invited Paper at the Int. Workshop on Plasma Technology (Ashkelon 1981).
14. I.P. Shkarovsky, T.W. Jonston and M.P. Bashinsky, The Particle Kinetics of Plasmas (Addison-Wesley Pub. Co., Reading, Massachusetts 1966).
15. Yu.L. Khait, Abstr. XIV Int. UPAP Conf. on Statistical Physics (Alberta, Canada 1980). Bull. Israel Phys. Soc. 27, I-13 (1981); 28, B-6 (1982).
16. Yu.L. Khait, J. de Phys. (Paris) C-941 (1980).
17. Yu.L. Khait, Proc. 6th Symp. on Plasma Chemistry (Montreal 1983).

18. F.B. Vursel, Yu.L. Khait, G.V. Lysov, L.S. Polak, E.N. Tchervochkin, Proc. All-Union Conf. on Low Temperature Plasma and Generators (Alma-Ata, U.S.S.R. 1970). Chimiya Vysokich Energii (Sov. High Energy Chemistry) 5, 105 (1971); 5, 112 (1971).
19. Yu.L. Khait, Non-Stationary Heat and Mass Transfer, Doctor of Science Thesis (USSR Academy of Sciences 1972, Moscow) chap. 9. Pre-print N 01, Plasma Processes Lab., Petrochem. Synthesis Institute, USSR Academy of Sciences, Moscow 1971).
20. Yu.L. Khait and O. Biblarz, J. Appl. Phys. 50 (1979), p. 4692.
21. P.Glandsdorf and I. Prigogine, Thermodynamic Theory of Structures, Stability and Fluctuations (Wiley, N.Y. 1971).
22. G.Nicolis and I. Prigogine, Self-Organization in Non-Equilibrium Systems (Wiley, N.Y. 1977).
23. I. Prigogine, From Being to Becoming (W.H. Freeman and Co., San Francisco 1980).
24. H.Haken, Sinergetics. The Introduction: Non-Equilibruim Phase Transitions and Self-Organization in Physics, Chemistry and Biology (Springer-Verlag, Berlin 1977).
25. A.M. Tiring, Phil. Trans. Roy. Soc. B 237, 37 (1952).
26. J. Neumann, Theory of Self-Reqpoducing Automata (Urbana, Univ. of Illinois Press 1966).
27. Yu.L. Khait, Physics Reports (in print).
28. Yu.L. Khait, Physica 103 A, 1 (1980).
29. Yu.L. Khait, Abstr. ESCAPMIG (Essen, W.Germany 1978). Abstr. 1978 Annual Meeting of Austrian Phys. Soc. (Innsbruck 1978).

SURFACE TREATMENT OF THE GLASS FIBERS IN LOW PRESSURE MICROWAVE PLASMA

Ryszard Parosa
Institute of Telecommunication and Acoustics, Technical University of Wrocław, Wybrzeże Wyspiańskiego 27, 50-370 Wrocław, Poland.

ABSTRACT

In the method described here, oxygen and air plasmas were generated at pressure of 1 - 10 Torr in quartz tube placed inside various kinds of microwave cavieties. Cavieties were supplied by 10 to 500 W of microwave power / f=2.45 GHz /. Processed fiber was fastly moved across the plasma region by special driving system. Experimentally the optimal process conditions, i.e. treatment time and gas pressure, were found. Moreover, a special construction of a "long" plasma reactor for industrial application of the process was worked out.

INTRODUCTION

In 1980 at the Institute of Telecommunication and Acoustics of Technical University of Wrocław an experimental study has started on the glass fiber / and glass cord / surface cleaning process in the low pressure microwave plasma. Impurieties on the glass fiber surface / like paraffins and oils / are necessary during the fibers production process, but are undesirable afterwards. In the conventional cleaning technology the cleaned glass fibers are intensively heated by a stream of a hot air inside a special construction furnace having temperature of 620 K. Heating process in such a case must be led continuously during 24 - 74 hours, what makes this technology very energy-consuming. As a result of intensive heat treatment of glass fibers or glass textures the impurity content on the fibers surface decreases up to a level of less than 1 % of the fibers weight, and the fiber tenacity is of the order of 40 % - 70 % lower in comparison with glass fibers before treatment.

In the proposed method the cleaning process is realised by using low pressure microwave plasma. Essential for this process is the rapid oxidation of the paraffins and oils on the glass fiber / or cord / surface, so that the air or oxygen plasma must be applied. Chemical reaction products / mainly CO_x / are in the gaseous state, and these products are easily removed outside the plasma region by a small gas flow.

EXPERIMENTS

In order to assure the required experimental conditions, the special experimental set was constructed [1]. This set of apparatus was equipped with the microwave line, vacuum chamber with pump, and the driving system that moved the fiber across the plasma region / see fig.1 /.

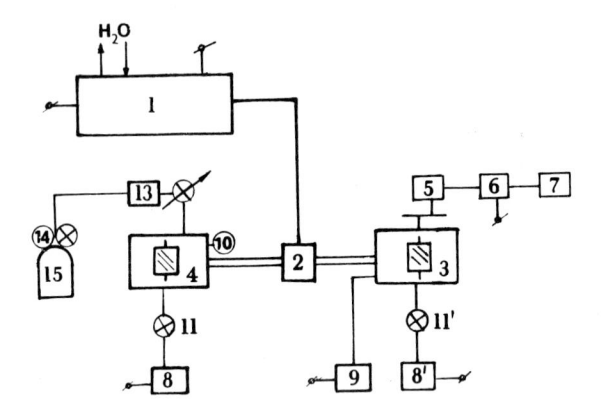

FIG. 1. Experimental set of apparatus.
1.-microwave line, 2.-discharge cavity, 3 and 4.-vacuum cavities with driving system, 5.-electric motor, 6.and 7.-power supply, 8.-vacuum pump, 9.-vacuummeter, 10.-manometer, 11.and 12.-vacuum valve, 13.-flowmeter, 14.-valve, and 15.-gas cylinder.

Microwave line was composed of: CW magnetron with power output of 2 kW / f= 2.45 GHz /, water cooled adjustable attenuator, coaxial slotted line with detector and discharge cavity. In experiments three kinds of microwave discharge cavieties have been tested; the rectangular waveguide cavity with TE_{10n} mode, the cylindrical cavity with TM_{011} mode, and coaxial resonator similar to Fehsenfeld at all. /2/ construction. Best results were obtained by applying the coaxial discharge cavity, and therefore only this microwave discharge structure will be here described. The simplified design of the coaxial resonator is shown in fig. 2.
The microwave power regulated in the range of 5 W to 2000 W was transmitted into the cavity by coaxial line and moveable antenna coupler. Resonance frequency of the cavity was adjusted by the matching element. Microwave breakdown inside a quartz tube crossing the cavity occurs, when the resonance frequency of the cavity is equal to the magnetron frequency, and the gas pressure inside a quartz tube is fairly low.
  In the experiments carried out the typical operation conditions were as follows:
- microwave power transmitted into cavity was of the order of 20 - 30 W / microwave power was regulated by magnetron current control cirquit and by water-cooled attenuator /,
- plasmagenious gas pressure was in the range of 0.5 - 4. hPa / 0.38 - 3. Torr /,
- travel spead of the glass fibers across the plasma region was in the range of 0.5 - 10. m/min.
Experiments show, that the optimal pressure for the process is about 1.3 hPa / 1 Torr /. If pressure is larger, the fiber tenacity decrease rapidly. In the case of very low pressure / $p < 1$ hPa /, the process efficiency is low; for example for

treatment time 0.4 sec the impurity content was 0.56 %, when the pressure was 0.67 hPa and 0.19 % when the pressure increases up to 1.3 hPa.

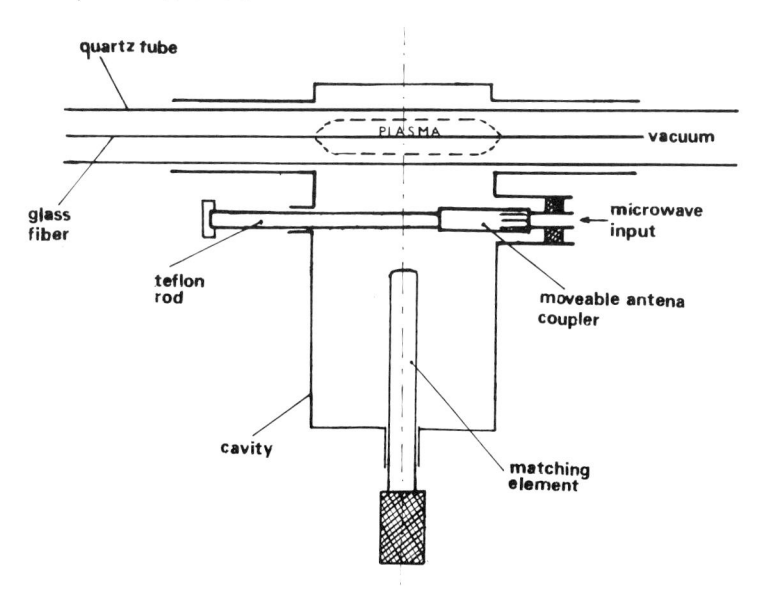

FIG. 2. The simplified design of the coaxial discharge resonator.

Very important parameter of the studied process is the treatment time / i.e. the time of the fiber being inside the plasma /. In the fig. 3. the measured dependence of the impurity content on the fiber surface as a function of treatment time is shown.

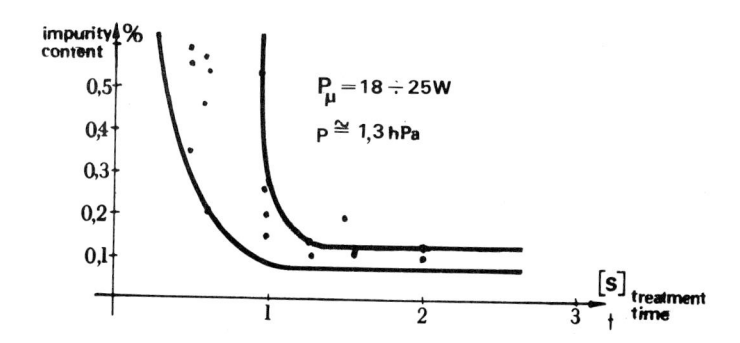

$P_\mu = 18 \div 25W$

$p \cong 1,3\ hPa$

FIG. 3. The impurity content as a function of the treatment time.

Taking into account the large number of experiments one may conclude, that usually required in the industrial practice the impurity content level of 0.1 - 0.3 % may be easily achived, if the plasma treatment time is of the order of 0.9 - 1.5 sec. Moreover, as it has been stated, the microwave power level transmitted into the cavity has an insignificant influence on the process efficiency, if the microwave power is of the order of 20 - 30 W. Also experiments carried out with different gases / i.e. with oxygen and air / have given similar results, but one may expect that the oxygen flow into discharge region must be large enough to assure the effective oxidation of the paraffins and oils. Here it should be pointed out, that the small gas flow inside the quartz tube is essential for removing the CO_2 and another chemical reaction products from the plasma region.

Very low energy-consumption as compare with the conventional methods is the most important parameter of the described plasma process - see table 1.

TABLE I. Energy-consumption level for various methods of the glass fiber / and glass cord / surface cleaning treatment.
88888888

Method	Energy consumption kWh / m^2 of fiber	
	final impurity content 0.2 %	final impurity content 0.1 %
Thermal / T=620 K /; treatment time - 48 - 74 hours [3]	0.2	0.63
By using arc plasmatron / T 8000 K /; treatment time - 0.6 - 2 sec [3]	2.6	8.6
By using plasma glow discharge; treatment time - 120 sec [3]	1.08	-
Microwave plasma treatment; treatment time - 1. - 1.5 sec	1.3×10^{-2}	$6. \times 10^{-2}$

The above results fully confirmed industrial utility of the microwave plasma method, pointing out to necessity of undertaking further work on industrial microwave plasma device.

LONG PLASMA REACTOR

In order to increase the amount of fibers treated, it was necessary to construct the microwave discharge cavity with a long plasma / 0.5 - 1. m /. In this case the fiber must be very fast moved through the long plasma region, faster than in the case of a previous plasma because we wanted to behave the same time of a plasma treatment, greatly increasing the amount of the fibers treated. This new plasma reactor is shown in the fig. 4.

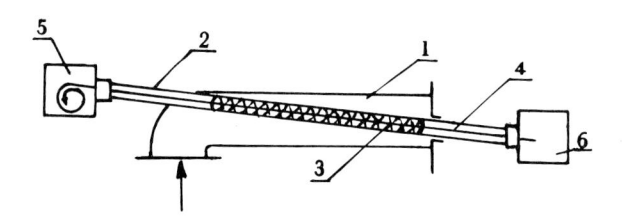

FIG. 4. Simplified draft of a long plasma reactor.
1.-rectangular waveguide, 2.-quartz tube, 3.-plasma, 4.-glass fiber, 5. and 6.-vacuum chambers with driving system.

It is essential for our construction that the microwave device before breakdown behave as a good microwave resonator with mode TE_{10n} helping to start gas discharge inside a quartz tube. When the discharge is started, generated plasma heavily loads microwave resonance cavity, decreasing Q factor and in this way making electric field distribution more homogenous.
Important implications from this solution are as follows:
- decrease of a power needed for discharge ignition,
- effective absorption of a microwave power by plasma,
- homogenous distribution of some plasma parameters along discharge tube.

Typical operational parameters of the long plasma reactor were:
- microwave power absorbed by the plasma was of the order of 300 - 600 W,
- gas pressure inside the discharge tube was in the range of 1. - 3. hPa,
- travel spead of the glass fibers through the plasma region was in the range of 30 - 40 m/min,
- lenth of the plasma column was about 65 cm.

In order to eliminate the possibility of the dissociation process of the CO_2 in the plasma region, the gas flow inside the discharge tube was increased up to order of 1. - 8. l/h. When the gas flow is lower, the dissociation process products / mainly carbon / are fastly sedimented on the quartz tube surface. This effect is very unprofitable because of intensive absorption of the microwave energy by carbon sedimented on the quartz tube, what leads to the intensive heating of the quartz and decreases the power absorbed by plasma.

CONCLUSIONS

Preliminary results point evidently out to a possibility of the industrial scale use of the worked out process. Especially important would be develop of this process for textures, and now our group in Technical University of Wrocław tries to solve this problem. In particular, the most important problems we are solving now are conected with assurance of the homogenious plasma treatment on full width of the textures, and the question how to eliminate the microwave leakage from the discharge cavity.

Expected advantages of the microwave plasma treatment of the glass textures / and fibers / are:
- very low energy consumption,
- comparatively low cost of the apparatus,
- high productivity.

REFERENCES

1. M.J. Kloza, R. Parosa, E. Reszke, Universal microwave set o of apparatus for investigation of plasma. Research Report of Institute of Telecommunication and Acoustics Technical University of Wrocław, No I-28/17/83.
2. Fehsenfeld F.C., K.M. Evenson, H.P. Broida, Microwave Discharge Cavieties Operating at 2450 MHz. The Review of Scientific Instruments, vol.36, nr. 3, March 1965.
3. Rakowski W., K. Bartos, J. Zawadzki, Research Raport, Institute of Textile Industry, No UT/380/81/bc, 1981.

A MASS SPECTROMETRIC SYSTEM FOR THE STUDY OF TRANSIENT PLASMA SPECIES IN
THIN FILM DEPOSITION

N.P. JOHNSON,* A.P. WEBB,** and D.J. FABIAN***
*Department of Metallurgy and Materials, University of Strathclyde,
Glasgow G1 1XN, Scotland; **Department of Electronics and Electrical
Engineering, University of Glasgow, Glasgow, Scotland; ***Department
of Physics, Simon Fraser University, Burnaby, B.C. Canada (On leave
from University of Strathclyde, Glasgow, Scotland).

ABSTRACT

A system is described for mass spectrometric detection
of transient gaseous species involved in reactive plasma
deposition of materials. The equipment comprises a three
stage differentially pumped UHV quadrupole mass spectrometer
chamber, which permits modulated molecular beam sampling over
a short path-length, direct from the plasma at 0.1-1.0 torr
pressure. Operation of the system and optimum conditions
for maximum signal-detection are detailed, and preliminary
results for species formed in a silane-argon high-power rf
discharge are reported. Spectra mostly agree with those
obtained by Turban and Catherine using a lower power rf
plasma, although some evidence is observed for the formation
of increased SiH species at higher power.

INTRODUCTION

From the earliest reports [1] on plasma or glow-discharge deposition
of thin films to the latest work current today [2], it has been recognised
that the fullest development of plasma-processing techniques would depend
on correct evaluation of the plasma parameters as well as on identification
of the gaseous species present and an understanding of the reaction
mechanisms involved.

One of the novel features in applications of plasma processing has been
the use of chemical equilibrium shifts in the gas phase of halocarbons for
the control of etch selectivity between SiO_2 and Si surfaces [3]. However,
the reaction mechanism in neither the etching nor the deposition of silicon
is yet fully understood, and to-date several research groups are extensively
investigating the fluorine-silicon reaction [4-6] for etching, and the
hydrogen-silicon reaction for thin-film deposition [7-8].

The energy available for a given chemical reaction in a glow discharge
depends on the extent of dissociation within the plasma and on the neutral
gas temperature. The degree of dissociation must depend on the particular
chemical species, and on the type of plasma, but generally it is accepted that
plasmas involve an overall dissociation not greater than 30% [9]; thus, for
instance, a relatively high temperature is normally used during growth of
inorganic films. The trend to lower substrate temperatures for deposition
has been slow, although deposition of silicon at low substrate temperatures
has been achieved with accompanying control of film properties [10].

From several viewpoints development of low-temperature deposition would
be advantageous; low melting materials such as plastics are, for example, used
as encapsulants for electronic components. Their surfaces can be modified by
plasma treatment reducing both outgassing and permeability to moisture and
corrosive gases. Coating of plastics has been reported, by vaporization [11]
as well as by ion plating [12,13] but there are film-to-substrate adhesion

problems associated with these methods. No previous work, other than preliminary reports by the present authors [14,15], appears to have been recorded on glow discharge or plasma deposition of inorganic films on plastics.

Various approaches have been used to lower the substrate deposition temperature. One example is the use of an endothermic chemical reaction to deposit the films by low-pressure glow-discharge chemical transport [16]. Ideally the method requires the degree of dissociation to be as high as possible without the neutral gas temperature being enhanced. This has been achieved by several investigators [17] for the production of ozone, but the full potential of these techniques has not been realized in thin film production except for some recently reported work by Sokolowski et al [18].

Overall it is generally accepted that in plasma processing there is far too little known of the fundamental mechanisms of etching and deposition of solids. Quantitive analyses of thin film properties grown under plasma conditions are wide spread, but it is the reaction that occurs between gaseous species which determine the nature and physical and chemical properties of a deposited material. If optimum conditions for deposition are to be established, and films are to be grown with control over their properties, a detailed knowledge of the gas-phase species involved is essential.

Clearly representative sampling is of primary importance. However, conclusions drawn from mass-spectrometric analysis of effluent gases must be reached with utmost caution. Recombination, deposition and desorption are just some of the likely processes that can occur between the glow discharge deposition region and the sampling position. On the other hand, plasma deposition conditions during thin-film deposition can be established in a suitable mass spectrometric sampling system and the detected gaseous species are more realistic than effluent gas studies. Then, by monitoring and controlling these species immediately prior to deposition, it should be possible to grow thin films in a more controlled manner than hitherto.

Many of the species in the plasma are short-lived transients and mass spectrometry, in the form of molecular or modulated beam analysis, lends itself ideally to the task of their detection and identification. Modulation of the sampled molecular beam permits separation of the sample mass spectrum from the extraneous background spectrum.

Molecular-beam sampling [19] and modulated-beam mass spectrometry [20] for the detection of transient species have been available to the scientific community for 30 years or more. However, considering their widespread potential in interdisciplinary applications, very little use has been made of this sensitive technique in reactive plasma sampling; although studies have been reported for methane [21], and more recently the method has been described in some detail [22-24]. The delay in their application to plasma sampling has probably been because of the associated problems of sampling from relatively high (plasma) pressures into the ultra-low pressures (UHV) required in the analyzer chamber of a high-resolution mass spectrometer. We describe in this report equipment employing specially designed 3-stage differential pumping and modulated-beam sampling with as short as possible path length from a plasma tube to the ion source chamber of a quadrupole mass spectrometer.

EXPERIMENTAL

The design of equipment for plasma sampling centres around minimizing the path-length between the plasma or glow discharge and the analyzer region (i.e. the quadrupole ion-source in our case). In principle a representative molecular beam is extracted from the stable glow discharge, operating under deposition conditions. The beam is modulated by a vibrating reed and the species arrive at the ionizer periodically. The detector system can then be tuned in phase with the periodicity of the modulated beam by means of a lock-in amplifier, thus separating the sample signal from the background signal.

The equipment employed in the present investigation (figure 1) comprises of an Extra-Nuclear Laboratories quadrupole mass spectrometer with a cross-beam ionizer and external vibrating reed beam-modulator (supplied by Spectrum Scientific Ltd.). The quadrupole system is mounted vertically in a 3-stage differentially pumped stainless steel vacuum chamber (constructed by Leisk Engineering Company). The vacuum is maintained by either turbomolecular pump or diffusion pump with water-cooled baffle, plus titanium sublimination pump, at the third stage, and by diffstacks on the first and second stages. The 3-stage chamber can be baked to 150°C. The quadrupole operates in an environment eight to ten orders of magnitude lower in pressure than the reactor tube, with a path length of <70 mm between the two.

Alignment of the quadrupole ion source with the beam sampling orifice is achieved with a micromanipulator mounting that permits x-y-z adjustment of the quadrupole, together with x-y adjustment of the reactor tube by means of a sliding teflon 'O' ring seal.

The quartz plasma reactor is constructed from 50mm diameter tubing in the form of a tee. A short (30mm) side-arm envelopes the plasma sampling orifice ($\sim$250µm diameter) which is formed at the tip of a quartz 'thimble' set into the side of the reactor tube, and which permits a representative molecular beam from the plasma to be extracted into the mass spectrometer analyzer. Gas flow and pressure in the reactor tube are monitored by flow controllers and a capacitance manometer. The reactor tube is evacuated by rotary pump (560 ℓ min^{-1}) and diffusion pump; also a cold finger trap, downstream of the plasma tube, condenses effluent gases which can be thermally

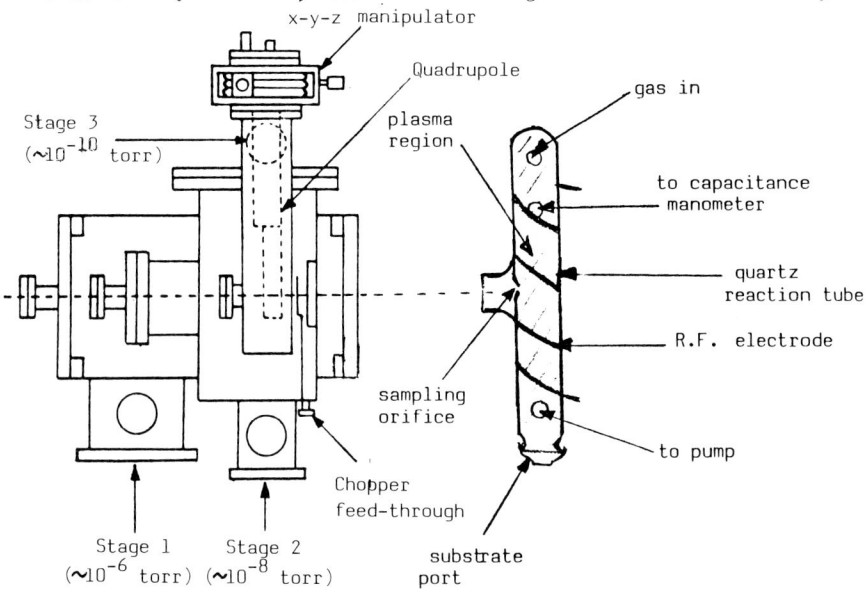

Fig. 1 Equipment for mass spectrometric sampling of plasma species.

regenerated for comparison of spectra. The discharge is excited by an rf generator (13.56 M Hz; supplied by Plasma Therm Ltd.) coupled inductively to the plasma through a matching network and copper coil around the reactor tube.

RESULTS

The beam chopper gives a suitably modulated argon-ion beam sampled from the discharge tube, while spectra with silane present have yet to be elucidated. The preliminary results that we report here were therefore obtained for an unmodulated beam, sampled from a silane-argon discharge with the rf generator operating up to 1.5kW power. The flow controllers were operated in a constant total pressure mode at 0.36 torr. with flow rates of argon and silane respectively 7±1 and 1.5±0.5 scm^{3}m^{-1}, giving a residence time of ~1 sec in the plasma reactor tube. Power input to the plasma was estimated from forward and reflected power at the generator.

Typical spectra in the mass range 25-30 amu are shown in figure 2 for rf power input to the discharge of up to 1.5kW. A decrease in peak intensity with increasing power occurs for masses 29-32. Relative changes in ion current with rf power are shown in figure 3. The 'cracking pattern' for silane gas at 90eV electron-beam energy, with zero rf power to the tube, agrees well with that reported by Turban et al[25]; except for a rather larger mass 29 peak than is observed by these investigators, which may be a large N_2H^+ background in our case.

With increasing rf power the background mass peaks 25-27 (see mass 25 illustrated in figure 3) remain constant, indicating that the rf field does not perturb the detection electronics; while overall decreases in intensity of mass peaks 29-31 (incl. 28, not shown) indicate depletion of silane in the discharge. Relative intensities appear to indicate increased formation of SiH species in the discharge, probably resulting from increased electron density at higher power. Quantitative data and correlation with film properties have yet to be obtained.

Thin ($Si_x H_y Ar_z$) films were deposited on glass substrates which varied in temperature from 60-80°C. Their optical spectra were measured with a double beam optical spectrometer. Freshly deposited films were bright yellow in transmission with a broad absorption at 3.2eV (figure 4) but were found to age over a two-week period with the peak absorption shifting to the red, changing the colour of the films from yellow to bronze.

These observations compare favourably with those of Catherine and Turban [7] for $Si_x C_y H_z$ films. They found the peak absorption to vary from 2.8 to 2.0eV depending on the silicon content. No measurements have to-date been made by the present authors to explain either the peak position or the ageing effects; but following Catherine and Turban both are probably related to the inherent amorphous nature of the films and to their silicon-argon content and it is probable that the gradual colour change is caused by the slow release of argon from the films. Adhesion of the films to the substrates was such that 10% HF was necessary to remove it from the inside wall of the reactor.

With higher gas pressures in the reaction tube gas-phase polymerisation results, as observed by Catherine and Turban. This yellow powder is presumably polymerised $Si_x H_y Ar_z$.

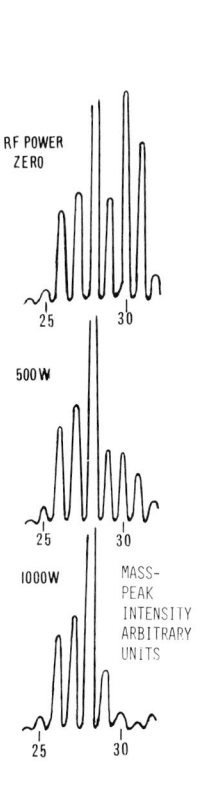

Fig.2 Representative Mass Spectra for SiH_4-Ar system.

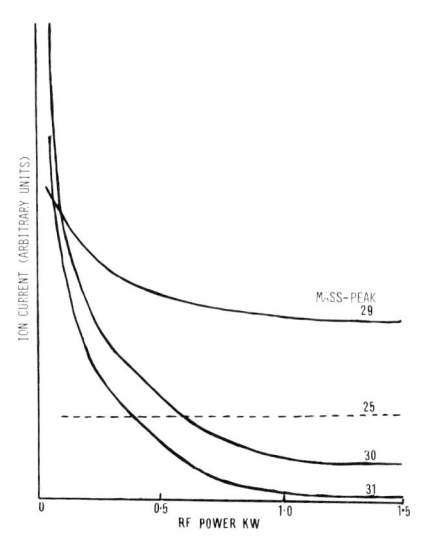

Fig.3 Variation of mass-peak intensity with rf power.

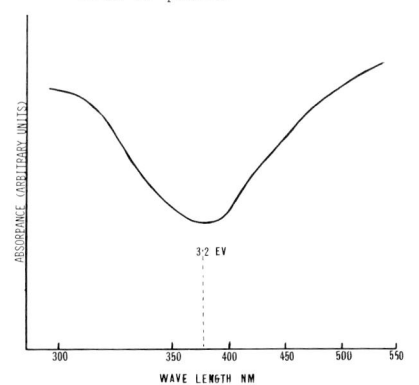

Fig.4 Visible light absorption in typical Si-H-Ar deposited film.

CONCLUSIONS

Mass spectrometric sampling with a truly collision-free path has yet to be achieved but early results appear promising. It is important to note that equipment used to sample a plasma or to make a diagnostic evaluation can easily disturb the discharge to be measured. It follows that glow discharge sampling can be highly dependent on reactor geometry. On the other hand if parameters can be adjusted to give ion, electron and neutral particle densities that are consistent from one discharge system to another then reaction mechanisms determined in a laboratory discharge should apply to those pertaining in a processing plasma. Such considerations are embodied in the 'similarity principle' [28] that underlies any attempt to obtain scaling laws.

REFERENCES

1. R.J. Joyce, H.F. Sterling and J.H. Alexander, Thin Solid Films 1, (1967/68) 481.
2. S.R. Ovshinsky, Proc.Int'l. Ion Engineering Congress ISIAT-IPAT '83, (Kyoto, Japan 1983) Ed.T. Tagaki, Vol.II, 817.
3. R.A.H. Heinecke, Sol.St. Elect. 18, (1975) 1146.
4. J.W. Coburn and E. Kay, Proc. 7th Int.Vac.Congr. and 3rd Int.Conf. Solid Surfaces (Vienna, 1977) 1257.
5. D.L. Flamm, V.W. Donnelly and J.A. Mucha, J.Appl.Phys. 52, (1981) 3633.
6. D.L. Flamm, V.W. Donnelly and D. Ibbotson, J.Vac.Sci.Tech. B1,(1983) 23.
7. G. Turban, Y. Catherine and B. Grolleau, Thin Solid Films 77 (1981) 287.
8. J. Wagner and S. Veprek, Plasma Chem. and Plasma Proc. 2, (1981) 95.
9. S. Veprek, Pure amd Appl.Chem. 48, (1976) 163; and Appl.Phys. 7,(1975) 271.
10. S. Veprek, Z. Iqbal, H.R. Oswald and A.P. Webb, J.Phys.C.Sol.St.Phys. 14, (1981) 295.
11. L. Holland, Vacuum 3 (1953) 330.
12. J.N. Avaritsiotis and R.P. Howson, Thin Solid Films 65, (1980) 101.
13. J.N. Avaritsiotis, R.P. Howson, M.I. Ridge and C.A. Bishop, Trans.Inst. Met.Finish (G.B.), 57, (1979) 65.
14. A.P. Webb and D.J. Fabian in Proc. 2nd European Conf. on Sol.St.Chem. (Netherlands, 1982); Elsevier Sci.Publ.Co. "Solid State Chemistry 1982"; Studies in Inorg.Chem. 3, (1983) 427.
15. A.P. Webb and D.J. Fabian, Proc. of the Int.Conf. on Ion Assisted Surface Treatments Techniques and Applications (Warwick, 1982) 20.1 (Publ. by Metals Soc.).
16. S. Veprek, Topics in Current Chem. 56, (1975) 139.
17. See for example Proc. 6th Int'l. Symp.Plasma Chemistry (ISPC6), Montreal 1983 (also ISPC5, Edinburgh 1981 and ISPC4, Zurich 1979).
18. M. Sokolowski et al, Thin Sol.Films 80, (1980) 249.
19. See for example: D.J. Fabian and W.A. Bryce, in 7th Int'l. Symp. on Combustion, Butterworths (London, 1958) 150.
20. S.N. Foner and R.L. Hudson, J.Chem.Phys. 21, (1953) 1374.
21. G. Smolinsky and M.J. Vasile, Int'l. J.Mass Spectry. Ion Phys. 16, (1975) 137.
22. W.L. Fite, Int'l.J. Mass Spec. Ion Phys. 16, (1975) 109.
23. M. Kaufman, Pure and Appl. Chem. 48, (1976) 155.
24. J.J. Wagner and W.W. Brandt, Plasma Chem. Proc. 1, (1981) 1.
25. G. Turban, Y. Catherine and B. Grolleau, Thin Sol. Films 67, (1980) 309
26. Y. Catherine and G. Turban, Thin Sol. Filma 60, (1979) 193.
27. G. Turban, Y. Catherine and B. Grolleau, Thin Sol. Films 60 (1979)158.
28. G. Francis, Ionization Phenomena in Gases, Butterworths, London, 1960.

ACKNOWLEDGEMENTS

This work was supported by the U.K. Science and Engineering Research Council and by the University of Strathclyde Research and Development Fund. The work was also carried out with the support of Standard Telecommunication Laboratories Ltd. and the authors wish to thank the Directors of STL Ltd. for their permission to publish this paper. They are grateful also to B. Hanks of Leisk Engineering Co. for help with design of the equipment, to N. Bridger of Spectrum Scientific Co. for help with the experiments, and to S. Veprek for generous advice at the design stage.

NOVEL RF-PLASMA SYSTEM FOR THE SYNTHESIS OF ULTRAFINE, ULTRAPURE SiC AND Si_3N_4

Gerald J. Vogt, Charles M. Hollabaugh, Donald E. Hull, Lawrence R. Newkirk, and John J. Petrovic
Materials Science and Technology Division, Los Alamos National Laboratory, Los Alamos, New Mexico 87545

ABSTRACT

A novel high-temperature plasma tube has been developed that overcomes the meltdown problem of the conventional water- and gas-cooled quartz plasma tubes commonly used. The key feature of this system is the placement of heavy-walled, water-cooled copper fingers inside a quartz mantle to shield the mantle from the intense radiation of the plasma. The copper fingers act as transformers to couple the plasma to the applied rf field.
This system has been used to produce ultrafine, ultra-pure silicon carbide powder by reaction of silane and methane. Powder of β-SiC has been obtained with a BET surface area of >160 m^2/g and a particle size range of 10 to 20 nm as measured by TEM. Likewise, powder of silicon nitride has been synthesized by reaction of silane and ammonia in the plasma. The resulting powder is approximately 50% Si_3N_4 with a mixture of α- and β-polymorphic forms. Boron carbide has also been successfully synthesized from diborane and methane.

INTRODUCTION

Plasma chemical synthesis has been utilized in recent years to prepare ultrapure, ultrafine refractory powders. There is considerable interest in using submicron ceramic powder to obtain dense, fine grained ceramic bodies with improved properties. By taking advantage of the surface energy available from the high surface area of these powders, it is expected that ceramic powders can be consolidated to near theoretical density at lower temperatures and without the conventional sintering additives. Because densification aids typically degrade the high temperature mechanical properties of structural ceramics, the ability to consolidate ultrafine powders without these aids can potentially lead to superior structural ceramic materials for high temperature and hostile environment applications.
Hamblyn and Ruben [1] have reviewed early synthesis work for numerous refractory and ceramic materials in an rf plasma. In several investigations [2-4] ultrafine β-SiC has been prepared in a thermal plasma by the pyrolysis of organochlorosilanes. Silicon carbide has also been produced by the thermal reaction of SiO or SiO_2 [5,6] with methane and by the reaction of a silicon halide and a halogenated hydrocarbon [7,8]. Ultrafine Si_3N_4 powders have been made in an arc plasma from $SiCl_4$ and NH_3 [9,10]. Submicron B_4C has been synthesized in an rf plasma from a $BCl_3/H_2/CH_4$ mixture [11].
This paper describes our progress toward synthesizing SiC, Si_3N_4, and B_4C in a novel rf-plasma reactor developed at Los Alamos. Our efforts are part of a Structural Ceramics Program at Los Alamos for the development of new structural ceramic materials. The plasma powders are obtained by the following reaction schemes:

$$SiH_4(g) + CH_4(g) \longrightarrow SiC(s) + 4H_2(g)$$

$$3SiH_4(g) + 4NH_3(g) \longrightarrow Si_3N_4(s) + 12H_2(g)$$

$$2B_2H_6(g) + CH_4(g) \longrightarrow B_4C(s) + 8H_2(g)$$

The ultrapure, ultrafine powders are prepared by injecting the reactants into the plasma jet of an argon plasma. The rf plasma provides a very clean environment for chemical synthesis with little contamination from the reactor, in comparison to an arc plasma where electrode erosion can lead to significant contamination. Moreover, our rf-plasma system can be readily operated for many hours with a stable plasma without the meltdown hazard of conventional water- and gas-cooled quartz plasma tubes.

EXPERIMENTAL APPARATUS AND PROCEDURE

The rf-plasma system, shown in Fig. 1, consists of a plasma tube and associated rf power generator, a reaction and quenching chamber, and a powder collection system as described previously [12,13]. The plasma tube is a novel design utilizing the concept of a transformer. Several heavy-walled, water-cooled copper fingers are positioned inside a quartz mantle, encircling the plasma zone. They transmit power from the primary rf coil to the plasma by an induced voltage and current. By means of an

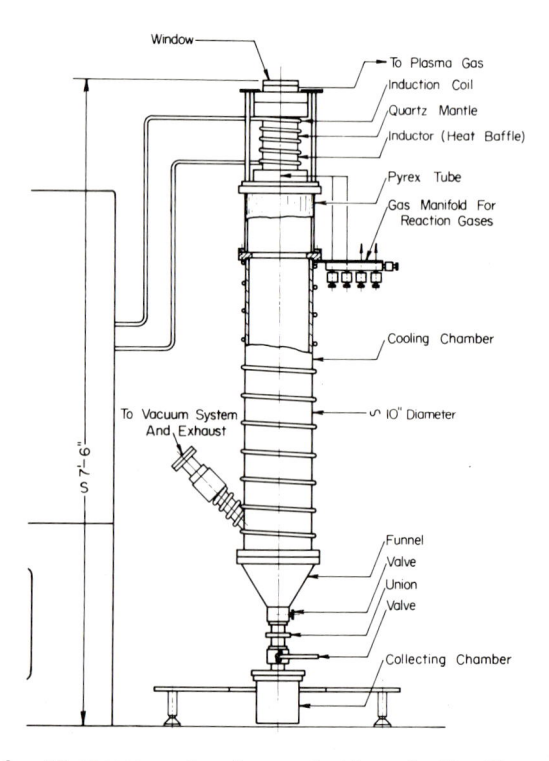

Fig. 1. RF plasma system for production of ultrafine powder.

interlocking chevron cross section, the fingers shield the mantle from the intense radiation and heat of the plasma and prevent the hazardous arcing, induced by the ionizing radiation, that can occur between the coil turns and between the coil and the mantle.

A tangentially injected argon stream stabilizes and centers the plasma, providing a tail flame out of the bottom of the plasma tube. Reactant gases are introduced into the tail flame in either a radial or axial mode. In the radial mode the reactants are injected through four ports orthogonally arranged around the lower end of the plasma tube. In the axial mode the reactants are fed into the tail flame through a water-cooled probe that is positioned on the centerline of the plasma tube and is passed through the plasma generation zone from above.

Below the reaction zone the quenching chamber is water-cooled in order to rapidly quench the hot gas/powder mixture exiting the tail flame. The entrained powder is collected in two cyclone separators connected in series. Although the mean particle size (10 to 20 nm) is too small to be separated by a cyclone, there is sufficient agglomeration to permit efficient separation and collection. Over 4 to 6 hours operating time about 45% of the powder collects on the wall of the reaction/quenching chamber, about 45% is separated and collected in the cyclones, and the remainder is lost in the exhaust. A production rate of 100 g/hr has been achieved, a rate limited by the gas injection system rather than the plasma capacity.

The rf plasma tube has been operated at frequencies between 200 kHz to 3.0 MHz at power levels from 15 to 50 kW. The powders collected in the cyclones were characterized by elemental analysis, x-ray diffraction, and electron microscopy.

RESULTS AND DISCUSSION

Silicon Carbide

Silicon carbide synthesis was carried out by radial injection of silane and methane into the tail flame of the plasma. The composition of the ultrafine powder was varied over a wide range from excess silicon to excess carbon. This was accomplished by adjusting the feed gas composition and rf power. Table I summarizes the gas flows and power levels for several experimental runs. The plasma argon was used to sustain the plasma and to provide the tail flame. A process argon stream of 4.5 slpm (standard liter per minute) served as a diluent for the silane and methane mixture. In runs 1 to 13 the reactant feed was also diluted with a hydrogen flow of 5.3 slpm.

In runs 1, 2, 4, and 6 variations in the silane flow have affected the overall stoichiometry of the powder. In each case β-SiC was present as the major constituent. The change in composition was clearly reflected in the color of the powders. Run 1 was a dark tan, run 2 black, and run 4 dark gray.

In run 5 the power level and plasma argon flow were increased, while the other gas flows were identical to run 4. The carbon-to-silicon ratio of the powder increased to 1.86, nearly double that found in run 4. This observation suggests that a higher power level may increase the carbon activity in the gas, providing a greater carbon level in the resulting powder.

In the remaining runs the plasma argon, silane, and methane flows were increased, but the methane-to-silane ratio was held constant at 3.6. Runs 7 and 8 at power levels of 16 and 18 kW produced silicon-rich powders. By comparison, the next three runs at 520 kHz and 20 kW yielded carbon-rich powders. This difference in the powder stoichiometry we attribute to the overall operational characteristics of the plasma reactor design.

The hydrogen diluent in the feed gas appears to lower significantly the

TABLE I. Process parameters and powder composition.

Run No.	Frequency (kHz)	Power (kW)	Plasma Ar (slpm)	CH_4 (slpm)	SiH_4 (slpm)	Powder C/Si Ratio
1	580	15	22	1.0	0.65	<1.00
2	580	15	22	1.0	0.21	>1.00
4	706	19	22	1.0	0.41	1.08
5	706	33	34	1.0	0.41	1.86
6	706	19	22	1.0	0.35	0.97
7	626	16	34	2.3	0.65	0.86
8	448	18	34	2.3	0.65	0.87
13	520	20	34	2.3	0.65	1.22
16	520	20	34	5.9	1.64	1.12
17	520	20	34	5.9	1.64	1.16

temperature of the plasma tail flame. This is illustrated by observing the change in the tail flame when only hydrogen and process argon are radially injected. The introduction of only half the amount of hydrogen used in the experiments will quench the visible emission of the tail flame, a clear indication of a significant cooling of the flame temperature. Similarly, a variation in the rf power will greatly affect the reactions in the tail flame by subsequent temperature changes. In runs 16 and 17 the hydrogen diluent was not included, anticipating that higher flow rates of silane and methane could be used with the greater tail flame temperature. This appears to be the case, as the flow rate of each gas was incremented by a factor of 2.5 over those used in run 13 with little change in stoichiometry.

The particle size range for the powders was determined by transmission electron microscopy. In Fig. 2 a photomicrograph of a typical particle sample shows a size range of 10 to 20 nm. This size range is consistent with mean particle diameters of 11.4, 12.1, and 18.5 nm derived from BET surface area measurements of 164, 154, and 102 m^2/g for select powders. The chain-like agglomeration in Fig. 2 is characteristic of our powders and of particles in this size range [14]. At present the nature and strength of the interparticle bond is not known, but the agglomeration will certainly be detrimental for the successful densification of the powders if the bonds cannot be broken.

X-ray diffraction results indicate that the silicon carbide powder consists mainly of the beta phase (cubic) with a lesser amount of the alpha polytypes. The cubic structure is preferred for structural ceramic applications. The coherently diffracting crystal size is estimated from the diffraction broadening to be between 3.0 and 7.5 nm, suggesting that the SiC particles are polycrystalline. The preferential formation of the beta phase can lead to some insight on the formation conditions of the powder. Cubic SiC formation is favored by a low temperature (<2000 K) and by the presence of excess silicon [15]. Temperatures greater than 2200 K will generally favor the stable alpha polytypes. Silicon carbide sublimes and decomposes at elevated temperatures (>2255 K) under atmospheric pressure [16]. We believe that during nucleation and initial growth conditions exist such that the activity of silicon is greater than that of carbon and an excess of silicon in the growing particles stabilizes the beta structure. These conditions must be momentary as the residence time at the formation temperature is certainly short. In order to form near stoichiometric silicon carbide, the activity of carbon relative to silicon must seemingly rise as the particles grow, allowing the carbon content of the powder to increase. It is possible, therefore, to have excess silicon during the initial stage of particle nucleation and growth and to have excess carbon during the final stage at a lower temperature.

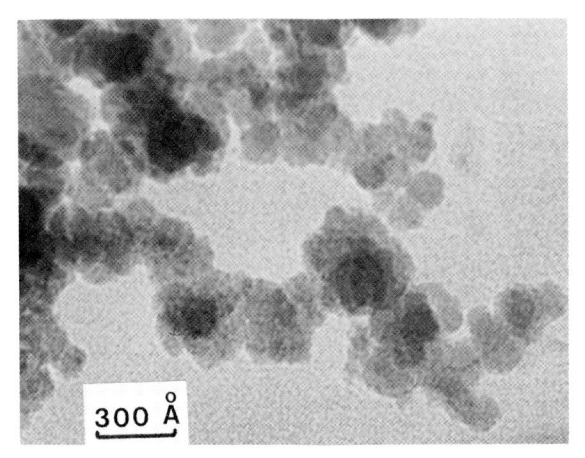

Fig. 2. TEM of silicon carbide powder (Run #21).

Spectrographic analysis of the plasma powders demonstrates that we have indeed produced ultrapure silicon carbide. Table II lists the trace impurities of the Los Alamos powders compared to an H.C. Starck Company Grade B-10 beta silicon carbide. With the exception of copper in one analysis, the impurity levels of the plasma powder are well below those of the high quality commercial powder. The analysis of the commercial powder was performed at Los Alamos. Elements not given in Table II are below detectable limits.

As with most high-surface materials prepared from chemically active reactants, oxygen contamination is a problem. The content of these powders as sampled from the process and maintained in a closed argon atmosphere is less than 0.5% by weight. Varying exposures to air can contribute to oxygen levels as great as 6% by weight. For example, the oxygen level in a high-surface (160 m^2/g) powder stored in an ordinary glass jar in air was 4.0% by weight after 16 days and 5.7% after 110 days. Without special care powders stored in an argon-atmosphere glove box (<10 ppm O_2) can reach a contamination level of 1% by weight. Consequently, special handling procedures, plus a degassing by heating under vacuum to remove volatile O_2-sensitive by-products, are required to assure a low oxygen content.

Silicon Nitride

The silicon nitride work was performed at a rf frequency of 450 kHz and at a power level of 35 kW. Our initial attempts to prepare Si_3N_4 used nitrogen as the nitriding agent in nitrogen-to-silicon atomic ratios of 7 to 35. These attempts failed to produce silicon nitride either by injecting the SiH_4/N_2 mixtures in a radial mode or by injecting the N_2 with the plasma argon stream. In each case the powder was essentially crystalline silicon.

Later we were able to obtain silicon nitride by using ammonia as the nitriding agent. With a large excess of ammonia the SiH_4/NH_3 mixture was injected axially into the tail flame. The resulting powder was highly crystalline with about 50% by mole silicon and nearly equal fractions of both α- and β-Si_3N_4.

The apparent difficulty in synthesizing silicon nitride may be related to the extreme chemical stability of diatomic nitrogen. The formation of

TABLE II. Impurities in the plasma SiC powder compared to commercial beta SiC powder.

Element[a]	Plasma powder (ppm)	Commerical powder[b] (ppm)
B	5-100	200
Mg	3-10	8
Al	10-50	800
Ca	1-50	50
Ti	3-20	50
Cr	<10	20
Fe	<10	300
Ni	<10	10
Cu	1-20,70	40
Zn	<30	<30
W	<300	1000

[a] Elements not shown are below detectable limits.
[b] H.C. Starck Company, Grade B-10.

this stable diatomic from the decomposed N_2 and NH_3 can occur at temperatures far too high (>2200 K) to permit the formation of the nitride. The competing formation of N_2 would deplete the available active nitrogen species before they could react with the active silicon species to produce Si_3N_4 below 2200 K. One approach to enhancing the synthesis of Si_3N_4 is to utilize a large excess of ammonia to increase the concentration of active species, recognizing that a large fraction of the ammonia will be lost as nitrogen. A more appealing approach is to lower the decomposition temperature of the ammonia so that the formation of silicon nitride can occur simultaneously. In this case a maximum usage is made of the available active nitrogen. A lower reaction temperature can be obtained by injecting the SiH_4/NH_3 mixture into a cooler region of the tail flame or by adding hydrogen to the argon plasma for a greater enthalpy.

Boron Carbide

We have demonstrated the feasibility of synthesizing boron carbide in our plasma system. The CH_4/B_2H_6 mixture was injected radially into the plasma tail, operated at 650 kHz and 23 kW. The resulting ultrafine powders were largely amorphous, but did contain a small fraction of crystalline material identified as hexagonal B_4C by x-ray powder diffraction. Diborane was chosen as the boron source due to the incompatibility of our plasma reactor with chlorine. Future work will use boron trichloride in a new halogen-compatible reactor.

CONCLUSIONS

Ultrafine, ultrapure silicon carbide powder has been synthesized by the injection of silane and methane into the tail flame of an rf argon plasma. Because the powders are highly agglomerated, further research and development must be conducted to yield the desired ultrafine material needed for our structural ceramic applications. Special handling for oxygen sensitive powder is essential.

Plasma synthesis of B_4C and particularly Si_3N_4 shows great promise for further research and development at Los Alamos.

ACKNOWLEDGEMENTS

This work was supported in part by the Department of Energy through the Division of Materials Science, Office of Basic Energy Sciences.
The authors wish to thank D.L. Rohr and P. Martin for their electron microscopy studies and R.B. Roof for his x-ray diffraction studies.

REFERENCES

1. S.M.L. Hamblyn and B.G. Reuben, Advances in Inorganic Chemistry and Radiochemistry 17, 89-114 (1975).
2. O. De Pous, F. Mollard and B. Lux, Proc. of the 3rd Int. Sym. on Plasma Chemistry 3, Paper S.4.7. (1977).
3. R.M. Salinger, Ind. Eng. Chem. Prod. Res. Develop. 11, 230-231 (1972).
4. J. Cateloup and A. Mocellin, Special Ceramics 6, 209 (1975).
5. I.G. Sayce and B. Selton, Special Ceramics 5, 157 (1972).
6. P.C. Kong, T.T. Huang, and E. Pfender, Proc. of the 6th Int. Sym. on Plasma Chemistry 1, 219-224 (1983).
7. F.G. Stroke, U.S. Patent 4,133,689 (1979).
8. F.G. Stroke, U.S. Patent 4,295,890 (1981).
9. G. Perugini, Proc. of the 4th Int. Sym. on Plasma Chemistry 2, 779-785 (1979).
10. T. Yoshida, H. Endo, K. Saito, and K. Akashi, Proc. of the 6th Int. Sym. on Plasma Chemistry 1, 225-230 (1983).
11. I.M. MacKinnon and B.G. Reuben, J. Electrochem. Soc. 122, 806-811 (1975).
12. C.M. Hollabaugh, D.E. Hull, L.R. Newkirk, and J.J. Petrovic, submitted to J. Mater. Sci. (1983).
13. C.M. Hollabaugh, D.E. Hull, L.R. Newkirk, and J.J. Petrovic, Proc. of Int. Conf. on Ultrastructure Processing of Ceramics, Glasses and Composites (1983).
14. J.R. Stephens and B.K. Kothari, Moon and Planets 19, 139-152 (1978).
15. P.T.B. Shaffer, Mat. Res. Bull. 4, S13-S24 (1969); in "Silicon Carbide 1968."
16. Engineering Property Data On Selected Ceramics. Volume II. Carbides (Metals and Ceramics Information Center, Battelle, Columbus, Ohio, 1979) p. 5.2.3-1.

USE OF OPTICAL EMISSION SPECTROSCOPY AS A DIAGNOSTIC TECHNIQUE FOR PLASMA
DEPOSITION OF HYDROGENATED AMORPHOUS SILICON AND CARBON

F. J. KAMPAS
Division of Metallurgy and Materials Science, Brookhaven National
Laboratory, Upton, New York 11973

ABSTRACT

Optical emission intensities have been measured as a function of composition for silane-argon and silane-hydrogen mixtures used in the deposition of hydrogenated amorphous silicon. It was found that changes in silane fraction have a large effect on the electron concentration and energy distribution in the discharge.

INTRODUCTION

Deposition of hydrogenated amorphous silicon (a-Si:H) and hydrogenated amorphous carbon (a-C:H) by the glow-discharge technique requires an optimization process involving a large number of parameters including deposition system geometry, frequency and power of electrical excitation, process gas composition, flow rate, substrate temperature, etc. These parameters determine the composition of the discharge which includes electrons, ions, reactive fragments of molecules, photons, and stable molecules. The discharge composition and potential distribution determine the flux onto the substrate of species that result in the deposited films. Optimization of deposition parameters is facilitated by an understanding of the basic mechanisms involved in film growth. The use of plasma diagnostics to determine the effect of deposition parameters on the discharge composition simplifies modeling of discharge processes and film growth mechanisms. Several diagnostic techniques are available, including the use of Langmuir probes, mass spectrometry, optical emission spectroscopy, and laser induced fluorescence. Results obtained using optical emission spectrosopy will be presented in this paper.

EXPERIMENTAL

Glow discharges in two capactively coupled rf systems were studied. Silane-argon mixtures at 0.1 Torr pressure, 10 watts power, were studied in a system with two 8.9 cm diameter electrodes 3.5 cm apart. Silane-hydrogen mixtures at 1.0 Torr pressure, 30 watts power, were studied in a larger system with 25 cm electrodes 3.5 cm apart. Both systems had metal chambers which were connected to one electrode. Gas flow was controlled with modified Veeco PV-10 piezoelectric valves and measured with Tylan FM-360 flow meters, using MKS 257 flow controllers. Total pressure was measured with MKS 222 capacitive pressure transducers. The discharges were excited by Plasma-Therm HFS-500E 13.56 MHz generators.
Light emitted by the glow discharges was dispersed by a Spex model 1400 0.75 m monochromator and detected with a thermoelectrically cooled Hamamatsu R955P photomultiplier tube (PMT). The emission intensities have been determined using a Spex PC-1 photometer, and more recently with a system consisting of a Photochemical Research Associates (PRA) model 1763 preamplifier, a PRA model 1762 amplifier/discriminator, and a Canberra model 2071 counter/timer. Count rates from the counter/timer were transferred to an Apple II Plus computer over an IEEE-488 interface bus. Pulse trains generated by a Cyborg model 91A "Isaac" laboratory interface unit connected

to the Apple were sent to the monochromator stepper motor driver unit to advance the monochromator. In the study of the silane-argon system, a 0.2 m Jobin-Yvon monochromator, a R911 Hamamatsu PMT, and a Keithley model 480 picoammeter were used simultaneously with the Spex monochromator and photometer in order to measure the intensities of two emission lines at the same time. The optics were aligned to collect light from the space near the electrode connected to the rf "hot lead" (the cathode). Time-resolved emission spectra were obtained using a Canberra model 2044 time-to-amplitude converter (TAC). The discriminated PMT output from the PRA model 1762 was connected to the start pulse input of the TAC. The rf voltage applied to the electrode of the deposition system was attenuated and processed by a second PRA model 1762 amplifier/discriminator and applied to the stop pulse input of the TAC. The output of the TAC was connected to an Nuclear Data model 62 multichannel analyzer (MCA). The time-resolved emisson spectrum was transferred from the MCA to the Apple II Plus computer over an RS232 serial interface line.

BACKGROUND

In this section processes occuring in the discharge relevant to deposition and optical emission will be described and the mathematical formalism used to analyze the data will be presented. Emitting excited states in glow discharges can be produced by excitation (reaction 1) or dissociative excitation (reaction 2),

$$e^- + X \longrightarrow e^- + X^* \qquad (1)$$

$$e^- + PM \longrightarrow e^- + Y^* + Z \qquad (2)$$

in which a parent molecule PM is broken up, leaving one of the fragments in an excited state. Other processes which take place in the discharge are dissociation (reaction 3), ionization (reaction 4), and dissociative ionization (reaction 5).

$$e^- + PM \longrightarrow e^- + Y + Z \qquad (3)$$

$$e^- + X \longrightarrow 2e^- + X^+ \qquad (4)$$

$$e^- + PM \longrightarrow 2e^- + Y^+ + Z \qquad (5)$$

For excitation and ionization the rate is given by

$$R = k\, n_e\, [X] \qquad (6)$$

whereas the expression for dissociative excitation and dissociative ionization is

$$R = k\, n_e\, [PM]. \qquad (7)$$

The electron concentration is n_e and the rate constant k is determined by the electron energy distribution $f(E)$ and the cross section for the process $\sigma(E)$ [1].

$$k = (2/m_e)^{1/2} \int E^{1/2}\, f(E)\, \sigma(E)\, dE \qquad (8)$$

Strongly emitting excited states have lifetimes of the order of tens of nanoseconds. Consequently the emission intensities are proportional to the generation rate for the excited states as long as the pressure in the discharge is around 1 Torr or less. Therefore the emission intensities are given by Eq. 9 for excitation and by Eq. 10 for dissociative excitation [2].

$$I_x = C_x k_x n_e [X] \qquad (9)$$

$$I_y = C_y k_y n_e [PM] \qquad (10)$$

The quantity C is independent of deposition conditions. Deposition occurs from reactive fragments and ions which are produced by dissociative excitation and dissociative ionization of the parent molecule. Another possible source of the film precursors is ion-molecule reactions [1]

$$W^+ + PM \longrightarrow X^+ + Y + Z \qquad (11)$$

which are very fast, being essentially diffusion controlled. Since the ion concentration is equal to the electron concentration in the plasma, the overall deposition rate can be written in the following way:

$$D = C_d k_d n_e [PM] \qquad (12)$$

where k_d is the sum of the rate constants for the various processes which produce the reactive fragments which grow the film.

Studies of silane discharges show that emission from Si and SiH result from dissociative excitation of SiH_4 [3,4]. Emission from H is a result of dissociative excitation of SiH_4 and also H_2 [4]. However, emission from H_2 is solely due to excitation of H_2 [4]. Dissociative excitation of CH_4 used in the deposition of a-C:H produces emission fom C, CH, and H [5].

RESULTS AND DISCUSSION

Emission intensity results from a previously published study [2] of the argon-silane system will be presented in this section and compared with the silane-hydrogen system. Also, the time resolved optical emission from the silane-hydrogen system will be described.

From Eqs. 9, 10, and 12, it is clear that the quantities $I_x/[X]$, $I_y/[PM]$ (normalized emission intensities), and D/[PM] (the normalized deposition rate) are proportional to the product of the electron concentration n_e and the appropriate rate constant, k_x, k_y, and k_d. These quantities are shown versus silane fraction for silane-argon mixtures in Fig. 1 and for silane-hydrogen mixtures in Fig. 2. The dependence of the deposition rate on silane fraction was not measured for the silane-argon system. However, the concentration of H_2 in the discharge is a measure of the deposition rate since H_2 is eliminated during a-Si:H film growth. The H_2 concentration was estimated from the intensity of H_2 emission, using the method of Coburn and Chen to find the excitation rate for H_2 from the Ar emission intensity [6]. Therefore the plot of $[H_2]/[SiH_4]$ in Fig. 1 is equivalent to DEP. RATE/$[SiH_4]$ in Fig. 2.

In comparing Figs. 1 and 2, one should remember that the measurements were done in different deposition systems at different total pressures. In both cases the normalized emission intensities and the normalized deposition rate decreased with increasing silane fraction. They decreased with a power law dependence for the silane-argon mixture, but decreased exponentially for the silane-hydrogen mixture. The bend upwards in the normalized atomic hydrogen emission in Fig. 2 for silane fractions above 60% is probably due to a contribution to the emission from the dissociative excitation of silane.

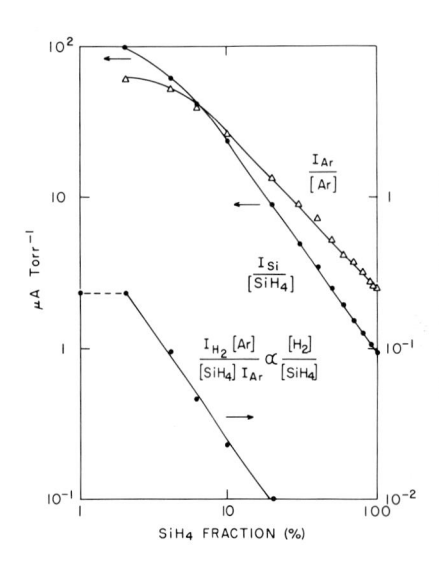

Fig. 1. Normalized emission intensities for a silane-argon mixture.

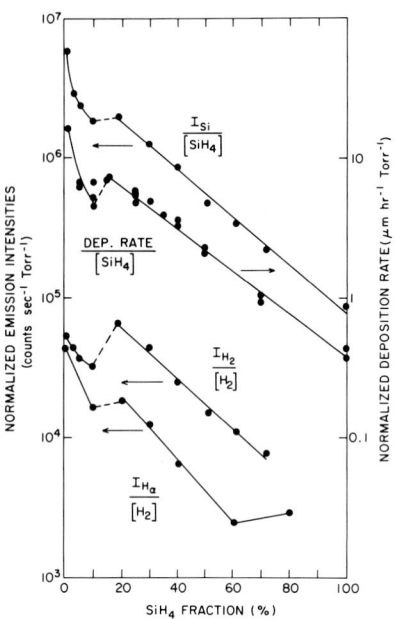

Fig. 2. Normalized emission intensities and normalized deposition rate for a silane-hydrogen mixture.

By taking the ratio of two normalized emission intensities it is possible to eliminate the electron concentration and find the ratio of the excitation rate constants. For example $(I_{AR}/[AR])/(I_{Si}/[SiH_4])$ is proportional to k_{Ar}/k_{Si}. The dependence of this ratio on silane fraction gives some information on the dependence of the electron energy distribution on silane fraction. This can be seen by consideration of a simple model which assumes that the electron energy distribution is Maxwellian and that the cross sections for excitation increase linearly above a threshold, E_t [2]. In this model the ratio of two excitation rate constants is given approximately by a fairly simple expression:

$$k_x/k_y \propto \exp[-1.5(E_{tx} - E_{ty})/E_{av}] \quad (13)$$

where E_{tx} and E_{ty} are the thresholds for excitation and E_{av} is the average electron energy. If the two excitation thresholds are the same, Eq. 13 states that the excitation rate constants remain in proportion as deposition conditions are changed. This is the assumption used by Coburn and Chen for finding ground state concentrations from emission intensities [6].

The excitation thresholds can be estimated from the minimum energy required for the excitation. This is 13.5 eV for the argon emission in Fig. 1 (the 750.4 nm line) and 9.5 eV for the Si line at 288.2 nm [2]. The fact that the normalized Si emission intensity decreased more rapidly with silane fraction than did the normalized argon emission implies that the average electron energy increased with silane fraction in the silane-argon mixture. Therefore the rate constants increased and the electron concentration decreased with silane fraction.

The excitation threshold for the H 656 nm (Balmer alpha) emission in Fig. 2 is 16.1 eV, which is greater than the threshold for the Si 288.2 nm emission or the H_2 continuum emission (11.9 eV) [2]. The fact that the normalized H emission decreased more rapidly with silane fraction than did the normalized Si or H_2 emission implies that the average electron energy decreased with silane fraction for the silane-hydrogen mixture.

The normalized H_2 production rate for the silane-argon mixtures and the normalized deposition rate for the silane-hydrogen mixtures both had silane fraction dependences similar to that of the normalized emission from Si. This implies that the film precursors are formed by electron impact dissociation of silane molecules, rather than ion-molecule reactions, since the rate constant for ion-molecule reactions should have a very different dependence on silane fraction than the rate constant for dissociative excitation.

Time-resolved emission spectra were taken of silane-hydrogen mixtures in order to gain some insight into the mechanisms of the discharge. de Rosny et al. [7] have reported that Si emission from pure silane rf discharges is sharply peaked in time and propagates from the electrodes into the discharge. A similar situation was found in silane-hydrogen mixtures, as shown in Fig. 3.

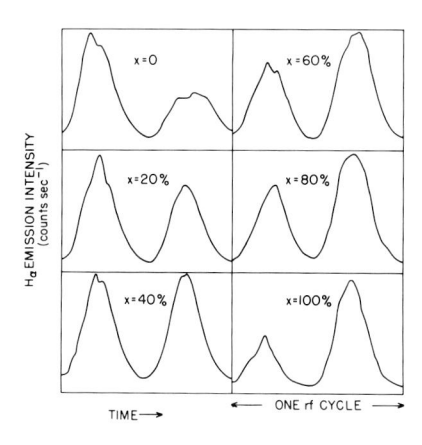

Fig. 3. Time-resolved emission from silane-hydrogen mixtures as a function of silane fraction, x.

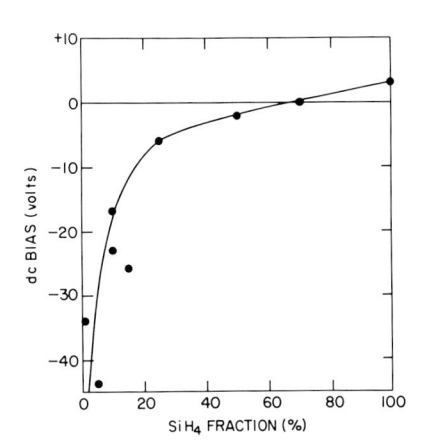

Fig. 4. Potential difference between rf electrode and grounded electrode versus silane fraction for silane-hydrogen discharges.

Two peaks are seen for each silane fraction, probably representing excitation peaks propagating from the two electrodes. The intensity of both peaks decreases with increasing silane fraction (the spectra in Fig. 3 have been normalized). de Rosny et al. stated that the peaks in optical emission were not due to secondary electron emission from the electrodes because they

were too sharply peaked in time. However, it is hard to imagine any other
mechanism. Furthermore, the relative intensities of the two peaks is
correlated with the voltage difference between the two electrodes, which is
shown in Fig. 4. At low silane fractions the "cathode" was negative in
comparison with the grounded electrode, and there was an obvious dark space
in the center of the discharge. Therefore the "earlier" peaks in Fig. 4
would seem to be due to secondary electrons from that electrode. As the
silane fraction was increased, the discharge became dimmer, the dark space
became less obvious, the voltage difference between the electrodes decreased
and changed sign, and the relative intensities of the two peaks in the
time-resolved emission reversed. This correlation is a strong indication
that the structure in the time-resolved emission is due to secondary
electrons. If that is the case, then the dependence of E_{av} on silane
fraction which was deduced from the emission intensities probably represents
a dependence of the plasma potential on silane fraction. Probe studies of
silane-hydrogen mixtures could confirm this hypothesis. Kushner's results
on rf discharges in fluorocarbon-oxygen mixtures studied by Langmuir probes
[8] appear to be qualitatively similar to those described here for
silane-hydrogen mixtures.

CONCLUSIONS

The optical emission intensities of both the silane-argon mixtures and
the silane-hydrogen mixtures show that the discharge electron concentration
and energy distribution are strong functions of the gas mixture composition.
The dependence of the deposition rate of a-Si:H from the discharge can be
correlated with the emission intensities. The time-resolved emission
studies imply that secondary electrons emitted from the electrodes play an
important role in the excitations responsible for the optical emission and
the deposition process as well.

ACKNOWLEDGMENTS

Loans of equipment by Drs. R. Corderman, S. Heald, and L. Snead are
gratefully acknowledged. The research was performed under the auspices of
the U.S. Department of Energy, Division of Materials Sciences, Office of
Basic Energy Sciences under Contract No. DE-AC02-76CH00016.

REFERENCES

1. I. Haller, J. Vac. Sci. Technol. A $\underline{1}$, 1376 (1983)
2. F.J. Kampas, J. Appl. Phys. $\underline{54}$, 2276 (1983)
3. F.J. Kampas and R.W. Griffith, J. Appl. Phys. $\underline{52}$, 1285 (1981)
4. J. Perrin and J.P.M. Schmitt, Chem. Phys. $\underline{67}$, 167 (1982)
5. D.E. Donohue, J.A. Schiavone, and R.S. Freund, J. Chem. Phys. $\underline{67}$, 769 (1977)
6. J.W. Coburn and M. Chen, J. Appl. Phys. $\underline{51}$, 3134 (1980)
7. G. de Rosny, E.R. Mosburg, Jr., J.R. Abelson, G. Devaud, and R.C. Kerns, J. Appl. Phys. $\underline{54}$, 2272 (1983)
8. M.J. Kushner, J. Appl. Phys. $\underline{53}$, 2939 (1982)

PLASMA ARC CARBIDE COATINGS ON TITANIUM

R. D. Shull, P. A. Boyer, L. K. Ives, and K. J. Bhansali
National Bureau of Standards, Metallurgy Division,
Washington DC 20234

ABSTRACT

The plasma transferred arc (PTA) process has been traditionally used to deposit wear resistant coatings on iron base alloy substrates, but has not been employed to coat lightweight alloys due to processing problems. In the current study, use of the PTA process to deposit TiC, WC, and Cr_3C_2 coatings on titanium substrates has been explored. The resistance of these coatings to dry abrasive wear has been measured and compared to that of the base metal. The variation in wear resistance of these coatings is discussed in terms of the carbide particle size and the microstructure of the deposit. A comparison is also made to coatings prepared by a laser surface melting and carbide particle injection process.

INTRODUCTION

For many applications the use of lightweight alloys based on aluminum or titanium can result in significant improvement in operating efficiency and performance. However, under conditions involving exposure to abrasive wear, the resistance of these materials is in general not sufficient. For iron and nickel based alloys, this problem has been solved [1,2] by hardfacing, that is, application of a wear resistant coating, especially by a welding process (the plasma transferred arc, PTA, technique being a notable example [3]). In addition to extending the life of metal parts subjected to severe abrasive service (e.g., in rock crushing and pulverizing equipment), hardfacing has also proved successful in controlling wear under light duty applications where close dimensional tolerances must be maintained (e.g., in control valves). A program was initiated to study the application of hardfacing technology to the surface modification of lightweight metals. Among the most wear resistant materials are the hard refractory carbides: TiC, WC, and Cr_3C_2. Recently, good adherent coatings of such materials on lightweight alloys have been prepared by laser surface melting [4]. In the present investigation conventional PTA hardfacing equipment was utilized to apply hard carbide coatings to CP titanium. During the study, the dependence between process parameters, microstructure, and abrasive wear performance was evaluated.

EXPERIMENTAL PROCEDURE

Surface coatings covering a 5 cm square area were deposited onto a 1.6 mm thick substrate of commerically pure (CP) titanium by the PTA process. The PTA unit used was capable of simultaneously feeding two separate powders or powder mixtures. In this work approximately equal volume concentrations of a binder (100/325 mesh 5456 aluminum alloy) and of a hard carbide (100/325 mesh titanium carbide, tungsten carbide, or chromium carbide) were applied. In some cases carbide powders were applied without the aluminum alloy binder. For each type of coating two plasma power levels, 120 and 170 amps at a potential difference of 25 volts, were used. Adjustment of the power level determines whether the injected powder is either partially or completely melted in the plasma before impinging on the partially melted top surface of the workpiece. Following the deposition, parallel slices at opposite sides of the deposit were taken for metallographic observation; and the middle portion was surface ground flat (surface roughness, 0.5 - 2.0 μm arithmetic average) with a diamond grinding wheel for wear testing.

The abrasive wear resistance of each coating was determined by the dry sand/rubber wheel test method. With this method abrasive wear is produced by 50-70 mesh silica sand rubbed against the specimen surface with a rotating rubber rimmed wheel. A detailed description of the test machine, materials and procedure is given in ASTM Standard G65-81. The conditions employed in this investigation were as follows: load, 134N; sand flow rate, 300 g/min.; test environment, air at 20-24 °C and relative humidity of 24-56 percent. Each test consisted of several increments of 350m (or 700m in some cases) sliding distance. Wear loss was measured by weight loss after each increment. Changes in the coating wear resistance could therefore be detected. In general the preferred means of reporting wear is in terms of volume loss since the serviceability of a machine component is usually limited by dimensional changes. Conversion from weight loss to volume loss was restricted here because of uncertainty concerning coating composition. The complicated shape of many of the wear scars also rendered direct measurement of scar dimensions impractical. Of the coating materials used, WC would be affected most on the basis of a weight loss comparison due to its much higher density.

RESULTS AND DISCUSSION

Mass loss as a function of sliding distance is shown in Fig. 1 for both the Ti substrate and the different coatings examined. In each case a steady state linear relationship between sliding distance and wear mass loss was found; and a straight line has been drawn representing a linear least squares fit to the data. In all cases except one (specimen G-12) the intercept of this line with the mass loss axis occurs at a positive value.

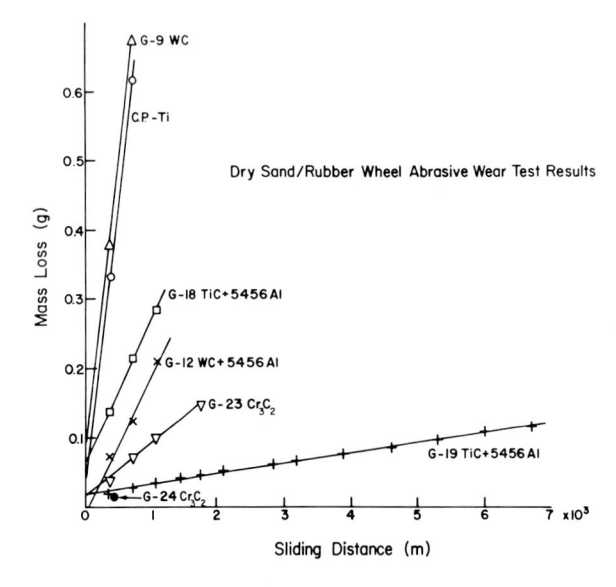

Fig. 1. Wear Mass Loss vs. Sliding Distance for PTA coatings.

This is a reflection of the high rate of wear which occurs during the initial contact of the cylindrical wheel with the flat specimen surface. For composite coatings with brittle particles, fracture of the exposed particles during the preparatory surface grinding may also result in a reduced initial wear resistance, as will also the rapid wear of the soft matrix phase which occurs before the hard particles stand out in relief. For specimen G-12 (Cr_3C_2 coating) a negative intercept is indicated. This was brought about by an accelerating wear rate caused by penetration of the coating layer during the test.

The wear rate values characteristic of each specimen are given by the slopes of the linear least squares lines. These values and the ratio of the wear rate of the Ti base metal to the wear rate of each coating are listed in Table 1. It can be seen that there are significant differences in wear rates of the various coatings. The reasons for these differences were in general explained on the basis of examinations of wear scar topography and coating microstructure revealed on metallographically polished cross sections prepared through the specimen surfaces. Specimen G-19 (TiC plus 5456 Al binder), applied under conditions where moderate melting of the substrate occurred, was significantly more wear resistant than the other

TABLE 1
ABRASIVE WEAR RESULTS ON PTA CARBIDE COATED TITANIUM

SPECIMEN NO.	COATING	WEAR RATE (G/M)	WEAR RATIO SUBSTRATE: COATING
C.P. Ti	(NONE)	8.25×10^{-4}	1
G - 19	TiC + 5456 Al	1.48×10^{-5}	56
G - 18	TiC + 5456 Al (DEEP MELT)	2.09×10^{-4}	3.9
G - 15	TiC	3.86×10^{-4}	2.1
G - 12	WC + 5456 Al (DEEP MELT)	2.00×10^{-4}	4.1
G - 9	WC	1.07×10^{-3}	0.77
G - 23	Cr_3C_2	7.55×10^{-5}	11
G - 24	Cr_3C_2	5.10×10^{-5}	16
	Ti-6 Al-4V*	4.13×10^{-4}	1
	TiC (50 μm) LASER MELT INJECTED*	3.43×10^{-5}	12

*DATA FROM REF. 4 EXPRESSED IN TERMS OF MASS LOSS/UNIT SLIDING DISTANCE.

coatings. Figure 2 shows a cross section through this coating: note the high concentration of TiC particles which do not appear to have been dissolved or noticeably altered during the deposition process. There are no cracks or voids in the coating interface (where intermixing between the Ti substrate and the 5456 Al binder has occurred) to indicate poor bonding. Figure 3 shows the corresponding worn surface. Matrix metal has been preferentially abraded, leaving the TiC particles raised above the surface. This is not unexpected since the TiC particles are significantly harder than both the matrix and the abrading sand particles. The exposed TiC paticles appear to be smooth and rounded, indicating that the wear process was very gradual without large scale fracture or crushing. The extent to which matrix material between the TiC particles was worn away is limited by the relatively large size (nominally 250 μm) of the abrading sand particles compared to the spacing (~ 60 μm) between TiC particles. In contrast, the wear rate of specimen G-18 (also coated with TiC plus 5456 Al binder) is significantly higher. In this case deposition was carried out at the higher (170 A) of the two current levels used. Appreciable melting of both the Ti substrate and deposit occurred; and the unmelted TiC particles remaining

Fig. 2. Cross section of specimen G-19, etched with Knoll's reagent.

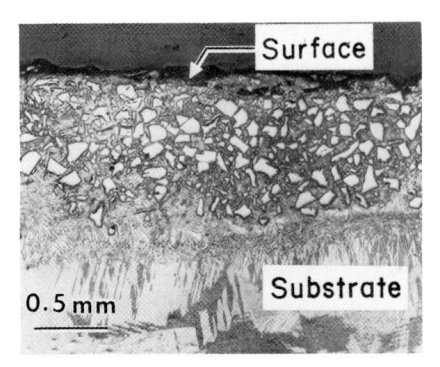

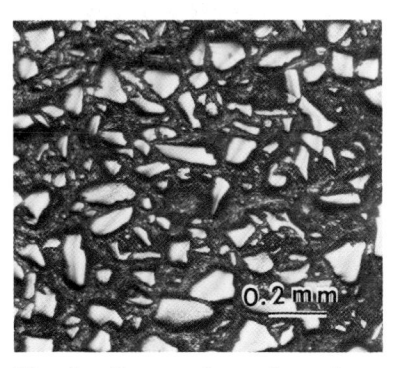

Fig. 3. Worn surface of specimen G-19.

were dispersed irregularly throughout a relatively large volume of metal (Fig. 4). The very rough topography of the worn surface (Fig. 5) results from the smaller wear loss in the regions of high TiC particle accumulation.

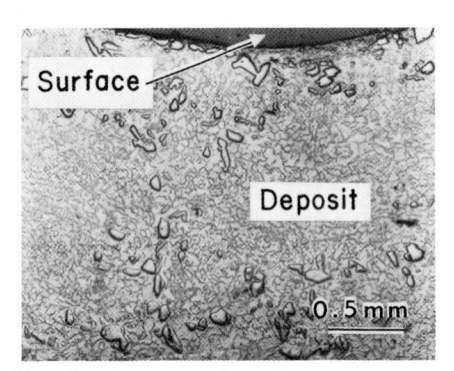

Fig. 4. Cross section of specimen G-18, etched with Knoll's reagent.

Fig. 5. Worn surface of specimen G-18.

A rather interesting wear effect was observed for one of the WC deposits (without a binder): specimen G-9 (Fig. 1). This deposit was again one of those prepared using the high 170A plasma current. In this case, however, the carbide particles did not melt. Instead, their high density (15.6 g/cm^3) resulted in their sinking deep into the low density (4.5 g/cm^3) molten region of the substrate, leaving only titanium at the surface. Consequently, the wear rate was approximately the same as the original substrate. In the case of specimen G-12 (WC plus 5456 Al) an adherent coating with a high concentration of WC was obtained. However, after surface grinding, only a thin layer was left which was penetrated very early in the wear test. Therefore, the data in Fig. 1 underestimates the true wear resistance of this coating. The Cr$_3$C$_2$ particle coatings (specimens G-23 and G-24) were applied without a binder. Under the conditions of application (120 amps) complete melting of the Cr$_3$C$_2$ occurred, except for a few particles at the surface. A cross section through the coating is shown in Fig. 6. The relative hardness of the different regions is indicated by the Knoop indentations. The Cr$_3$C$_2$ particles at the surface

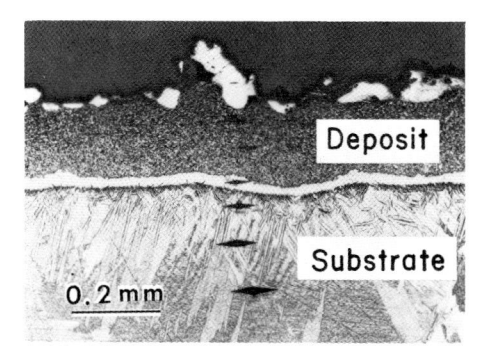

Fig. 6. Cross section through specimen G-24, showing undissolved Cr_3C_2 particles (top) and the Ti substrate (bottom), etched with Knoll's reagent.

are substantially harder than the other regions; note the smaller indentation dimensions. Farther from the surface there is a broad zone consisting of small spherical particles (probably precipitated Cr_3C_2 particles) in a dark etched matrix. A thin, apparently single phase layer is next, followed by an alloyed region in the substrate that exhibits what appears to be an α+β titanium structure. The wear rate of the resultant coating was lower by a factor of ~10 than the substrate.

As indicated in Table 1 the most wear resistant coating (56 times greater than the Ti substrate) was given by a high concentration of TiC particles in a 5456 Al alloy binder (specimen G-19). Dissolved and then apparently precipitated Cr_3C_2 coatings (specimens G-23 and G-24) also possessed low wear rates – having a wear resistance of 11-16 times greater than the Ti substrate. For comparsion, the results of Ayers, et al. [4] on TiC coated Ti-6Al-4V are also shown. From these data, it may be seen that the coatings of Ayers, et al. possessed an effective wear resistance ratio (compared to CP titanium) of 24, about one-half that of the slowly cooled TiC coating (G-19) of the present study. To put these wear rates into perspective, the abrasive wear resistance of D2 tool steel (61 HRC hardness) is approximately one-third that of the G-19 TiC coating of the present study.

CONCLUSION

It has been shown that the PTA process can be used to apply hard carbide coatings to titanium, with or without a binder phase; and that such coatings can provide more than a 50-fold increase in resistance to abrasive wear over that of CP-grade titanium.

ACKNOWLEDGEMENT:

The authors acknowledge Mr. E. L. Long of the Oak Ridge National Laboratory for partial support of this work under DOE Contract NO. DE-AT05-830R21322.

REFERENCES

1. American Welding Society, Welding Handbook, Section 5, pp. 94-58 (1973).
2. K. J. Bhansali, "Wear Coefficient of Hard-Surfacing Materials, Wear Control Handbook, pp. 373-383, (1980). American Society for Mechanical Engineers
3. Hardfacing, Metals Handbook, Vol 6, pp. 771-803 (1983). American Society for Metals.
4. J. Ayers, L. K. Ives, F. Matanzo, and A. W. Ruff, Proceedings on Wear of Materials, ASME, Reston, VA, (1983), p. 265.

AUTHOR INDEX

Apelian, D.: 91; 187; 197
Bhansali, K.J.: 297
Boulos, M.I.: 53; 207
Boyer, P.A.: 297
Bradley, E.R.: 235
Cheney, R.F.: 163; 225
Choi, H.K.: 77
Correa, S.M.: 197
Coudert, J.F.: 37
Dallaire, S.: 207
Fabian, D.J.: 277
Fauchais, P.: 37
Fiedler, H.C.: 173
Gauvin, W.H.: 77
Giacobbe, F.W.: 133
Harry, J.E.: 245
Heberlein, J.V.R.: 101
Herman, H.: 181
Hollabaugh, C.M.: 283
Houck, D.L.: 225
Hull, D.E.: 283
Ives, L.K.: 297
Johnson, N.P.: 277
Jurewicz, J.: 207
Kaczmarek, R.: 207
Kampas, F.J.: 291
Khait, Yu.L.: 263
Knight, R.: 245
Knoll, R.W.: 235
Kravchenko, S.K.: 255

Laktyushina, T.V.: 255
Lee, Y.C.: 141
Lesinski, J.: 37
Littlewood, K.: 127
MacCrone, R.K.: 181
Moore, J.J.: 117; 151
Newkirk, L.R.: 283
Paliwal, M.: 187; 197
Parosa, R.: 271
Peterson, L.G.: 217
Petrovic, J.J.: 283
Pfender, E.: 13; 141
Reid, K.J.: 117; 151
Robert, W.: 207
Roman, W.C.: 61
Schilling, W.F.: 217
Schmerling, D.W.: 133
Shankar, N.R.: 181
Shull, R.D.: 297
Sivertsen, J.M.: 117
Smith, R.W.: 217
Szekely, J.: 1
Tylko, J.K.: 151
Vardelle, A.: 37
Vardelle, M.: 37
Vashkevich, V.A.: 255
Vogt, G.J.: 283
Webb, A.P.: 277
Wei, D.: 197
Yasko, O.I.: 255

SUBJECT INDEX

Acetylene, 108; 127
Aerodynamic heating effect, 199
Agglomerates, 163; 225
Air plasma spraying, 177
Al_2O_3, 141
Arc characteristics, 23
Arc column, 16; 156
Arc discharges, 255
Arc gas heaters, 28
Argon plasma jet, 127
Blast furnace firing, 110
Boron carbide, 288
Boundary layer, 143
Carbide coatings, 297
Carbon, 133
Carbon dioxide, 133
Carbon monoxide, 133
Catalytic, 143
Centrifugal furnace, 64
Chemical synthesis, 101; 110; 283
Chromite, 118; 157
Chromium carbide, 225
Circuit breakers, 101
Coal, 126
Coatings, 207; 225; 235; 297
Cobalt, 163
Cold plasma, 61
Compressibility effects, 198
Consolidation, 217
Continuum flow, 200
Continuum mechanics, 99; 197
Cooling rate, 92; 123; 163
Copper reduction, 158
Copper smelting, 158
Crystallographic texture, 240
D.C. arc furnaces, 245
D.C. plasma furnaces, 245
D.C. plasma jets, 56
Dendrite fragments, 219
Deposition, 291
Deposits, 187; 219
Diagnostics, 37; 291
Dielectric constant, 182
Dielectric measurements, 181
Direct smelting, 152
Discharge cavity, 272
Discharge lamps, 101
Discharge parameters, 256
Discharge resonator, 273
Drag coefficients, 199
Electric arcs, 255
Electrical discharge, 61; 245
Electrical furnaces, 77
Electrical stability, 24
Electrode-stabilized arcs, 20
Electromagnetic interaction, 245
Electron temperature, 13; 61; 146
Enthalpy distribution, 31

Epitaxial grain growth, 220
EPR spectra, 181
Equilibrium Plasma, 1; 13
Expanded conical plasma, 153
Expanded precessive plasma, 153
Extended plasma arc flash reactor, 70
Extractive metallurgy, 77
Falling film reactor, 67
Fe-C metallics, 120
Fe-Cr-C metallics, 121
Ferro-alloy production, 111
Ferrochrome, 83
Flow regimes, 198
Forced convection stabilized arcs, 21
Free molecular flow, 145; 200
Gas discharges, 263
Gas heating, 101
Gas temperature, 61
GDR plasma furnace, 69
Glow discharges, 1; 277; 291
Growth defects, 240
Hardsurfacing, 163
Heavy particle temperature, 13; 146
Heat transfer, 59; 141; 197
Heat transfer coefficient, 200
Heat transfer mechanisms, 142
High intensity arc, 16
Hollow electrode melting, 71
Hydrogenated amorphous silicon, 291
In-flight metal extraction, 151
In-flight plasma reactions, 245
In-flight plasma reduction, 117
Interdendritic fracture, 219
Ion bombardment, 235
Jet-Kote^R coatings, 225
Lattice spacings, 238
Linde plasmarc furnace, 67
Local thermodynamic equilibrium, 13
Low-pressure microwave plasma, 271
Low-pressure plasma
  deposition, 94; 187; 217
Low-pressure plasma jet, 197
Macrosegregation, 91
Magnetically spun arc torch, 32
Magnetically stabilized arcs, 21
Mass spectrometric system, 277
Metal-ceramic system, 207
Metal melting, 245
Metallurgical processing, 101
Metallurgical reduction, 245
Metastable metallic alloys, 117
Metastable phase, 173
Metastable states, 181
MHD equations, 4
Microprocessor control, 207
Microsegregation, 91
Microwave cavities, 271
Molybdenum, 86; 163

Momentum transfer, 197
Monocrystal growing, 74
Multiple arc discharges, 245
Multiple electrode plasma reactor, 248
Multiple plasma source reactor, 72
Nichrome, 225
Nickel base superalloys, 217
Niobium, 87
Nitrogen fixation, 108
Non-catalytic, 143
Non-continuum effects, 99; 197
Non-equilibrium, 263
Non-equilibrium plasma, 1; 13
Nuclear waste processing, 73
Optical emission spectroscopy, 291
Particle melting, 197
Particle-plasma interactions, 6
Particle trajectory, 197
Particle velocity, 198
Plasma arc, 62
Plasma etching, 263
Plasma furnaces, 68; 77
Plasma gas heating, 152
Plasma gas temperature, 197
Plasma generation, 1; 13
Plasma generators, 102
Plasma induction furnace, 68
Plasma jet reactor, 133
Plasma jet velocities, 198
Plasma melting, 165
Plasma processing, 1; 53; 102
Plasma reactors, 63; 70; 133; 151; 271
Plasma sintering process, 141
Plasma smelting, 152
Plasma spray processing, 191
Plasma sprayed alumina, 173; 181
Plasma sprayed coatings, 181; 225
Plasma spraying, 37; 94
Plasma synthesis, 106; 163
Plasma torch, 28
Plasma transferred arc, 297
Plasmadust process, 79
Plasmared process, 79
Plasmasmelt process, 79
PMRS process, 163
Powder feeder, 207
Powder fragmentation, 190
Powder melting, 187
Powder metallurgy, 217
Progressive casting reactor, 72
PTA process, 297
Pyrolysis, 126
Rapid solidification, 91; 187; 217
Rapid solidification processes, 92
Rapidly solidified powders, 163
Reactive plasma deposition, 277
Recombination, 143
Resistivity, 183

RF discharges, 13; 277
RF plasma, 57; 141; 283
RF sputter deposition, 235
RF thermal plasma torch, 65
Scrap agglomeration, 73
Self-stabilized arcs, 21
Shockwave plasma reactor, 117; 153
Silicon carbide, 283
Silicon nitride, 283; 287
Spray drying, 163
Sputter-depositing, 235
Sputtering, 263
Stabilization, 62
Stabilized arcs, 257
Substrate bias, 235
Supersonic flow, 199
Surface conditions, 143
Surface treatment, 74; 271
Synthesis processing, 101
Taconite, 118; 157
Tantalum, 87
Temperature, 5; 13; 40; 54; 61; 221
Thermal conductivity, 173
Thermal plasmas, 3; 13; 61
Thermal RF discharges, 25
Titanium, 85
Torch design, 101
Toxic wastes, 110
Transferred arc, 94
Transferred arc furnaces, 152
Transferred arc plasmas, 78
Transferred arc reactor, 77
Transient plasma species, 277
Tungsten, 163
Tungsten carbide, 163
Vanadium, 87
Velocity, 5
Velocity distribution, 31
Velocity profiles, 197
Vortex flow, 257
Vortex-stabilized arcs, 19; 103
Wall-stabilized arcs, 18; 102
Wear resistance, 298
Wear testing, 226
Yttria, 235
Zirconia, 235
Zirconium, 84